T0075355

ADVANCED COMPUTER SCIENCE APPLICATIONS

Recent Trends in AI, Machine Learning, and Network Security

AAP Research Notes on Optimization & Decision-Making Theories

ADVANCED COMPUTER SCIENCE APPLICATIONS

Recent Trends in AI, Machine Learning, and Network Security

Edited by

Karan Singh, PhD

Latha Banda, PhD

Manisha Manjul, PhD

First edition published 2024

Apple Academic Press Inc.
1265 Goldenrod Circle, NE,
Palm Bay, FL 32905 USA

760 Laurentian Drive, Unit 19,
Burlington, ON L7N 0A4, CANADA

CRC Press
6000 Broken Sound Parkway NW,
Suite 300, Boca Raton, FL 33487-2742 USA

4 Park Square, Milton Park,
Abingdon, Oxon, OX14 4RN UK

Apple Academic Press exclusively co-publishes with CRC Press, an imprint of Taylor & Francis Group, LLC

Library and Archives Canada Cataloguing in Publication

CIP data on file with Canada Library and Archives

Library of Congress Cataloging-in-Publication Data

CIP data on file with US Library of Congress

ISBN: 978-1-77491-239-3 (hbk)
ISBN: 978-1-77491-240-9 (pbk)
ISBN: 978-1-00336-906-6 (ebk)

AAP RESEARCH NOTES ON OPTIMIZATION AND DECISION-MAKING THEORIES

SERIES EDITORS:

Dr. Prasenjit Chatterjee
Department of Mechanical Engineering, MCKV Institute of Engineering, Howrah, West Bengal, India
E-Mail: dr.prasenjitchatterjee6@gmail.com / prasenjit2007@gmail.com

Dr. Dragan Pamucar
University of Defence, Military academy, Department of Logistics, Belgrade, Serbia; E-Mail: dpamucar@gmail.com

Dr. Morteza Yazdani
Department of Business & Management, Universidad Loyola Andalucia, Seville, Spain; E-Mail: morteza_yazdani21@yahoo.com

Dr. Anjali Awasthi
Associate Professor and Graduate Program Director (M.Eng.), Concordia Institute for Information Systems Engineering, Concordia, Canada
E-Mail: anjali.awasthi@concordia.ca

Most real-world search and optimization problems naturally involve multiple criteria as objectives. Different solutions may produce trade-offs (conflicting scenarios) among different objectives. A solution that is better with respect to one objective may be a compromising one for other objectives. This compels one to choose a solution that is optimal with respect to only one objective. Due to such constraints, multi-objective optimization problems (MOPs) are difficult to solve since the objectives usually conflict with each other. It is usually hard to find an optimal solution that satisfies all objectives from the mathematical point of view. In addition, it is quite common that the criteria of real-world MOPs encompass uncertain information, which becomes quite a challenging task for a decision maker to select the criteria. Also, the complexities involved in designing mathematical models increase. Considering, planning, and appropriate decision-making require the use of analytical methods that examine trade-offs; consider multiple scientific, political, economic, ecological, and social dimensions; and reduce possible conflicts in an optimizing framework. Among all these, real-world multi-criteria decision-making (MCDM) problems related to engineering optimizations are categorically important and are quite often encountered with a wide range of applicability.

MCDM problems are basically a fundamental issue in various fields, including applied mathematics, computer science, engineering, management, and operations research. MCDM models provide a useful way for modeling various real-world problems and are extensively used in many different types of systems, including, but not limited to, communications, mechanics, electronics, manufacturing, business management, logistics, supply chain, energy, urban development, waste management, and so forth.

In the aforementioned cases, modeling of multiple criteria problems often becomes more complex if the associated parameters are uncertain and imprecise in nature. Impreciseness or uncertainty exists within the parameters due to imperfect knowledge of information, measurement uncertainty, sampling uncertainty, mathematical modeling uncertainty, etc. Theories like probability theory, fuzzy set theory, type-2 fuzzy set theory, rough set, grey theory, neutrosophic uncertainty theory available in the existing literature deal with such uncertainties. Nevertheless, the uncertain multi-criteria characteristics in such problems are not explored in depth, and a lot can be achieved in this direction. Hence, different mathematical models of real-life multi-criteria optimization problems can be developed on various uncertain frameworks with special emphasis on sustainability, manufacturing, communications, biomedical, electronics, materials, energy, agriculture, environmental engineering, strategic management, flood risk management, supply chain, waste management, transportations, economics, and industrial engineering problems, to name a few.

Coverage & Approach:

The primary endeavor of this series is to introduce and explore contemporary research developments in a variety of rapidly growing decision-making areas. The volumes will deal with the following topics:

- Crisp MCDM models
- Rough set theory in MCDM
- Fuzzy MCDM
- Neutrosophic MCDM models
- Grey set theory

- Mathematical programming in MCDM
- Big data in MCDM
- Soft computing techniques
- Modelling in engineering applications
- Modeling in economic issues
- Waste management
- Agricultural practice
- Material selection
- Renewable energy planning
- Industry 4.0
- Sustainability
- Supply chain management
- Environmental policies
- Manufacturing processes planning
- Transportation and logistics
- Strategic management
- Natural resource management
- Biomedical applications
- Future studies and technology foresight
- MCDM in governance and planning
- MCDM and social issues
- MCDM in flood risk management
- New trends in multi-criteria evaluation
- Multi-criteria analysis in circular economy
- Multi-criteria evaluation for urban and regional planning
- Integrated MCDM approaches for modeling relevant applications and real-life problems

Types of volumes:

This series reports on current trends and advances in optimization and decision-making theories in a wider range of domains for academic and research institutes along with industrial organizations. The series will cover the following types of volumes:

- Authored volumes
- Edited volumes
- Conference proceedings
- Short research (thesis-based) books
- Monographs

Features of the volumes will include recent trends, model extensions, developments, real-time examples, case studies, and applications. The volumes aim to serve as valuable resources for undergraduate, postgraduate and doctoral students, as well as for researchers and professionals working in a wider range of areas.

CURRENT & FORTHCOMING BOOKS IN THE SERIES

Multi-Criteria Decision-Making Techniques in Waste Management: A Case Study of India
Editors: Suchismita Satapathy, Debesh Mishra, and Prasenjit Chaterjee

Applications of Artificial Intelligence in Business and Finance: Modern Trends
Editors: Vikas Garg, Shalini Aggarwal, Pooja Tiwari, and Prasenjit Chatterjee

Advances in Data Science and Computing Technology: Methodology and Applications
Editors: Suman Ghosal, Amitava Choudhury, Vikram Kr. Saxena, Arindam Biswas, and Prasenjit Chatterjee

Precision Agriculture for Sustainability: Use of Smart Sensors, Actuators, and Decision Support Systems
Editors: Narendra Khatri, Ajay Kumar Vyas, Celestine Iwendi, and Prasenjit Chatterjee

Attacks on Artificial Intelligence: The New Facets of Cyber Ecospace
Editors: Prasenjit Chatterjee, Kukatlapalli Pradeep Kumar, Vinay Jha Pillai, and Boppuru Rudra Prathap

Advanced Computer Science Applications: Recent Trends in AI, Machine Learning, and Network Security
Editors: Karan Singh, Latha Banda, and Manisha Manjul

Decision-Making Models and Applications in Manufacturing Environments
Editors: Pushpdant Jain, Kumar Abhishek, and Prasenjit Chatterjee

ABOUT THE EDITORS

Karan Singh, PhD
Assistant Professor, School of Computer and Systems Sciences, Jawaharlal Nehru University, New Delhi, India

Karan Singh, PhD, is Assistant Professor in the School of Computer and Systems Sciences at Jawaharlal Nehru University, New Delhi, India. His areas of interest include multicast communications, information security, cryptography, security issues in wireless sensor networks, intelligent vehicular networks, Internet of Things, and cyber-physical systems. He has published many journal articles and conference papers. He has also attended and many international conferences and workshops and has delivered invited talks and lectures. He is a senior member of IEEE and is an IEEE MGM awardee.

Latha Banda, PhD
Department of CSE, ABES Engineering College, Ghaziabad, Uttar Pradesh, India

Latha Banda, PhD, is Associate Professor of Computer Sciences and Engineering at ABES Engineering College, Ghaziabad, U.P., India. She was previously an Associate Professor at Sharda University and Lingaya's University, India. Her areas of interest include artificial intelligence, recommender systems, machine learning, network security, and soft computing. She has published several journal articles and conference papers and has attended and participated at international conferences and workshops. She is a member of the Association for Computing Machinery, Institute of Electrical and Electronics Engineers, and Institution of Electronics and Telecommunication Engineers.

Dr. Manisha Manjul
Department of Computer Science, G. B. Pant DESU
Okhla I campus, New Delhi

Manisha Manjul, PhD, is working with the Department of Computer Engineering at G. B. Pant DESU Okhla-I Campus, New Delhi, India. She has published several papers in Scopus-indexed journals as well as several conference papers. She organizes workshops, conferences, and faculty development programs. Her primary research interests are in computer network, network security, multicast communication, and object-oriented programming. She is a life member of the Computer Society of India. She received her Engineering degree (Computer Science & Engineering) from KNIT, Sultanpur, UP, India. She received her MTech (Computer Science & Engineering) from NIT Jalandhar, PB, India, and PhD (Computer Science & Engineering) from Gautam Buddha University, UP, India. She worked at Gautam Buddha University, UP, India.

CONTENTS

CONTRIBUTORS

Naveen Aggarwal
CSE, UIET, Panjab University, Chandigarh, India

Latha Banda
ABES Engineering College, Ghaziabad, Uttar Pradesh, India

Anshul Bhardwaj
Dronacharya College of Engineering, Gurgaon, India

B. R. Chandavarkar
Department of Computer Science and Engineering, National Institute of Technology Karnataka, Surathkal, Karnataka, India

Chiranjoy Chattopadhya
Department of Computer Science and Engineering, Indian Institute of Technology Jodhpur, Jodhpur, Rajasthan, India

Muskaan Chopra
CSE, CCET, Panjab University, Chandigarh, India

J. Deebika
School of Computing, SASTRA Deemed University, Thanjavur, Tamil Nadu, India

Sayali Deshpande
Department of Computer Engineering, Pune Institute of Computer Technology, Pune, Maharashtra, India

Indu Dohare
School of Computer and Systems Sciences, Jawaharlal Nehru University, New Delhi, India

Shishir Gangwar
Department of Computer Science and Engineering, National Institute of Technology Karnataka, Surathkal, India

Devendra Gautam
ABES Engineering College, Ghaziabad, Uttar Pradesh, India

K. Geetha
School of Computing, SASTRA Deemed University, Thanjavur, Tamil Nadu, India

Shruti Gite
Electronics and Telecommunication Engineering, Mumbai University, Pillai HOC College of Engineering and Technology, Rasayani, India

Barnini Goswami
Department of Computer Science and Engineering, Krishna Engineering College, Ghaziabad, India

Anshul Gupta
CSE, CCET, Panjab University, Chandigarh, India

Kajal Gupta
Department of Computer Science and Engineering, Krishna Engineering College, Ghaziabad, India

Sumeet Gupta
School of Electronics and Communication Engineering, Shri Mata Vaishno Devi University, Kakryal, Jammu & Kashmir, India

Sachin Kumar Gupta
School of Electronics and Communication Engineering, Shri Mata Vaishno Devi University, Kakryal, Jammu & Kashmir, India

Ankita Jaiswal
School of Computer and System Sciences, Jawaharlal Nehru University, New Delhi, India

Krupa N. Jariwala
Department of Computer Engineering, Sardar Vallabhbhai National Institute of Technology, Surat, Gujarat, India

Sneha Kamble
Department of Computer Science and Engineering, National Institute of Technology Karnataka, Surathkal, Karnataka, India

P. K. Kapur
Amity University, Amity Centre for Interdisciplinary Research, Noida, Uttar Pradesh, India

Pankaj Kashyap
School of Computer and System Sciences, Jawaharlal Nehru University, New Delhi, India

Navroop Kaur
Computer Science Engineering, ACET Amritsar, Punjab, India

Prateek Kembhavi
Department of Computer Science and Engineering, National Institute of Technology Karnataka, Surathkal, Karnataka, India

Rasika Gururaj Khade
Department of Computer Engineering, Sardar Vallabhbhai National Institute of Technology, Surat, Gujarat, India

Amina Khan
School of Electronics and Communication Engineering, Shri Mata Vaishno Devi University, Kakryal, Jammu & Kashmir, India

Santosh Khandal
Department of Computer Science and Engineering, Sharda University, Greater Noida, Uttar Pradesh, India

Deepak Kumar
Amity University, Amity Institute of Information Technology, Noida, Uttar Pradesh, India

Mukesh Kumar
School of Computer & Systems Sciences, Jawaharlal Nehru University, New Delhi, India

Sudhakar Kumar
CSE, CCET, Panjab University, Chandigarh, India

Sushil Kumar
School of Computer and System Sciences, Jawaharlal Nehru University, New Delhi, India

Shreya Majumdar
Department of Computer Science and Engineering, Krishna Engineering College, Ghaziabad, India

Achintya Kumar Pandey
Department of CSE, Greater Noida Institute of Technology, Greater Noida, India

Vinay Pathak
School of Computer and Systems Sciences, Jawaharlal Nehru University, New Delhi, India

Sneha Mishra
School of Computing Science and Engineering, Galgotias University, Greater Noida, Uttar Pradesh, India

Hari Mohan Rai
Faculty of Electronics and Communication, Krishna Engineering College, Ghaziabad, India

R. Rajaraman
Amity University, Amity Institute of Information Technology, Noida, Uttar Pradesh, India

Sailesh Rana
Department of Computer Science and Engineering, Sharda University, Greater Noida, Uttar Pradesh, India

Vindhya Guru Rao
School of Computing, SASTRA Deemed University, Thanjavur, Tamil Nadu, India

Gopal Singh Rawat
School of Computer and Systems Sciences, Jawaharlal Nehru University, New Delhi, India

Manyam Nandeesh Reddy
Department of CSE, School of Computing, SASTRA Deemed University, Thanjavur, Tamil Nadu, India

B. B. Sagar
Computer Science and Engineering, Birla Institute of Technology, Mesra, Ranchi, India

Deepti Sahu
Sharda University, Greater Noida, Uttar Pradesh, India

Tende Ivo Sake
Department of Computer Sciences, Sharda University and Engineering, Greater Noida, Uttar Pradesh, India

Prabhpreet Singh Sandhu
Sharda University, Greater Noida, Uttar Pradesh, India

Kanchan Sapkota
Department of Computer Science and Engineering, Sharda University, Greater Noida, Uttar Pradesh, India

N. Sasikaladevi
Department of CSE, School of Computing, SASTRA Deemed University, Thanjavur, Tamil Nadu, India

Mohd Shariq
School of Computer and Systems Sciences, Jawaharlal Nehru University, New Delhi, India

Pankaj Sharma
Department of Computer Science and Engineering, Sharda University, Greater Noida, Uttar Pradesh, India

Kollabathini Siddhardha
Centre for European Studies, School of international Studies, Jawaharlal Nehru University, New Delhi

Amarpreet Singh
Computer Science Engineering, ACET Amritsar, Punjab, India

Chandra Shekhar Singh
Dronacharya College of Engineering, Gurgaon, India

Karan Singh
School of Computer and Systems Sciences, Jawaharlal Nehru University, New Delhi, India

Sunil Kr. Singh
CSE, CCET, Panjab University, Chandigarh, India

Sunny Singh
Data Scientist, NextGenTechEdge Solutions Pvt Ltd., India

Mansi Subhedar
Electronics and Telecommunication Engineering, Mumbai University, Pillai HOC College of Engineering and Technology, Rasayani, India

Yugaraj Tamang
Department of Computer Science and Engineering, Sharda University, Greater Noida, Uttar Pradesh, India

Rajat Kishor Varshney
Department of CSE, Greater Noida Institute of Technology, Greater Noida, India

Satvik Vats
Computer Science and Engineering, Graphic Era Hill University, Dehradun, India

K. Velmurugan
Anjalai Ammal Mahalingam Engineering College, Kovilvenni, Thiruvarur, Tamil Nadu, India

Dileep Kumar Yadav
School of Computing Science and Engineering, Galgotias University, Greater Noida, Uttar Pradesh, India

Saneh Lata Yadav
School of Engineering and Technology, K. R. Mangalam University, Gurugram, Haryana, India

Saurabh Yadav
Department of Computer Science and Engineering, National Institute of Technology Karnataka, Surathkal, India

ABBREVIATIONS

AI	artificial intelligence
ANN	artificial neural networks
ARM	association rule mining
ASL	American Sign Language
ASP	automatic speculative parallelization
AUT	antenna under test
BoG	Board of Governors
BSL	British Sign Language
BSs	base stations
CA	certification authority
CC	correlation coefficient
CH	cluster heads
CNN	convolutional neural network
CRL	certificate revocation list
CRM	customer relationship management
CVS	cooperative vehicle safety
DAG	directed acyclic graph
DAO	DODAG advertisement object
DDoS	distributed denial of service
DEC	deterministic energy-efficient clustering protocol
DHCP	Dynamic Host Configuration Protocol
DIO	DODAG information object
DIS	DODAG information solicitation
DM	data mining
DOM	document entity model
DoS	denial of service
DPA	Distributed Parallel Apriori
DSRC	dedicated short-range communication
DST	Department of Science and Technology
DTCWT	dual-tree complex wavelet transforms
ECC	elliptic curve cryptosystem
ERP	enterprise resource planning
FANETs	flying ad hoc networks
FBR	final beacon rate

FFNN	feed-forward neural network
FM	frequency modulation
FMOG	fuzzy mixture of Gaussian
FND	first node die
GA	genetic algorithm
GIS	geographic information system
GMM	Gaussian mixture model
GPS	global positioning system
HBM	human body model
HCF	hop count filtering
HCI	human-computer interaction
HDFS	Hadoop distributed file system
HEED	hybrid energy-efficiency distributed clustering
HND	half node die
HSM	hardware security module
IBRAC	intelligent beacon rate adaptation component
ICTs	internet and communication technologies
ILP	instruction-level-parallelism
IoMT	internet of medical things
IoT	internet of things
ISL	Indian Sign Language
ITS	intelligent transportation system
KDD	knowledge discovery databases
KNN	k-nearest neighbor
LLN	low power and lossy networks
LoS	line-of-sight
LSTM	long short-term memory
L2R	learning to rank
MAC	media access control
MANETs	mobile ad hoc networks
MAP	mean average precision
MCA	multicore chip architecture
MCI	Medical Council of India
MD	malicious domain
MINLP	mixed-integer nonlinear programming
ML	machine learning
MoHFW	Ministry of Health and Family Welfare
MPI	message passing interface
MWA	malicious web application

NSL	Nepali Sign Language
OBU	on-board unit
OS	operating system
PKI	public key infrastructure
PNN	probabilistic neural network
POS	part of speech
PR	precision-recall
QoS	quality of service
RA	router advertisement
ReLU	rectified linear unit
RFA	random forest algorithm
RLC	run length encoding
RNN	recurrent neural network
RoI	region of interest
RPL	routing protocol for low-power and lossy networks
RSU	roadside unit
SAR	search and rescue
SAR	specific absorption rate
sDiDi	stochastic distance discriminant
SIEC	swarm information exchange component
SLAP	succinct and lightweight authentication protocol
SLR	sign language recognition
SMT	simultaneous multithreading
SPAN	security protocol animator
SQMC	self queue monitoring component
SSL	secured shell
TA	trusted authority
TBR	temporary beacon rate
TCP	transmission control protocol
TD	trustworthy domain
TOF	time of flight
TPD	temper proof device
UAVs	unmanned aerial vehicles
UE	user equipment
uRPF	unicast reverse path forwarding
VANETs	vehicular ad hoc networks
VCC	vertex chain code
V WA	weak web application
VWPS	ventilator weaning prediction system

WBAN	wireless body area networks
WSNs	wireless sensor networks
WWW	World Wide Web
XML	extensible markup language
XSS	cross-site scripting
ZC	Zaslavsky chaotic

PREFACE

Advanced Computer Science Applications: Recent Trends in AI, Machine Learning, and Network Security consist of three parts. The first part focuses on machine learning algorithms in security analytics, which discusses the creation and training of supervised machine learning models for prediction and binary categorization tasks, including logistic regression and linear regression. In this, the reader will learn the fundamentals of machine learning. This part also explains how to apply these methods to create practical security applications. It offers a comprehensive introduction to contemporary machine learning and security analytics, covering green computing concepts, quality of software ecosystems based on transactions, survey on static images and videos, UAV-enabled disaster management, application layer concepts, and deployment techniques in IoT.

The reader may learn the key ideas by the conclusion of this specialization and acquire the practical skills necessary to apply machine learning and security applications quickly and effectively to difficult real-world problems. For the new machine learning specialization, it is best to start with security applications to break into a security career or develop a related career in the same field.

The second part of the book focuses on AI and machine learning, which will aid in understanding of artificial intelligence by combining theory and experience with machine learning. The reader will learn about pattern recognition, classification using web usage mining, AI-assisted applications, various machine learning algorithms applied on different applications, etc.

The third part of our book covers network security applications. The digital sphere now encompasses a vast majority of our existence. We use the internet to conduct business, stay in touch with loved ones and colleagues, make purchases, find entertainment, and conduct study. Birth dates, Social Security (or other identification) numbers, health histories, credit histories, bank accounts, utility bills, and a variety of other details are just a few examples of the confidential data we keep online. Hackers and cybercriminals could access all of that information and then conduct all of those transactions. The danger of compromise increases as we devote more of our lives to the internet. Additionally, the Internet of Things (IoT) will continue to be important, which means that wireless networks will be used more frequently.

This will only broaden the threat environment and give criminals more ways and opportunities to commit crimes. This part covers the concepts of IoT, security early detections for COVID-19, multimetric geographical routing in VANETs, V2X communication in VANET, and optimization of congestion control scheme for VANETs.

We hope that the information conveyed in these chapters will give the reader have a firm grasp of the methods in the field of artificial intelligence, machine learning, and security.

PART I

Machine Learning Algorithms in Security Analytics

CHAPTER 1

SPECULATIVE PARALLELISM ON MULTICORE CHIP ARCHITECTURE STRENGTHEN GREEN COMPUTING CONCEPT: A SURVEY

SUDHAKAR KUMAR[1], SUNIL KR. SINGH[1], and NAVEEN AGGARWAL[2]

[1]*CSE, CCET, Panjab University, Chandigarh, India*

[2]*CSE, UIET, Panjab University, Chandigarh, India*

ABSTRACT

In the present scenario, nobody wants to compromise on computational speed when it comes to competitive business, transferring information quickly in a communication link, or even imparting education from a distance, and many more. These requirements lead to high carbon emissions in the environment from a computing machine. The high carbon emissions are one of the main reasons behind green computing. This problem can be solved by speculative parallelism operating procedure. Speculative parallelism is the process of parallel execution aggressively of sequential applications into multiple parallel threads without relying on a compiler in advance. This technique generates different fragments of threads without considering any data and control inter-thread dependence. So, this concept has a substantial fast speed in execution as compared with ILP with multicore chip architecture (MCA). But if any data and control inter-thread dependencies found during execution,

Advanced Computer Science Applications: Recent Trends in AI, Machine Learning, and Network Security. Karan Singh, PhD, Latha Banda, PhD & Manisha Manjul, PhD (Eds.)

the whole execution of the sequential work is aborted and re-executed. These aborts are the victim of total lower performance in this concept. One has to deal with various overheads to limits the frequency of aborts leading to high performance and less power consumption. Also, speculative parallelism has less chip's energy consumption hence promoting green computing. In this chapter, a survey is presented highlighting the working of speculative parallelism techniques, overheads due to mis-speculation of parallel threads, performance, and finally efficient power consumption, which will encourage green computing.

1.1 INTRODUCTION

The uniprocessor system can also be called as "Single Core on Single Chip System," which can handle only one process at a time. But with the help of instruction-level-parallelism (ILP)[1–3] many processes can be executed in a single core too. ILP can reorder, pipeline instructions, can do branch prediction, and even split them into microinstructions. Thus, it can scale up the performance at the same time. But, ILP has an inherent problem[2,3] due to many factors like memory latency, dependent instruction, limitation in register renaming, imperfect prediction of prefetch at the start of executions, and memory-address alias problem, which leads for the search of even better solutions.

The architecture is scaled up from "Single Core on Single Chip system"[4] to "Multiple Cores on Single Chip system (MCA)"[5] and from ILP to thread-level-parallelism as devised from many researchers in this field. Increasing the cores on a single chip will increase the transistor counts, which in result increase the chip's heat and can hit the power wall[6] of the die's area. So, the excessive usage of cores into a single chip is not proper solution in performance hungry competition. Next architecture transition is involvement of thread-level parallelism into MCA with different level of memory hierarchy onto same chip. Here, threads that are in the forms of different fragments of a sequential process can be allocated to multicore for running at the same time. This will increase the overall speedup of the system as now same multicore are not only able to handle different parallel requests but also capable to schedule those parallel threads of a sequential workload efficiently.

Converting a sequential workload to explicit (independent) parallel threads[7–9] is done mostly through two ways. Firstly, with the use of compiler, which detect the explicit threads automatically and second way is to parallelize it manually with proper mechanism of synchronization. Finding the explicit or independent threads, either through compiler or manually is not easy and error prone. The reasons behind this are inter-thread data and control dependencies. Also, compilers sometimes act very conservative whenever they cannot assure the absence of any type of dependence. Moreover, as per the Amdahl's law,[10] only few fractions of whole sequential workload are parallel and hence only those parallel parts can be allocated to different cores not whole program. So, overall speedup of the system can be improved to one level only.

Speculative parallelism[11–14] is a one type of operating procedure in which the implicit (speculative or dependent) threads[15] are parallelized automatically keeping in view of original sequential semantic of the program contrasting on the other hand programmers has to take extra overhead of partitioning them into explicit threads. It is an optimistic technique of multithreading where a sequential workload is divided into implicit threads fragments automatically at runtime assuming there will be no any dependencies among them and same is allocated to different cores for execution. But if there will be any dependency detected at latter stage, corrective action is taken by software or hardware technique. Also, upon finding cross-thread dependency violation, the results of evaluated victim thread that consumed erroneous datum are to be discarded and offending implicit thread has to be squashed also a fresh execution is performed after returning the system to a previously nonspeculative correct state through software or hardware or combination of both mechanisms. In the next section, different types of speculative parallelism are discussed.

1.2 SOURCE OF AUTOMATIC SPECULATIVE PARALLELISM

Automatic speculative parallelism can be understood by its different techniques. These types are broadly divided on the basis of source of speculation given below.

(a) Loop as a source of Automatic Speculative Parallelization (ASP).[16]
(b) Procedure or module as a source of ASP.[17]

Loop as a source of ASP: It is the most easy and obvious way of the automatic partitioning of a sequential workload into speculative parallel threads, where loop is considered as one of the sources of speculation and also one of the working models for automatic parallelization technique. Loops are divided into blocks of iteration to make different speculative threads and executed in parallel fashion keeping the sequential semantic. If there will be mis-speculation, that is, error in sequential semantic, offending threads have to be squashed off and again a fresh execution is lodged with correct value.

Procedure or module as a source of ASP: In this type of ASP, function or procedure or subroutine is considered as main source of speculation. Every procedure or subroutine and code after every subroutine will be new thread, which will execute in parallel speculatively in different cores. The mis-speculation[17,18] occurs due to unwanted dependency. These may be either data dependency or control dependency violations. Data dependency violation can be of different forms namely RAW (Read After Write), WAR (Write After Read), and WAW (Write After Write).[19]

RAW data dependency occurs when one block of code namely B_1 or predecessor thread and another block of code namely B_2 or successor thread shares a shared location namely L_1, and B_2 or successor thread reads to L_1 before B_1 or predecessor thread writes to same shared location. This dependency is called flow dependency also.

$$\textit{Predecessor Thread: } \boldsymbol{T_j} \leftarrow \boldsymbol{T_i} + \boldsymbol{T_k} \tag{1.1}$$

$$\textit{Successor Thread: } \boldsymbol{T_l} \leftarrow \boldsymbol{T_j} + \boldsymbol{T_k} \tag{1.2}$$

Where i, j, k, l.... are in chronological order.

In WAR conflict, B_2 or successor thread try to write to a shared location, L_1, before B_1 or predecessor thread reads to same L_1. This dependency is called antidependency also.

$$\textit{Predecessor Thread: } \boldsymbol{T_j} \leftarrow \boldsymbol{T_i} + \boldsymbol{T_k} \tag{1.3}$$

$$\textit{Successor Thread: } \boldsymbol{T_k} \leftarrow \boldsymbol{T_l} + \boldsymbol{T_m} \tag{1.4}$$

And in WAW data dependency, B_2 or successor thread try to writes to the same location L_1 before B_1 or predecessor thread writes.

$$\textit{Predecessor Thread: } \boldsymbol{T_j} \leftarrow \boldsymbol{T_i} + \boldsymbol{T_k} \tag{1.5}$$

$$\textit{Successor Thread: } \boldsymbol{T_j} \leftarrow \boldsymbol{T_l} + \boldsymbol{T_m} \tag{1.6}$$

Based on these data dependencies conflict another set of loops description can be defined, which is more suited to the ASP. The different researchers defined these loops broadly into Doall, Forall, and Doacross loops.[20]

Doall Loops[21]: In this, all different blocks of loop fragments have no data dependency whatsoever and hence its name Doall loops means execute all block of loops in parallel speculatively.

Forall Loops: In this, two or more blocks of loops have RAW conflict, that is, the value generated by one iteration (Write operation) of the loop is read by next or latter loop iteration. This kind of loops can be executed at the same time if the producer iteration (that produce any variable into a shared location L_1) execute earlier than the consumer iteration. Otherwise, synchronization method is required.

Doacross Loops[22,23]: In this, two or more blocks of loops have WAR conflict. If the result generated by producer iteration is known at compile time before consumer iteration read a value, then the blocks of iteration can be executed speculatively keeping in view of sequential semantics. If not, the offending threads of iteration have to squashed off and execution has to restart again according to ASP.

Either it is for all or doacross loops, ASP is able to extract parallelism either automatically or at lower performance with squashing for offending threads and restarting the execution again. Next type of violation occurs when a job is wrongly branched with a mis-prediction popularly known as Control dependence.[24]

1.3 WORKING MODEL OF SPECULATIVE PARALLELISM

The working model of this paradigm can be found in different related works.[17,25,26] In both techniques as a source of speculation, either a loop of iteration or a module or code after a subroutine is an operating unit called a thread in speculative parallelism. The whole sequential workload is divided into different threads namely T_1, T_2, ...; and T_n executing in parallel on different cores namely C_1, C_2, ………, and C_n based on above techniques as mentioned. The execution of the whole sequential program starts with a single thread T_1, which is called nonspeculative in nature. After reaching to new iterations or new procedure, a new thread is spawned called speculative thread, T_2. Here, T_2 may be dependent or independent on T_1. Again, if another new iteration or new procedure will encounter, a new and more speculative thread, that is, T_3 will be spawned and so on. Also, T_3 can be dependent or nondependent on T_2 and T_1 both and so on. Thus, in automatic speculative parallelism whole sequential workload will be divided into two or many parallel speculative threads without considering their data or control dependence. But if any data dependence violation will be found during run time execution resulting in mis-speculation, the victim threads along with all its successors will be squashed and then re-execution of the threads is launched preserving the correct sequential order. Here, Only, nonspeculative threads are allowed to commit not speculative threads. In other words, a speculative thread like T_2 or T_3 can commit only if its predecessor like T_1 has committed.

With the assumption of MCA of four cores namely C_1, C_2, C_3, and C_4 and different threads (T_1 to T_7), which can execute in ASP manner in sequential order (as shown in Fig. 1.1). So, T_1 is first nonspeculative predecessor threads to be executed on C_1 and T_2 to T_4 are speculative successor threads to be scheduled and executed on C_2 to C_4, respectively, in speculative parallel environment until all four cores are totally occupied. These threads are supposed to execute concurrently on different cores and commit in sequential total order from 1 to 7.

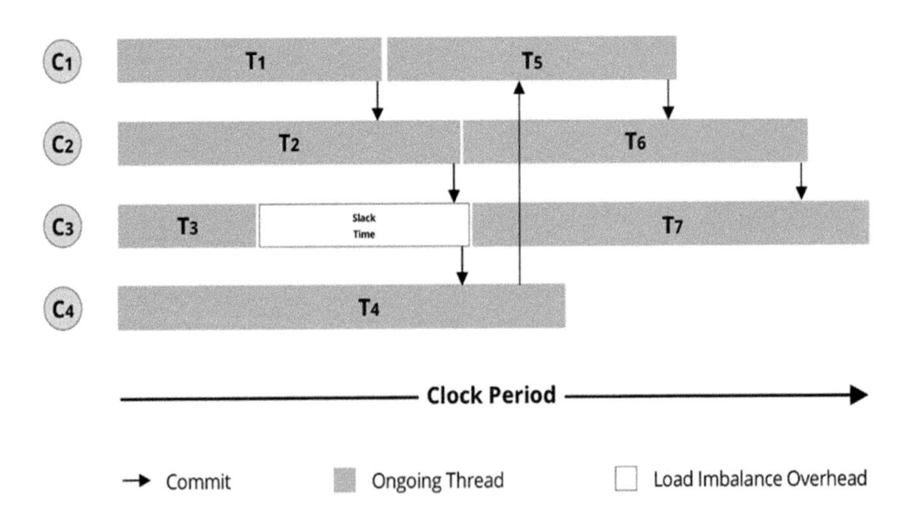

FIGURE 1.1 Successful speculations with four cores.

If there is no any dependence on predecessor thread, all successor threads will commit after nonspeculative threads (shown in Fig. 1.1) and all cores are free to take new sets of threads. So, finally, T_5 to T_7 will be allocated to cores from C_1 to C_4 again. The time between thread allocation to core and and waiting time for the commit of predecessor thread is slack time. In this case, the execution time(E_{Ti}) of thread does not only depends on its run time but also on the commit time (C_{Ti}) of previous allocated thread. So, commit time of successful speculation of *i*th thread with n-cores can be given by $\boldsymbol{C_{Ti} = max\{ C_{Ti-1}, C_{Ti-n} + E_{Ti} \}}$.

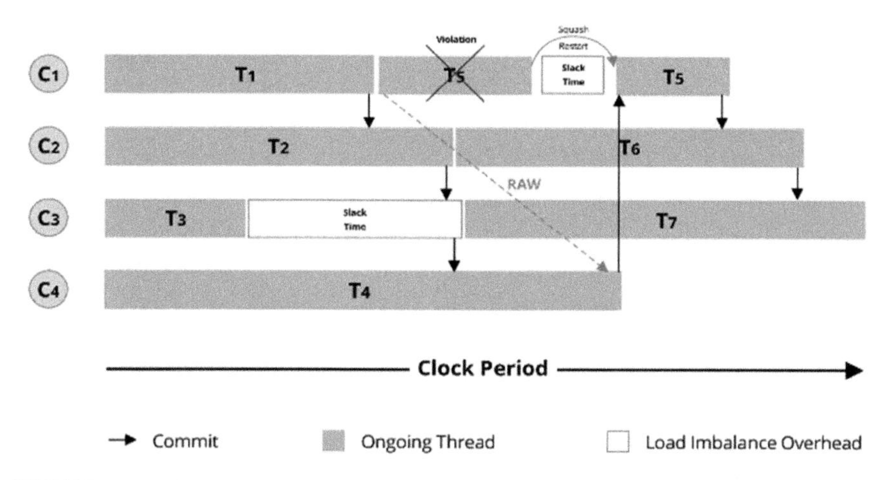

FIGURE 1.2 Mis-speculations with four cores.

But if there is mis-speculation, for example, between T_4 and T_5, a RAW dependency violation(shown in Fig. 1.2) can be detected and all these successor threads including offending threads will be squashed off as they might use hazardous data prior by the offending thread. Thus, T_5 will be reschedule to execute after the commit of T_4. A proper speculative hardware support is needed for the underlying detection of mis-speculation at run time.

1.4 OVERHEADS DUE TO MISSPECULATION

The overheads[27,28] can be defined as all unlike additional tasks that are required to perform with the intention to fix problems arisen due to mis-speculation in speculative parallelism. It can also be seen as all extra tasks that does not performed in execution of sequential instruction.

One type of compulsory overhead is due to thread start and commit. The thread start is actually extraction process of the threads from a sequential workload for speculative parallelism procedure. The commit operation is an exclusive process performed when a nonspeculative thread ends its execution as shown in Figure 1.1.

Other types can be roll-back overhead, squash overhead, load imbalance overhead, data versioning overhead, hardware overhead, communication and cache replacement overhead, which are not compulsory. Roll-back overhead is a result of data dependencies violation. The squash overhead is extra unwanted task performed by speculation mechanisms when all successor threads including offending threads are removed from the different cores as well as its data from speculative buffer after misspeculation as shown in Figure 1.2. The load imbalance overhead occurs when a speculative thread finishes its execution but wait for its predecessor thread(s) to become nonspeculative making the core unproductive as shown in Figures 1.1 and 1.2. Data versioning overhead occurs when different speculative tasks write to same variable which may generate different versions of the variable and thus these versions must be kept in buffer discretely in order to avoid data dependency violation. A special hardware support of cache hierarchy is required to impose data versioning and inspection of data dependence, resulting a hardware overhead. An intra-processor communication overhead happens where there is a need of cache replacement from L_1 cache or any memory components, which give rise to cache replacement overhead in order to re-execute the threads again with proper data in speculative buffer keeping the sequential semantic intact.

All mentioned overheads occur due to speculative parallelism procedure and can be minimized choosing proper and efficient speculation mechanisms protocols. This speculation mechanism protocol will be focused on avoiding minimum memory dependencies to the large possible extent with improved predictors and developing effective squashing techniques.

1.5 IMPACT OF MCA ON SPECULATIVE PARALLELISM

Increased cost, complicated operating system, and large main memory requirement lead to complex hardware and software in a multiprocessor. To elevates these problems in the multiprocessor system, the researchers moved to a better solution like multicore architecture. Unlike the "Single Core on Single Chip system" that is not of much use as per the many tasks is a concern, "Multiple cores on Single Chip" are devised as the solution and is a new trend architecture. This architecture is also called as MCA.[29–31] It is a special kind of tightly coupled multiprocessor system where all processors are on a chip or die. These different cores can execute multiple instructions that may reside in different parts of shared memory (multiple data).

1.6 IMPACT OF SIMULTANEOUS MULTITHREADING (SMT) ON SPECULATIVE PARALLELISM

The A thread can be thought of as the segment of a process or the whole process can be a thread. Actually, a process can be divided into more than one thread. Threads could be a part of the same program or different program. These threads can be allocated to different resources of a CPU or different CPUs for execution. Hence, these threads can execute concurrently and hence it will improve the resulting performance. This mechanism is known as "multithreading" or Thread Level Parallelism". If one thread encounters a long latency operation, other threads can utilize the CPU's resources efficiently for doing useful work. So, it has inherent quality for tolerating the latency caused by memory. Also, it reduces the context switch penalty. Coarse-grained, fine-grained, and SMT are three categories of multithreading basically. In coarse-grained multithreading, the switching between processes happens only during the costly stall of a process and not at every clock cycle. But by coarse-grained multithreading, the other processes may starve to get the CPU hence less throughput. In fine-grained multithreading, the CPU switches to different threads of different processes in round-robin fashion generally on every clock cycle causing the execution of each process in an interleaved manner. So, no process has to stall long at the start of execution but the execution of individual process delay. The third type of multithreading is SMT. The researchers also devised a solution by inducing the parallelisms on a coarser scale called simultaneous thread-level parallelism (SMT).[32] In simultaneous thread-level parallelism is a microarchitectural paradigm where as many as independent threads will be allowed to execute in the same core. SMT has a single copy of each resource available to a single core. It allows multiple instructions' threads to be fetched and execute simultaneously, thus, it eats resources more efficiently, and both instruction throughput and speedups are greater in the same pipeline.

With these two new architectures of multicore on single chip and SMT, researchers come up with the duo combination of multicore with SMT, where threads can be allocated to multi-cores which can be executed simultaneously. Now the number of SMT with the help of multicore can be as high as the programmer wants, thus this type of phenomenon can be termed as "hyperthreading"[33] by Intel Inc.

1.7 PERFORMANCE OF SPECULATIVE PARALLELISM

Despite of all mentioned overheads that occurs because of mis-speculation, the automatic speculative parallelism with MCA takes less execution time for a speculative loop in comparison of the sequential loop on a single processor. It is evident from the various research studies[34–38] related to automatic speculative parallelism that has clearly achieve significant speedups.

1.8 POWER CONSUMPTION BY SPECULATIVE PARALLELISM

Energy-efficient architecture means the architecture that consumes less power. Many researchers noticed that with mis-speculation overheads there is an increase in power consumption.[39,40] In the speculative parallelism paradigm, the power consumption wrongly understood by many researchers due to mis-speculation only.[41–47] There are many sources of energy consumption in this architecture which considered as energy hungry. These sources are due to the overheads occurred due to mis-speculation like thread roll-back, squash, data versioning, hardware, communication, and cache replacement overheads. But many researchers have shown that in spite of these overheads also there is a scope of energy-efficient as well as performance-oriented speculative parallelism optimizations. For example, communication latency can be decreased with little improvement of cache hierarchy and coherence protocol aimed for speculative buffer and data versioning with less thermal consumption. These architectures can be classified as energy-centric architecture which motivate to promote green computing concept.

1.9 CONCLUSIONS

Automatic speculative parallelism is capable of extracting parallel threads in hard-to-parallelize sequential workload aggressively without considering any data or control dependence violation analysis. This paradigm is skilled to enhance speedup excellently as compared with the sequential loop on a single processor. However, this operating procedure is highly susceptible to mis-speculation, which occur at run-time. Upon mis-speculation, nonspeculative threads have to abort its execution along with all its successor threads and re-executed again. This leads to various overheads in this concept of concurrent execution of threads hence decreasing the overall speedup and high thermal consumption. Nevertheless, this survey disproves the profess that speculative parallelism eats excessive energy. With slight impact in execution speed, there is a wide scope of energy-saving and performance-oriented optimizations. These energy-centric architectures will definitely encourage green computing and at the same it will catch the required speedup in real scenario.

KEYWORDS

- **speculative**
- **parallelism**
- **mis-speculation**
- **overheads**
- **speedup power efficiency and green computing**

REFERENCES

1. Rau, B. R.; Fisher, J. A. Instruction-Level Parallel Processing: History, Overview, and Perspective. *J. Supercomput.* **1993,** *7* (1), 9–50.
2. Wall, D. W. Limits of Instruction-Level Parallelism. *SIGARCH Comput. Archit. News* **1991,** *19* (2), 176–188.
3. Postiff, M. A.; Greene, D. A.; Tyson, G. S.; and Mudge, T. N.; The Limits of Instruction Level Parallelism in SPEC95 Applications. *SIGARCH Comput. Arch. News* **1999,** *27* (1), 31–34.
4. Yao, W.; Wang, D.; Zheng, W.; Guo, S. Architecture Design of a Single-Chip Multiprocessor. In *Current Trends in High Performance Computing and Its Applications*; 2005; pp 165–174.
5. Wang, Y.; An, H.; Liu, Z.; Li, L.; Huang, J. A Flexible Chip Multiprocessor Simulator Dedicated for Thread Level Speculation. In *2016 IEEE Trustcom/BigDataSE/ISPA*; 2016; pp 2127–2132.
6. Villa, O. et al. Scaling the Power Wall: A Path to Exascale. In *SC '14: Proceedings of the International Conference for High Performance Computing, Networking, Storage and Analysis*; 2014; pp 830–841.
7. Ungerer, T.; Robič, B.; Šilc, J. A Survey of Processors with Explicit Multithreading. *ACM Comput. Surv.* **2003,** *35* (1), 29–63.
8. Naishlos, D.; Nuzman, J.; Tseng, C.-W.; Vishkin, U. Towards a First Vertical Prototyping of an Extremely Fine-Grained Parallel Programming Approach. In *Proceedings of the Thirteenth Annual ACM Symposium on Parallel Algorithms and Architectures*, 2001; pp 93–102.
9. Vishkin, U.; Dascal, S.; Berkovich, E.; Nuzman, J. Explicit Multi-Threading (XMT) Bridging Models for Instruction Parallelism (Extended Abstract). In *Proceedings of*

the Tenth Annual ACM Symposium on Parallel Algorithms and Architectures, 1998; pp 140–151.

10. Hill, M. D.; Marty, M. R. Amdahl's Law in the Multicore Era. *Computer (Long. Beach. Calif).* **2008,** *41* (7), 33–38.
11. Prabhu, M. K.; Olukotun, K. Using Thread-Level Speculation to Simplify Manual Parallelization. In *Proceedings of the Ninth ACM SIGPLAN Symposium on Principles and Practice of Parallel Programming*, 2003; pp 1–12.
12. Blake, G.; Dreslinski, R. G.; Mudge, T.; Flautner, K. Evolution of Thread-Level Parallelism in Desktop Applications. In *Proceedings of the 37th Annual International Symposium on Computer Architecture*, 2010; pp 302–313.
13. Ioannou, N.; Cintra, M. Complementing User-Level Coarse-Grain Parallelism with Implicit Speculative Parallelism. In *Proceedings of the 44th Annual IEEE/ACM International Symposium on Microarchitecture*, 2011; pp 284–295.
14. Jeffrey, M. C.; Ying, V. A.; Subramanian, S.; Lee, H. R.; Emer, J.; Sanchez, D. Harmonizing Speculative and Non-Speculative Execution in Architectures for Ordered Parallelism. In *Proceedings of the 51st Annual IEEE/ACM International Symposium on Microarchitecture*, 2018; pp 217–230.
15. Ooi, C.-L.; Kim, S. W.; Park, I.; Eigenmann, R.; Falsafi, B.; Vijaykumar, T. N. Multiplex: Unifying Conventional and Speculative Thread-Level Parallelism on a Chip Multiprocessor. In *Proceedings of the 15th International Conference on Supercomputing*, 2001; pp 368–380.
16. Bhattacharyya, A.; Amaral, J. N. Automatic Speculative Parallelization of Loops Using Polyhedral Dependence Analysis. In *Proceedings of the First International Workshop on Code OptimiSation for MultI and Many Cores*, 2013.
17. Warg, F.; Stenstrom, P. Reducing Misspeculation Overhead for Module-Level Speculative Execution. In *Proceedings of the 2nd Conference on Computing Frontiers*, 2005; pp 289–298.
18. Bhattacharyya, A.; Amaral, J. N.; Finkel, H. Data-Dependence Profiling to Enable Safe Thread Level Speculation. In *Proceedings of the 25th Annual International Conference on Computer Science and Software Engineering*, 2015; pp 91–100.
19. Omar, R.; El-Mahdy, A.; Rohou, E. IR-Level Dynamic Data Dependence Using Abstract Interpretation Towards Speculative Parallelization. *IEEE Access* **2020,** *8*, 99910–99921.
20. Polychronopoulos, C. D.; Kuck, D. J. Guided Self-Scheduling: A Practical Scheduling Scheme for Parallel Supercomputers. *IEEE Trans. Comput.* **1987,** *C–36* (12), 1425–1439.
21. Kim, H.; Johnson, N. P.; Lee, J. W.; Mahlke, S. A.; August, D. I. Automatic Speculative DOALL for Clusters. In *Proceedings of the Tenth International Symposium on Code Generation and Optimization*, 2012; pp 94–103.
22. Chen, D. K.; Torrellas, J.; Yew, P. C. An Efficient Algorithm for the Run-Time Parallelization of DOACROSS Loops. In *Proceedings of the 1994 ACM/IEEE Conference on Supercomputing*, 1994; pp 518–527.
23. Cui, Y.; Liu, S.; Zou, N.; Wu, W. A Dynamic Parallel Strategy for DOACROSS Loops. In *Proceedings of the International Conference on High Performance Computing in Asia-Pacific Region*, 2018; pp 108–115.
24. Lam, M. S.; Wilson, R. P. Limits of Control Flow on Parallelism. In *Proceedings of the 19th Annual International Symposium on Computer Architecture*, 1992; pp 46–57.
25. Xekalakis, P.; Ioannou, N.; Cintra, M. Mixed Speculative Multithreaded Execution Models. *ACM Trans. Arch. Code Optim.* **2012,** *9*, 3.

26. Ying, V. A.; Jeffrey, M. C.; Sanchez, D. T4: Compiling Sequential Code for Effective Speculative Parallelization in Hardware. In *Proceedings of the ACM/IEEE 47th Annual International Symposium on Computer Architecture*, 2020; pp 159–172.
27. Radulović, M. B.; Tomašević, M. V.; Milutinović, V. M. Chapter One—Register-Level Communication in Speculative Chip Multiprocessors; Hurson, A., Ed., , Vol. 92; Elsevier, 2014; pp 1–66.
28. Radulovic, M. B.; Tomasevic, M. V. Towards an Improved Integrated Coherence and Speculation Protocol. In *EUROCON 2007—The International Conference on Computer as a Tool*, 2007; pp 405–412.
29. Lee, S.; Tuck, J. Automatic Parallelization of Fine-Grained Metafunctions on a Chip Multiprocessor. *ACM Trans. Arch. Code Optim.* **2013,** *10*, 4.
30. Hammond, L.; Willey, M.; Olukotun, K. Data Speculation Support for a Chip Multiprocessor. In *Proceedings of the Eighth International Conference on Architectural Support for Programming Languages and Operating Systems*, 1998; pp 58–69.
31. Krishnan, V.; Torrellas, J. Hardware and Software Support for Speculative Execution of Sequential Binaries on a Chip-Multiprocessor. In *Proceedings of the 12th International Conference on Supercomputing*, 1998; pp 85–92.
32. Eggers, S. J.; Emer, J. S.; Levy, H. M.; Lo, J. L.; Stamm, R. L.; Tullsen, D. M.; Simultaneous Multithreading: A Platform for Next-Generation Processors. *IEEE Micro* **1997,** *17* (5), 12–19.
33. Koufaty, D.; Marr, D. T. Hyperthreading Technology in the Netburst Microarchitecture. *IEEE Micro* **2003,** *23* (2), 56–65.
34. Olukotun, K.; Hammond, L.; Willey, M. Improving the Performance of Speculatively Parallel Applications on the Hydra CMP. In *Proceedings of the 13th International Conference on Supercomputing*, 1999; pp 21–30.
35. Prabhu, M. K.; Olukotun, K. Exposing Speculative Thread Parallelism in SPEC2000. In *Proceedings of the Tenth ACM SIGPLAN Symposium on Principles and Practice of Parallel Programming*, 2005; pp 142–152.
36. Prabhu, M. K.; Olukotun, K. Using Thread-Level Speculation to Simplify Manual Parallelization. In *Proceedings of the Ninth ACM SIGPLAN Symposium on Principles and Practice of Parallel Programming*, 2003; pp 1–12.
37. Packirisamy, V.; Luo, Y.; Hung, W-L.; Zhai, A. Yew, P-C.; Ngai, T-F. Efficiency of Thread-Level Speculation in SMT and CMP Architectures—Performance, Power and Thermal Perspective. In *2008 IEEE International Conference on Computer Design*, 2008; pp 286–293.
38. Bhattacharjee, A.; Martonosi, M. Thread Criticality Predictors for Dynamic Performance, Power, and Resource Management in Chip Multiprocessors. In *Proceedings of the 36th Annual International Symposium on Computer Architecture*, 2009; pp 290–301.
39. Co, M.; Weikle, D. A. B.; Skadron, K. Evaluating Trace Cache Energy Efficiency. *ACM Trans. Arch. Code Optim.* **2006,** *3* (4), 450–476.
40. Das, B.; Dalui, M.; Mondal, A.; Mandi, S.; Das, N.; Sikdar, B. K. Evaluation of Misspeculation Impact on Chip-Multiprocessors Power Overhead. In *Proceedings of the 2018 7th International Conference on Software and Computer Applications*, 2018; pp 129–133.
41. Cai, G. Z. N. Power-Sensitive Multithreaded Architecture. In *Proceedings of the 2000 IEEE International Conference on Computer Design: VLSI in Computers & Processors*, 2000; p 199.

42. Tanaka, Y.; Sato, T.; Koushiro, T. The Potential in Energy Efficiency of a Speculative Chip-Multiprocessor. In *Proceedings of the Sixteenth Annual ACM Symposium on Parallelism in Algorithms and Architectures*, 2004; pp 273–274.
43. Renau, J. et al. Thread-Level Speculation on a CMP Can Be Energy Efficient. In *Proceedings of the 19th Annual International Conference on Supercomputing*, 2005; pp 219–228.
44. Tuck, J.; Liu, W.; Torrellas, J. CAP: Criticality Analysis for Power-Efficient Speculative Multithreading. In *2007 25th International Conference on Computer Design*, 2007; pp 409–416.
45. Luo, Y.; Packirisamy, V.; Hsu, W.-C.; Zhai, A. Energy Efficient Speculative Threads: Dynamic Thread Allocation in Same-ISA Heterogeneous Multicore Systems. In *Proceedings of the 19th International Conference on Parallel Architectures and Compilation Techniques*, 2010; pp 453–464.
46. Li, P.; Guo, S. Energy Minimization on Thread-Level Speculation in Multicore Systems. In *2010 Ninth International Symposium on Parallel and Distributed Computing*, 2010; pp 125–132.
47. Luo, Y.; Hsu, W.-C.; Zhai, A. The Design and Implementation of Heterogeneous Multicore Systems for Energy-Efficient Speculative Thread Execution. *ACM Trans. Arch. Code Optim.* **2013,** *10*, 4.

CHAPTER 2

MEASURING PERCEIVED QUALITY OF SOFTWARE ECOSYSTEM BASED ON TRANSACTIONS IN CUSTOMER MANAGEMENT TOOLS

RAJARAMAN R.[1], KAPUR P. K.[2], DEEPAK KUMAR[1], and VELMURUGAN K.[3]

[1]*Amity University, Amity Institute of Information Technology, Noida, Uttar Pradesh, India*

[2]*Amity University, Amity Centre for Interdisciplinary Research, Noida, Uttar Pradesh, India*

[3]*Anjalai Ammal Mahalingam Engineering College, Kovilvenni, Thiruvarur, Tamilnadu, India*

ABSTRACT

Every software vendor uses tools for managing transactions with customer. perceived quality is customer satisfaction criteria with overall system quality. In software metrics terminology, general definition of perceived quality is the logical feel of the system lying between system assurance and product(s) reliability. We would like to propose a model for perceived quality as a customer point of view measured using Weibull distribution. Our study is not limited to defects but including all the transactions involved. We are considering intensity, time between requests, active number at any point

Advanced Computer Science Applications: Recent Trends in AI, Machine Learning, and Network Security. Karan Singh, PhD, Latha Banda, PhD & Manisha Manjul, PhD (Eds.)

of time, and timeline of each transactions. The distribution can reveal the quality perspective and focus areas for improving the perception.

2.1 INTRODUCTION

There are numerous studies about quality measurement and quality improvement strategies. Presence of reliability growth models based on the failure intensity of the system under test is highly quantitative and can be used easily within the purview of development teams. In this chapter, the study and proposed model provide a qualitative feel for the customer using quantitative factors readily available without taking additional. We will be applying Weibull distribution on the quantitative data. Three quality parameters are considered here. Each of the parameter can individually provide necessary metrics. A combination of metrics is created here for the perception measurement: open requests, time taken to close a transaction, time between transactions.

2.2 LITERATURE REVIEW

The research areas reviewed are focused mostly on existing metrics for software quality and statistical models on software reliability. Analyzing different defect-oriented models available for study and the application in the industry. Tools widely used in software supply chain and software maintenance are also noted.

Heiju et al.[1] proposed a combinatorial model for combining product quality, service quality, and experience quality. Brady[2] used the term perceived service quality. Yap[7] analyzed various type of normality tests. Our study commenced by analyzing normal and non-normal distributions for the available data. We are able to conclude through the articles available that requests are mostly mapping to non-normal distribution.

The articles on service quality, different models depicting the service quality, and customer satisfaction relationships are studied. Zang[3] proposed defect-based quality evolution metrics. There is a clear depiction of tangential differences between the measurements used for defect based product quality and communication-oriented service quality. The customer satisfaction and loyalty-powered quality adjustments can be seen in various studies across industrial sectors. The third most important factor we studied is about the academic references on the different tools used by industries for various interactions with customer.

2.3 ASSUMPTIONS AND DEPENDENCIES

Here is the list of assumptions and prerequisites we would like to enlist for research focus, which help us in determining the necessary parameters. We assume the vendor is managing customer interactions through a tool where we can deduce the data. If the software vendor maintains separate tools for Defect tracking and different tool for other interactions with the customer, we should be able to combine all the activities sorted by time.

Vendor might be delivering few or more software applications as a part of ecosystem. Vendor should consider customer as a central entity for calculating all the transactions. Excel spreadsheet formulae and R programming language are used for all the calculations and necessary curve plot. The calculations can also be derived without using this tool.

2.4 PERCEIVED QUALITY

The term perceive quality denotes in plain English term what is the perception of overall system quality provided by vendor. We consider perceived quality as critical part of customer satisfaction. It helps in determining the continuation of a business deal along with other factors like loyalty relationships, etc. Lalband[15] introduced software development technique for improving customer satisfaction. Our approach will be proposing a similar goal with more descriptive data received from transactions under study.

Though defects oriented models can give a better quantitative description of reliability, customer might be having a mixed/different opinion compared to what vendor has. Customers may or may not understand the derivations involved in calculating quality metrics, but they can easily tell you whether they are happy or unhappy with the system.

Any product quality strategies always involve the combination of calculations and customer targeted surveys or other form of interactions. Our attempt here is to pre-empt the customer feel before even we get into any surveys.

2.5 CUSTOMER RELATIONSHIP MANAGEMENT (CRM) AND ENTERPRISE RESOURCE PLANNING (ERP) TOOLS

Customer satisfaction is the primary goal of any business, be it a product or a service. The term satisfaction is very qualitative in nature and depends on the three major factors. The first factor is delivery, second factor is performance of the product or service, and third factor is how customer is equipped to handle their interactions. Many of the software vendors utilize Tools for managing customer interactions. The term "customer management" is used irrespective of type of tool. Two prominent categories of this tool are CRM and ERP. These tools are used for different purposes say—sales, marketing, etc. We will be focusing on the product-oriented transactions starting from an Order, followed by delivery and software maintenance. Figure 2.1 shows the transactions under consideration.

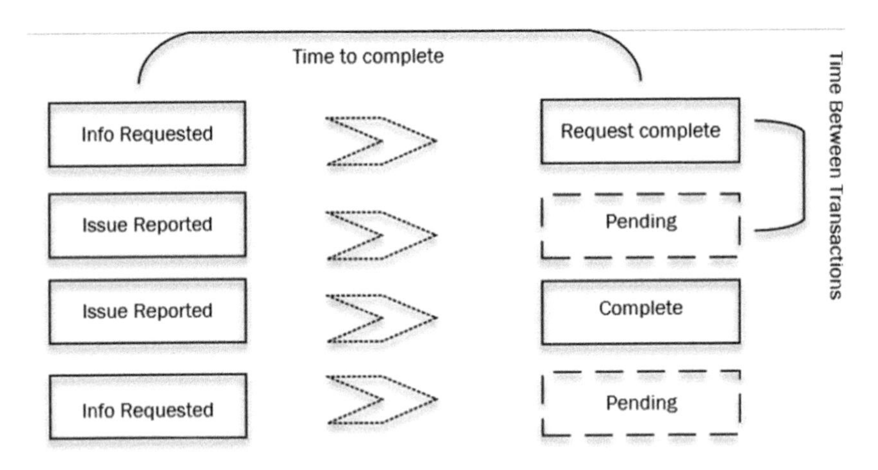

FIGURE 2.1 Customer transactions sample.

Ekatrina[12] introduced a detailed study on queues pending with customers and their relationship with impatience caused. Our study takes these scores into consideration.

2.6 MEASURING PERCEIVED QUALITY

The study of distributions applied in the field of software reliability with normal and non-normal distributions of data. Kapur et al.[10] proposed optimal strategies of time-to-market using multiattribute utility theories. There are numerous distribution driven and defect oriented models available for study. Emerging models are targeting Bayesian and Weibull-based approaches. Niveditha[4] proposed a two-parameter Weibull distribution for measuring six sigma quality based on lifetime test data. The proposal is to apply a two-parameter Weibull distribution for our approach. The primary reason for us choosing Weibull distribution over others due to its existential practice in wide array of industrial applications. We are considering centered Weibull process. A continuous random variable X is said to follow Weibull distribution with parameters η and β if its probability density function is given by

f(t)=βη(tη)β−1e−(tη)β

where

$f(t) \geq 0,\ t \geq \gamma$
$\beta > 0$
η= scale parameter, or characteristic life
β= shape parameter (or slope)

The mean and variance of Weibull distribution can be expressed as:

$$\text{Mean } \mu_w = \beta\Gamma\left(1+\frac{1}{k}\right)$$

$$\text{Variance } \sigma_w^2 = \beta^2\left[\Gamma\left(1+\frac{2}{k}\right)-\left\{1+\frac{1}{k}\right\}^2\right]$$

The reliability function of Weibull distribution is given by

f(*t*)=*βη*(*tη*)*β*−1e−(tη)β

The mean (also called *MTTF*) of the Weibull *pdf* is given by

MTTF =*η*·Γ(1*β*+1)

Our proposal is to classify the transactions into different parts and mapping them with the quality parameters. The equation based on the number of transactions and other parameters is as follows. At any given point of time.

$$\text{Request Intensity (R.I)} = \text{Load} \times \text{MTTR} / \text{MTBF}$$

Whereas load is the ratio of open transactions to closed transactions for a time period.

MTTR is the mean time to close a transaction, and MTBF is the mean time between transactions.

The primary reason for considering load transactional data is the direct proportion to positive feel. Table 2.1 illustrates the feel factor with respect to different parameter.

TABLE 2.1 Customer Feel Over Quality Parameters.

Customer feel over parameter values	More/high value	Less/low value
Number of open transactions	Not Happy	Happy
Time to close a request	Not Happy	Happy
Time between requests	Happy	Not Happy

Let N: Total number of requests for a given time period
N1: Average active requests at any given point of time.
N2:
T: Average time to close a request
T1: Average time to close request with some actions
T2: Average time to close a request without any action
B1: Average time between two requests.

Steps involved:

1. Getting the sample values of N, N1, N2, T, T1, T2, B1
2. Calculating k and β
3. Formulating Weibull distribution Table
4. Calculating MLE
5. Calculating tail probabilities
6. Target mean factor μw
7. Mapping tail probabilities to Satisfaction factor considering target mean as "customer is ok."
8. Repeating the steps with all the available data.

The equation that can be applied for finalizing the Weibull parameters is:

R.I = (No. of Open Txns.)/(Total Txns) × (Mean Time to close a Txns.) / (Mean Time Between Transactions).

Two curves are plotted using the data collected. The first curve is about number of pending requests in a weekly interval. Figure 2.2 depicts the simple box plot.

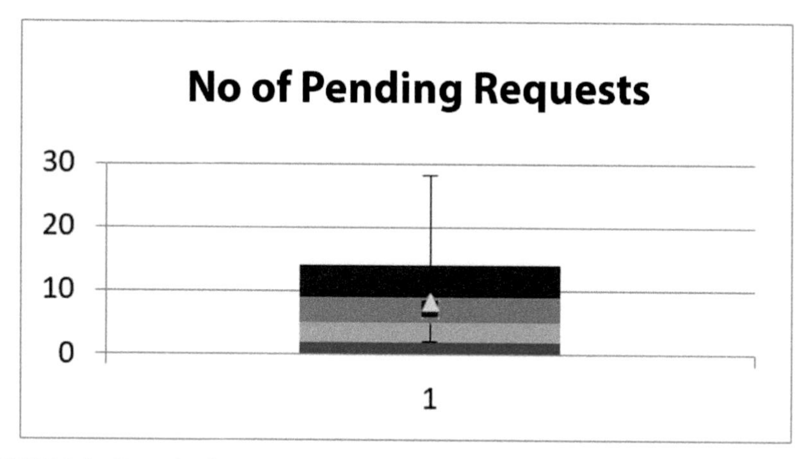

FIGURE 2.2 Box plot for number of pending requests.

The second curve is about the average time to complete a request week on week. Figure 2.3 depicts the box plot.

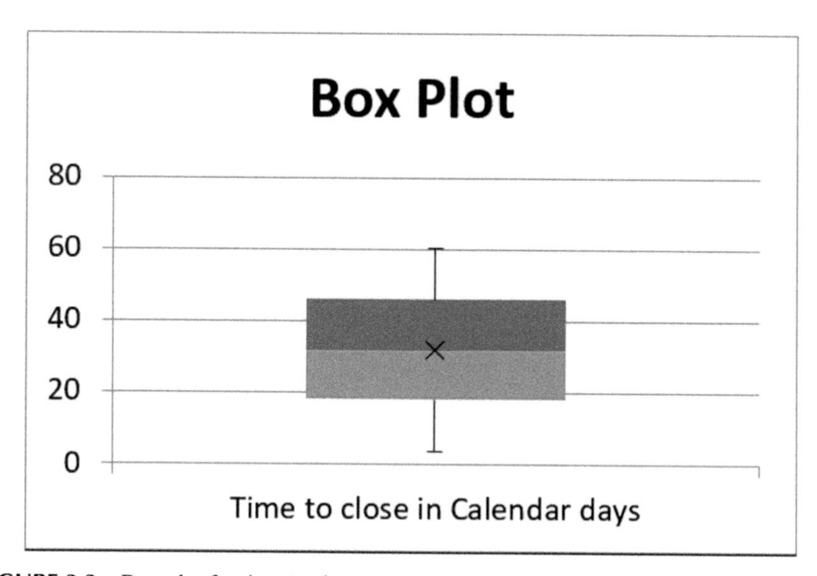

FIGURE 2.3 Box plot for time to close.

Figure 2.4 shows the normal fitting of time to complete.

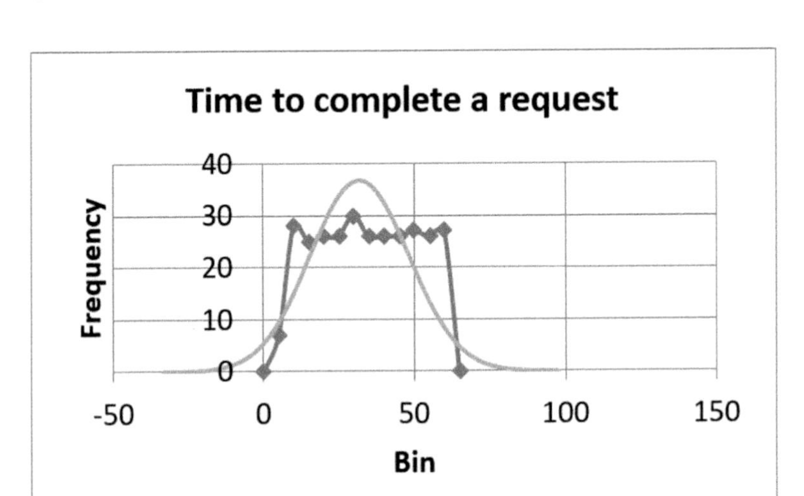

FIGURE 2.4 Normal distribution mapping for time to complete.

Figure 2.5 depicts the Weibull distribution from the sample A of 30 transactions from a customer A. Here the combination of open requests is certainly not decreasing.

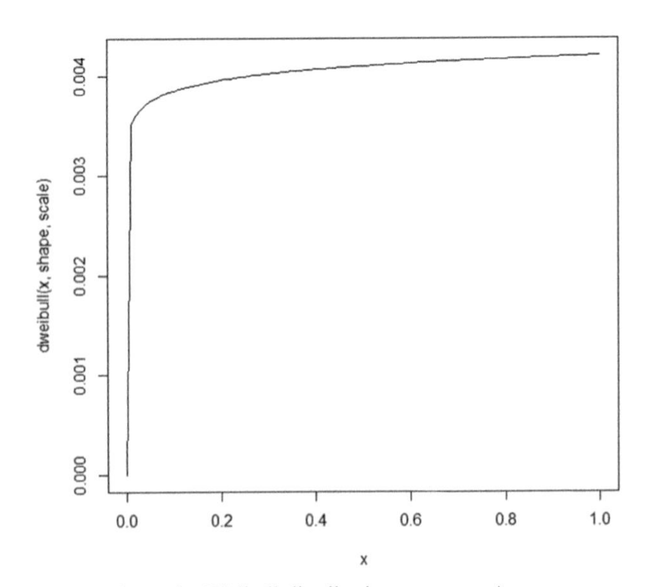

FIGURE 2.5 Request intensity Weibull distribution customer A.

Figure 2.6 shows the Weibull distribution from the sample B of 30 transactions from a customer B. We can conclude the intensity is reducing over the period of time.

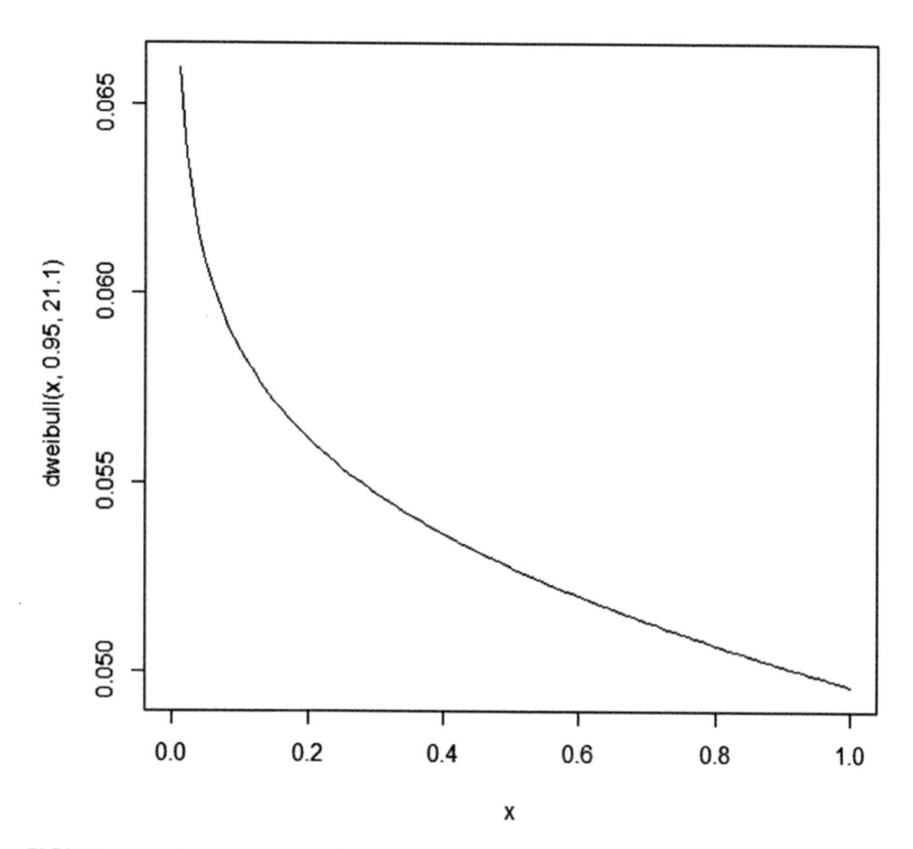

FIGURE 2.6 Request intensity Weibull distribution customer B.

Application of the request intensity calculation and formulation of a Weibull model is the last step in our Approach. Niveditha[4] described the procedure for obtaining six sigma metrics like DPMO (defects per million op) and EPGMO (extremely good per million opportunities). We infer similar practices here without referring to Six Sigma metrics but not matching them for direct calculations. The term positive perception can be equivalent of EPGMO and negative perception can be equivalent of DPMO.

2.7 APPLICATION AND USAGE

Perceived quality can be applied to determine the overall customer satisfaction. It can depict both product quality and service quality of the vendor.

What in general the customer like to see:

1. Faster resolution time
2. Quick response (not just automated emails)
3. Better training
4. Reduced dependencies

This can be calculated either from vendor side or customer side. We can easily convert the result into strategy for mapping customer specific needs as parametric or configuration requirements, training requirements, product improvements, etc.

The model is tested and illustrated using a real data set containing 300 transactions. If target μw is on right-tail region meaning the quality index is showing negative which means customer might not be happy with the vendor. If target μw is on left-tail region, the customer is happy with the progress and perceived quality can be termed "positive" or "good." The study focus on a simple three-way gauge with "Bad," "Ok," and "Good." The values can also be drilled down for any scalar range we need. Figure 2.7 depicts the gauge report for visualizing the perceived quality.

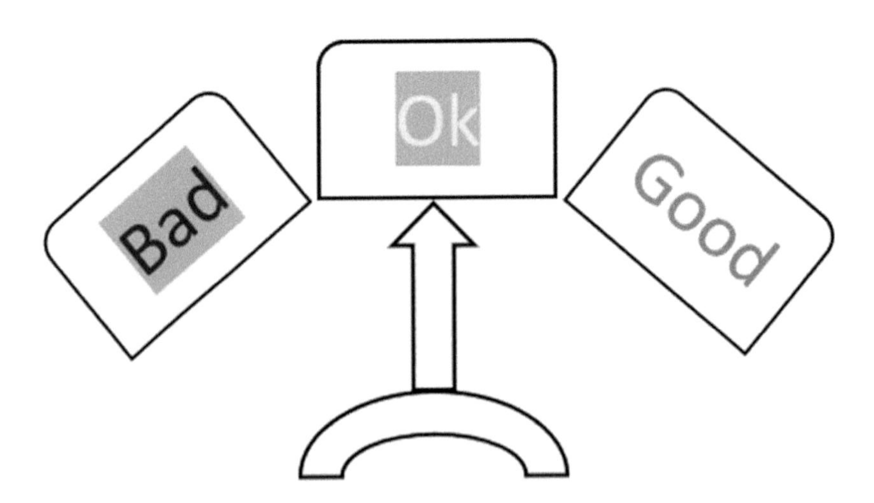

FIGURE 2.7 Perceived quality gauge report.

2.8 CONCLUSIONS

Perceived quality can be used along with existing metrics to determine the focus area. We can extend the calculations to drill down the data and derive the parameters at product level or component level. The quality management strategies can include perceived quality as a part of Business value analysis. The proposal suits well for the teams following time-bound measurements for deliveries like operational scrum teams. This can also suit Kanban teams provided the teams follow fixed queries for transactional data. The proposal can also be extended to identify the priority of customer expectations using decision-making models depending on the industry. Though the intention of the model is depiction of quality perception, We can also create a prediction of perception using the hazard function over the period of time.

KEYWORDS

- **reliability**
- **software quality**
- **customer relationship management (CRM)**
- **enterprise resource planning (ERP)**
- **weibull distribution**
- **software metrics**

REFERENCES

1. Hueiju,W. F. Relative Impacts from Product Quality, Service Quality, and Experience Quality on Customer Perceived Value and Intention to Shop for the Coffee Shop Market. *Total Qual. Manage. Busi. Excell.* **2009,** *20–11*, 1273–1285.
2. Brady, M. K.; Cronin. Some New Thoughts on Conceptualizing Perceived Service Quality: A Hierarchical Approach. *J. Market* **2001,** *65*, 37.
3. Zhang, H.; Kim, S. Monitoring Software Quality Evolution for Defects. In *IEEE Software* **2010,** *27* (4), 58–64.
4. Niveditha, A.; Joghee, R. Six Sigma Quality Evaluation of Life Test Data Based on Weibull Distribution. *Int. J. Qual. Reliab. Manage.* Aug **2020**.
5. Poth, A.; Sunyaev, A. Effective Quality Management: Value- and Risk-Based Software Quality Management. *IEEE Softw. 2014*, *31* (6), 79–85. DOI: 10.1109/MS.2013.138.
6. Singh, M.; Baranwal, G. Quality of Service (QoS) in Internet of Things. In *2018 3rd International Conference On Internet of Things: Smart Innovation and Usages (IoT-SIU)*; Bhimtal, 2018; pp 1–6, DOI: 10.1109/IoT-SIU.2018.8519862.

7. Yap, B. W.; Sim, C. H. Comparisons of Various Types of Normality Tests. *J. Stat. Comput. Simulation 2011*, *81* (12), 2141–2155.
8. Lanna, A.; Castro, T.; Alves, V.; Rodrigues, G.; Schobbens, P-Y.; Apel, S. Feature-Family-Based Reliability Analysis of Software Product Lines. In *Proceedings of the 23rd International Systems and Software Product Line Conference— Volume A* (*SPLC '19*). Association for Computing Machinery: New York, 2019; p 64.
9. Tickoo, A.; Kapur, P. K.; Shrivastava, A. K.; Khatri, S. K. Discrete-Time Framework for Determining Optimal Software Release and Patching Time. In *Quality, IT and Business Operations. Springer Proceedings in Business and Economics*; Kapur, P., Kumar, U., Verma, A., Eds.; Singapore, 2018. https://doi.org/10.1007/978–981–10–5577–5_11
10. Kapur, P. K.; Panwar, S.; Singh, O. et al. Joint Optimization of Software Time-to-Market and Testing Duration Using Multi-Attribute Utility Theory. *Ann Oper Res*. **2019**. https://doi.org/10.1007/s10479–019–03483-w
11. Singh, O.; Kapur, P. K.; Shrivastava, A. K. et al. Release Time Problem with Multiple Constraints. *Int. J. Syst. Assur. Eng. Manag.* **2015,** *6*, 83–91. https://doi.org/10.1007/s13198–014–0246–1
12. Evdokimova, E.; De Turck, K.; Fiems,D. Coupled Queues with Customer Impatience. *Perf. Eval.* **2018,** *118*, 33–47. ISSN 0166–5316. https://doi.org/10.1016/j.peva.2017.10.002.
13. Zerbino, P.; Aloini, D.; Dulmin, R.; Mininno, V. Big Data-enabled Customer Relationship Management: A Holistic Approach. *Inf. Process. Manage.* **2018,** *54* (5), 818–846. https://doi.org/10.1016/j.ipm.2017.10.005.
14. Pavlov, N.; Iliev,A.; Rahnev, A.; Kyurkchiev, N. On Some Nonstandard Software Reliability Models. *Dyn. Syst. App.* **2018,** *27* (4), 757–771.
15. Neelu, L.; Kavitha, D. Software Development Technique for the Betterment of End User Satisfaction Using Agile Methodology. *TEM J.* **2020,** *9* (3), 992–1002.
16. Durmic, N. Factors Influencing Project Success:A Qualitative Research. *TEM J.* **2020,** *9* (3), 1011–1020.
17. Gupta, V.; Kumar, D.; Kapur, P. K. Assessment of Quality Factors in Enterprise Application Integration. In International Conference on Reliability, Infocom Technologies and Optimization, 2015; pp 1–6.
18. Mallikarjuna, C.; Sudheer Babu, K.; Chitti Babu, P. A Report on the Analysis of Software Maintenance and Impact on Quality Factors. *Int. J. Eng. Sci. Res.—IJESR* **2014,** *05*, Article 01335.
19. Hui, Z.; Liu, X. Research on Software Reliability Growth Model Based on Gaussian New Distribution. In *3rd International Conference on Mechatronics and Intelligent Robotics*, 2019.
20. Pham, L.; Pham, H. Software Reliability Models with Time-Dependent Hazard Function Based on Bayesian Approach. In *IEEE Transactions On Systems, Man, and Cybernetics—Part A: Systems and Humans* **2000,** *30* (1).
21. Wooluru, Y.; Swamy, D. R.; Nagesh, P. Process Capability Estimation For Non-Normally Distributed Data Using Robust Methods—A Comparative Study. *Int. J. Qual. Res.* **2015,** *10* (2) 407–420.
22. Esaki, K. System Quality Requirement and Evaluation—Importance of Application of the ISO/IEC25000 series. *Global Perspect. Eng. Manage* **2013,** *2* (2), 52–59.
23. Rossi, B.; Russo, B.; Succi, G. Modelling Failures Occurrences of Open Source Software with Reliability Growth. *OSS* 2010, *IFIP AICT* **2010,** *319*, 268–280.

CHAPTER 3

MOVING OBJECT DETECTION IN VIDEO, CAPTURED BY STATIC CAMERA: A SURVEY

SNEHA MISHRA and DILEEP KUMAR YADAV

School of Computing Science and Engineering, Galgotias University, Greater Noida, Uttar Pradesh, India

ABSTRACT

This chapter presents survey regarding the study of object detection in videos or continuous sequence of frames captured by the static camera. Although many researches and good works have been performed in the area of computer vision, but this survey will make the clear image of comparison of object detection methods performed in prior researches. This chapter focuses on the real-time application areas of video surveillance and the concerning challenges in the respective areas in near future. All the background modeling techniques and their subcategories studied and implemented by various authors are mentioned in survey. This survey suggests the study of background modeling and background subtraction along with various other literature studies that justify the role of moving object detection in computer vision. It also depicts the major challenging issues available in real-time environment.

Advanced Computer Science Applications: Recent Trends in AI, Machine Learning, and Network Security. Karan Singh, PhD, Latha Banda, PhD & Manisha Manjul, PhD (Eds.)

3.1 INTRODUCTION

The gaining of high-level understanding from the images or videos is computer vision. It can automate the tasks, which the human eyes can do. Somehow, it is always not possible for human visuals to reach at every place to take a track of the work, like in deep mining activities and border security across the border, in any unfavorable climatic condition or unsound geographical areas. For such places, video surveillance is required to be performed. Object detection is the key application in computer vision. The object detection may further lead to many more applications of computer vision like object tracking, object recognition, activity detection, or behavioral analysis. In real-time scenarios, the captured video faces the following problems.

- Climatic changes such as the weather may be rainy, mist, dusty, foggy, or sunny.
- Illumination variation in which light may be dark or too bright.
- Background motion that may be due to swaying trees, sprinkling of water from fountain, or any other moving sources.
- Noise due to camera hardware.

Nowadays, manufacturing industries and agricultural department are also using video surveillance as there may be many human-made mistakes like inattentiveness, double vision, or blurred eyes. Detection, identification, motion, and simultaneous tracking of multiple objects or targets in coastal, navy, army, underwater, and indoor–outdoor video surveillance system is developing effectively in today's era. Various application areas of video surveillance may be manufacturing industry, sea surveillance, restricted areas, and leakage detections are tried to be described in Table 3.1.

TABLE 3.1 Real-Time Applications with Description.

Application areas	Description
Agriculture	To detect the type and size of the material or objects in agriculture is very important for their further classification, so as to detect the diseases or classification of breeds.
Thermal imaging	This gives the ability to detect in miscellaneous background. It provides ability to visualization in dark environment. Leakage detection in gas pipeline or underwater leakage and leakages in electric wire are also the application of thermal imaging.

TABLE 3.1 *(Continued)*

Application areas	Description
Security	In detecting the object that may be a crime suspect, the behavioral tracking of the suspect can give an alarm. The person climbing the fencing can be detected.
Manufacturing industry	The sorting and assembling in manufacturing industry is a very time taking task by human and again carries abnormalities and errors, object detection and tracking can ease the manufacturing world with less or no abnormalities.
Packaging industry	Packaging industry may include medicine packing and many retail objects packing, video surveillance may locate the object with no filling or partial filing of the packet. Human visuals may not be successful detectors in this industry.
Mining	Mining may be very deep under the earth where the human worker or expert is not possibly safe to go inside the deep penetrations for mining of coal or any other minerals, surveillance through camera is the best way to detect the object as much far it is.
Sea surveillance	Sea surveillance by visual cameras is to protect the ship or overseas borders, so the navy is alarmed before any danger.
Restricted zone	Restricted zones may be some deep zones, or dense forest zones where wild animals or terror activities are performed may be tracked by surveillance cameras.
Transportation	Traffic surveillance plays the major role in controlling the accidents and over speeding track of the vehicles. If the camera is mounted on the vehicle then it is much easy for the driver to get alarms as soon as the object, pedestrian, or any vehicle is near to the vehicle to save it from accident.
Army	Video surveillance in military purposes can reduce the offences like crimes in the nation, border security is major job of army that can be better controlled by border surveillance by thermal imaging or infrared imaging.

3.2 MOTIVATION

In this survey, the motivation is dynamic background as this requires a change in consecutive frames due to moving objects. Thus it needs to upgrade the background model every time and then detect the foreground. Even in indoor scenes where the environment is controlled, the shadow or light effects may be undesired. There are various real-time application areas of video surveillance like packaging.

Object detection in videos requires robust system rather it is simple to detect an object when the background is static. There are plenty of approaches by which object detection may be performed like optical flow, background subtraction, and filtering techniques. The background subtraction is the most used technique and so is to be concerned in this chapter also. The main goal is to detect the changes between consecutive frames. The idea behind this concept is that no information prior is necessary to classify the state of a pixel as foreground or background object. So the main objective is to resolve these challenging issues and detect the moving or stationary pixels of the scene.

3.3 LITERATURE WORK

In earlier decades, many researchers have applied nonadaptive methods of backgrounding, which have lots of drawbacks. These methods were required without any changes in the video scene and only for highly supervised and short-term applications. It required manual or physical re-initialization when error occurred. Adaptive back grounding averages the image over time. This is working in situations of moving objects (foreground).

Stauffer and Grimson introduced Gaussian Mixture Model (GMM),[1] his work attempts to model the value of each and every pixel as a solitary distribution as a Gaussian mixture. According to the persistence and variance of all (each), they identified the background and foreground colors.[1] Lee developed the statistical framework for GMM techniques, which increased the convergence speed of learning mechanism without compromising the stability of the model.[2]

Haque further proposed a technique to model every pixel independently by a mixture of k Gaussian distributions of the scene.[3] Lavanya worked effectively on the BGS-based scheme for critical background. The extraction of foreground pixel was based on Fisher's ratio-based threshold.[4]

Yadav has improved the GMM by introducing postprocessing.[5,6] Yadav further improvised in thermal video frames for moving object detection by applying statistical parameter based on background subtraction method by using different threshold Kullback–Leiber divergence based method. This work was divided into three respective stages—(i) Training phase (background modeling build by trimmed means based simple average method), (ii) Testing phase (generating KLD-based threshold values, differentiating background reference frame and the current frame, classifying the pixels), and (iii) Enhancement phase (applying image processing tools and morphological filters to improve detection quality).[7] Similarly, Sharma et al.[4] has looked into a new concept for moving objects detection under thermal environment where static camera has been used for capturing of video.

Allili introduced a new improvement by using Bayesian-based estimation and applied infinite Gaussian mixtures.[8]

Thierry Bouwman then worked on a new approach fuzzy mixture of Gaussian whose parameters were based on fuzzy c-mean algorithm. This focuses on précised estimation of parameters.[9] Thierry Bouwman in his survey proposed the statistical method for background modeling is the most used one; he classified the statistical background modeling methods according to three generations. This survey also classified the improvements as intrinsic improvements and extrinsic improvements.[9]

Zeng et al.[10] introduces a *Type-2* fuzzy set in *MOG*, which is called *T2-MOG* for uncertainty.[10] Several authors gave their contribution in different background modeling methods.

Steps involved in Background subtraction techniques are described as:

i. Background modeling
ii. Threshold generation
    - Fixed (constant) threshold
    - Dynamic or adaptive threshold
iii. Background subtraction
iv. Background maintenances (updation)
    - Background maintenance
    - Threshold updation or adaptive threshold generation or maintenance of threshold as per dynamic condition generated in the scene.

The background subtraction is the procedure of separation of foreground (moving object) with static information. For background subtraction, background modeling is the first step. Table 3.2 categorizes the various background modeling methods with their respective subcategories and respective authors who contributed in the related researches to provide an improved or innovative idea in background modeling. According to the study, every background modeling methods have some drawbacks and advantages over other method, which are described in Table 3.3 to compare the methods.

There are numerous amount of advantages and disadvantages of any method but a few and needful are summarized here. This may somehow contribute to the readers.

TABLE 3.2 Category Wise Literature.

Types of background modeling	Subcategories of background modeling	Authors
Basic background modeling	Histogram Mean Median Pixel intensity classification Pixel change classification	Lee[2] Mac Farlane[11], Zheng[12]
Statistical background modeling	Single Gaussian Mixture of Gaussian Single general Gaussian Mixture of general Gaussian Kernel Density Estimation Support Vector Machine Subspace learning applying PCA Subspace learning applying ICA Subspace learning applying INMF	C. Wren[13] C. Stauffer[1] H. Kim[14] M. Allili[8] A. Elgammal[15] H. Lin[16] N. Oliver[17] D. Tsai[18] S. Bucak[19]
Fuzzy backgrounding modeling	Type-2 Fuzzy Gaussian mixture model Fuzzy mixture of Gaussian (FMOG)	J. Zeng[10] T. Bouwman[9]
Background estimation	Kalman filter Wiener filter Chebychev filter	S. Messelodi[21] K. Toyama[20] R. Chang[22]

TABLE 3.3 Benefits and Limitations.

Type of background modeling	Advantages	Limitations
Basic background modeling	Easiest method to perform. It performs well in static background.	It does not work in dynamic Background.
Statistical background modeling	Most used to their robustness to the critical situation. This method gave better performance.	It only presents a difficult variable enhancement method.

TABLE 3.3 *(Continued)*

Type of background modeling	Advantages	Limitations
Fuzzy background modeling	FMOG is more robust than mixture of Gaussian.	Works only for dynamic backgrounds (waving Vegetation).
Background estimation	It has less trouble with frequently changing light (illumination variation), handles better than mixture of Gaussian. Gives more accuracy but fewer computations.	Probably not suitable for most applications. Very less computation is carried.

In computer vision, moving object detection is the initial step to be done. On basis of studying a lot of literature, there are too many investigating fields contributing a big amount of publications. In Table 3.4, these many investigating fields with their respective authors responsible for the representation of their publications are indicated. This will eventually help the readers to carry forward any of the approach of background modeling that works for most robust system and real-time application.

TABLE 3.4 Techniques Wise Overview.

Techniques	Contributions	Improvements
Fuzzy logic based	Zeng et al.[10] T. Bouwman[9]	• Suitable for infrared videos. • More robust in case of dynamic backgrounds such as waving vegetation.
Gaussian based	Wren et al.[14], Stauffer[1], Kim[15], Allili[8] Medioni[24], Zhao[25]	• Gives good results in presence of gradual lightening changes and reflections(shadows). • Intrinsic and extrinsic improvements.
Kernel density estimation based	Elgammal[16], Pahalawatta[26] Orten[27], Tanaka[28]	• Deals with multimodal backgrounds like waving trees and water rippling fast changes. • Many works adopted to reduce computation time.
Optical flow based	Sotirios et al.[32], Lucas et al.[33]	• Great work in object identification with high accuracy.[32] • Analysis of optic flow may be performed on the sequence of frames to determine a explanation of motion.

TABLE 3.4 *(Continued)*

Techniques	Contributions	Improvements
Subspace learning	Oliver et al.[17], Tsai and Lai[19], Bucak et al.[20]	• A noticeable improvement in presence of multimodal background. • An intrinsic improvement in presence of shadows.
Deep learning	Krishevesky et al.[29], Simonyan et al.[30], Tianming et al.[31]	• Background subtraction and convolution neural network combined for anomaly detection in much abnormal objects surveillance. • Remarkable achievements in image recognition, object detection and classification.

3.4 ROLE OF MACHINE LEARNING, ARTIFICIAL INTELLIGENCE, AND DEEP LEARNING FOR OBJECT DETECTION IN COMPUTER VISION

Over a decade, background subtraction has been very effectively and frequently used technique for object detection. As discussed in above sections, many background modeling methods are and were in existence to detect the objects for further applications like object tracking, classification, extraction, or recognition. With the constant development around artificial intelligence and deep learning, computers are been trained to build background models to detect the moving objects. Table 3.5 describes various techniques and applications of ML, Deep learning, and AI in computer vision.

TABLE 3.5 Methodology and Application-oriented Study.

Approach	Methodology & Application	Details
Machine Learning in Computer Vision	Support Vector Machines	Support vector machine classifier can classify the particular points of input set in two possible classes. It is a nonparametric method. Thus, it has the pliability to consider complex functions and at the same time being tough to overfitting.
	Neural networks	The organization of layers in neural network is in layers where each unit gets the inputs only from the next higher layer node.

TABLE 3.5 *(Continued)*

Approach	Methodology & Application	Details
		Representation of any continuous function is possible with arbitrary accuracy.
	Deep learning	Deep Learning Networks are more complex means of layer interconnection. They need resilient computational training power. Following are the main architectures types: • Convolutional neural networks • Unsupervisioned pretrained networks • Recursive Neural network. • Recurrent Neural network
	Applications	Machine learning helps computers to understand what they see. It can be used in wide variety of areas like below : • Agriculture (detecting diseases in crops based on visual symptom) • Visual Character Recognition/Generating subtitles for videos • Surveillance etc.
Deep Learning in Computer Vision	Convolutional Neural Networks.	There are three types of neural layers in CNN that are convolution layers, fully connected layers, and pooling layers. Every layer of it converts the input to an output Volume of neuron activation.
	Applications	1) Object Detection 2) Face Recognition 3) Action and Activity Recognition 4) Human Pose Estimation 5) Self-driving Cars
Artificial Intelligence in Computer Vision	Fuzzy logic	To perform image processing by fuzzy logic, there are following three steps: • Fuzzification of image: Used to change the values of a particular data set or image • Fuzzy clustering: Used to modify the membership values once image data are transformed from fuzzification • Defuzzification: Decoding of the results

TABLE 3.5 *(Continued)*

	Genetic algorithm	There are various steps of Genetic algorithms: • initial population • fitness function • selection • crossover • mutation Typically use the mechanisms of crossover and mutation and work toward obtaining global optimum.
	Applications	Can be used in wide variety of fields like Natural Sciences, Biological Sciences, Industry, Management and Engineering, etc.

3.5 CHALLENGING ISSUES

Moving object detection from video sequences is not an easy task. There are lots of difficulties or challenges faced by the camera to capture the sequences. Thierry Bouwman in his survey presented all the challenges in detail.[23] Some of the challenges faced by the camera are briefed as follows:

- *Illumination Variation*: Lightening situations of the target scene might always not accordingly. The appearance of the object changes due to motion of light sources like weather conditions—sunny, foggy, rainy, mist, windy in outdoor scenes, on and off of light in indoor scenes, reflection from sun or any other bright source.
- *Occlusion*: The target might get occluded by some other objects. Occlusion may be partial or complete, if partial part is occluded or hidden then it is partial occlusion but if full object is hidden then it is called complete occlusion. The pedestrian may be occluded by trees. This is a challenge to detect the object if it is occluded by the background either partially or completely.
- *Complex background*: In outdoor scenes, nature may be variable every now and then such as waving trees, sprinkles of water from fountain, and movement of clouds are highly variable so these movements are nonperiodic and not in control of camera and thus challenging for background modeling.

- *Shadows*: Occurrence of shadow is part of illumination variation; it may be of different size at various times of the day, which is again a complicated task to detect object.
- *Camouflage*: The camouflage is somehow related to partial occlusion but camouflage foreground is subsumed by the background. This causes difficulty to categorize background and foreground.
- *Bootstrapping*: This is a scenario when background frames are unavailable this becomes critical situation to model the background.
- *Foreground Aperture*: This is the condition where foreground region has the uniform color region as that of the background; this increases the possibility of false negative.

3.6 CONTRIBUTION

This chapter contributes to draw a comparison of various techniques of background modeling, their advantages and limitations. The study of existing research work gave idea of various steps used in background subtraction. Background subtraction has four steps—background modeling, threshold generation, background subtraction, and background maintenances. The threshold generation may be fixed and dynamic. Background maintenance is threshold maintenance as per the dynamic conditions generated in the scene. Several application areas in the real-time, which keenly substitute computer vision object detection technique as the key role instead of human are the major attention of the paper. The challenges that come in the way of object detection are well described in this survey.

3.7 CONCLUSIONS AND FUTURE SCOPE

The background modeling is the basic and most important task in background subtraction technique for object detection. The survey covers all the advantages and limitations of background modeling methods, based on these limitations an improved background subtraction method with some improvised technique will be developed for object detection. The future will try to overcome the challenges discussed in the survey by applying the better aspect of background modeling technique.

KEYWORDS

- **object detection**
- **static camera**
- **background subtraction**
- **support vector machines (SVM)**
- **machine learning (ML)**
- **deep learning (DL)**

REFERENCES

1. Stauffer, C.; Grimson, W. Adaptive Background Mixture Models for Real-Time Tracking. In *Proceedings IEEE Conference on Computer Vision and Pattern Recognition*; CVPR, 1999; pp 246–252.
2. Lee. B.; Hedley, M. Background Estimation for Video Surveillance. *Image Vis. Comput. New Zealand* **2002,** 315–320.
3. Haque, M.; Murshed, M.; Paul, M. On Stable Dynamic Background Generation Technique Using Gaussian Mixture Models for Robust Object Detection. In *5th Int. Conference on Advanced Video and Signal Based Surveillance*, IEEE, 2008; pp 41–48.
4. Sharma, L.; Yadav, D. K.; Singh, A. *Fisher's Linear Discriminant Ratio Based Threshold for Moving Human Detection in Thermal Video*; Elsevier, Infrared Physics and Technology, 2016.
5. Yadav, D. K.; Sharma, L.; Bharti, S. K. Moving Object Detection in Real-Time Visual Surveillance Using Background Subtraction Technique. In *14th International Conference on Hybrid Intelligent Systems*; IEEE, 2014; pp 79–84.
6. Yadav, D. K. Efficient Method for Moving Object Detection in Cluttered Background Using Gaussian Mixture Model. In *International Conference on Advances in Computing, Communications and Informatics (ICACCI)*, 2014; pp 943–948. DOI: 10.1109/ICACCI.2014.6968502.
7. Yadav, D. K.; Singh, K. *A Combined Approach of Kullback–Leibler Divergence and Background Subtraction for Moving Object Detection in Thermal Video*, Vol. 76; Elsevier, Infrared Physics and Technology, 2016; pp 21–31. ISSN 1350-4495.
8. Allili, M.; Bouguila, N.; Ziou, D. *A Robust Video Foreground Segmentation By Using Generalized Gaussian Mixture Modeling*; CRV, 2007; pp 503–509.
9. El Baf, F.; Bouwmans, T.; Vachon, B. Type- 2 Fuzzy Mixture of Gaussians Model: Application to Background Modeling. *Int. Symp. Visual Comput.* Dec **2008,** 772–781.
10. Zeng, J.; Xie, L.; Liu, Z. Type-2 Fuzzy Gaussian Mixture Models. *Pattern Recogn.* Dec **2008,** *41* (12), 3636–3643.
11. McFarlane, N.; Schofield, C. Segmentation and Tracking of Piglets in Images. *Br. Mach. Vis. App.* **1995,** 187–193.
12. Zheng, J.; Wang, Y. *Extracting Roadway Background Image: A Mode Based Approach*; Transportation Research Board, TRB, 2006.

13. Wren, C.; Azarbayejani, A. Pfinder: Real-Time Tracking of the Human Body. *IEEE Trans. Pattern Analys. Mach. Intell.* **1997,** *19* (7), 780–785.
14. Kim, H.; Sakamoto, R.; Kitahara, I.; Toriyama, T.; Kogure, K. Background Subtraction Using Generalized Gaussian Family Model. *IET Electron. Lett.* Jan **2008,** *44* (3), 189–190.
15. Elgammal and Davis, L. Non-Parametric Model for Background Subtraction. In *6th European Conference on Computer Vision, ECCV 2000*; 2000; pp 751–767.
16. Lin, H.; Liu, T.; Chuang, J. A Probabilistic SVM Approach for Background Scene Initialization, ICIP 2002; September 2002.
17. Oliver, N.; Rosario, B.; Pentland, A. A Bayesian Computer Vision System for Modeling Human Interactions. *Int. Conf. Vis. Syst.* Jan **1999,** *22* (8), 841–843.
18. Tsai, D.; Lai, C. Independent Component Analysis Based Background Subtraction for Indoor Surveillance. *IEEE Trans. Image Process.* Jan **2009,** *18* (1), 158–167.
19. Bucak, S.; Gunsel, B. Incremental Subspace Learning and Generating Sparse Representations Via Non-negative Matrix Factorization. *Pattern Recogn.* **2009,** *42* (5), 788–797.
20. Toyama, K.; Krumm, J. Wallflower. Principles and Practice of Background Maintenance. *Int. Conf. Comput. Vis. ICCV 1999*; 1999; pp 255–261.
21. Messelodi, S.; Modena, C. A Kalman Filter Based Background Updating Algorithm Robust to Sharp Illumination Changes. *Int. Conf. Image Analys. Process. ICIAP 2005* **2005,** *3617*, 163–170.
22. Chang, R.; Ghandi, T.; Trivedi, M. Vision Modules for a Multi Sensory Bridge Monitoring Approach. *IEEE Conference on Intelligent Transportation Systems, ITS 2004*; 2004; pp 971–976.
23. Yazdi, M.; Bouwmans, T. New Trends on Moving Object Detection in Video Images Captured by a Moving Camera: A Survey. *Comput. Sci. Rev.* **2018,** *28*, 157–177.
24. Francois, G. M. Adaptive Color Background Modeling for Real–Time Segmentation of Video Streams. *Int. Conf. Imag. Sci. Syst. Technol.* June **1999,** 227–232.
25. Zhao, M.; Li, N.; Chen, C. Robust Automatic Video Object Segmentation Technique. In *International Conference on Image Processing*, September, 2002.
26. Pahalawatta, P.; Depalov, D.; Pappas, T.; Katsaggelos, A. Detection, Classification, and Collaborative Tracking of Multiple Targets Using Video Sensors. In *International Workshop on Information Processing for Sensor Networks* Apr 2003; 529–544.
27. Orten, B.; Soysal, M.; Alatan, A. *Person Identification of Surveillance Video by Combining Mpeg-7 Experts*; WIAMIS, 2005.
28. Tanaka, T.; Shimada, A.; Arita, D.; Taniguchi, R. Object Segmentation Under Varying Illumination Based on Combinational Background Modeling. In *Proceeding of the 4th Joint Workshop on Machine Perception and Robotics*, 2008.
29. Krizhevsky, I.; Sutskever, G. E. H. ImageNet Classification with Deep Convolutional Neural Networks. *Adv. Neural Inf. Process. Syst.* **2012,** *1*, 1097–1105.
30. Simonyan, K.; Zisserman, A. Very Deep Convolutional Networks for Large-Scale Image Recognition, arXiv:1409.1556, 2014.
31. Yu, T.; Yang, J.; Lu, W. Combining Background Subtraction and Convolutional Neural Network for Anomaly Detection in Pumping-Unit Surveillance, May 2019.
32. Diamantas, S.; Alexis, K. Optical Flow Based Background Subtraction with a Moving Camera: Application to Autonomous Driving, 2018.

33. Lucas, B. D.; Kanade, T. An Iterative Image Registration Technique with an Application to Stereo Vision. *Proc. 7th Int. Joint Conf. Artif. Intell. (IJCAI)* Aug **1981,** *24–28*, 674–679.
34. Schrof, F.; Kalenichenko, D.; Philbin, J. FaceNet: A Unifed Embedding for Face Recognition and Clustering. In *Proceedings of the IEEE Conference on Computer Vision and Pattern Recognition (CVPR '15)*. IEEE: Boston, MA, June 2015; pp 815–823.
35. Taigman, Y.; Yang, M.; Ranzato, M.; Wolf, L. DeepFace: Closing the Gap to Human-Level Performance in Face Verification. In *Proceedings of the IEEE Conference on Computer Vision and Pattern Recognition (CVPR '14*); Columbus, OH, June 2014; pp 1701–1708.
36. Venkata Prasad, V. A.; Rayia, C. S.; Naveena, C.; Satpute, V. R. Object's Action Detection Using GMM Algorithm for Smart Visual Surveillance System. In *International Conference on Robotics and Smart Manufacturing*, 2018.
37. Massoli, F. V.; Carrara, F.; Amato, G.; Falchi, F. Detection of Face Recognition Adversarial Attacks. *Comput. Vis. Image Understand.* **2021,** *202.* ISSN 1077–3142, https://doi.org/10.1016/j.cviu.2020.103103.
38. Salazar, E. B.; Escudero, B. N.; Nathaly, S. Rea Minango Omnidirectional Transport System for Classification and Quality Control Using Artificial Vision. In *Conference: Proceedings of the* 2019, *3rd International Conference on Virtual and Augmented Reality Simulations—ICVARS*, 2019. DOI: 10.1145/3332305.3332321

CHAPTER 4

UAV-ENABLED DISASTER MANAGEMENT: APPLICATIONS, OPEN ISSUES, AND CHALLENGES

AMINA KHAN, SUMEET GUPTA, and SACHIN KUMAR GUPTA

School of Electronics and Communication Engineering, Shri Mata Vaishno Devi University, Kakryal, Jammu & Kashmir, India

ABSTRACT

One of the 21st century's most important innovations has been recognized as Unmanned Aerial Vehicles (UAVs). The accelerated growth of UAVs and their application in several fields open up a new vision for their use in natural and man-made disaster management. In UAV-based disaster management applications, UAV applications not only analyze the region impacted but also help to create a communication network between rescue teams and disaster survivors and the nearest mobile networks. The implementation of the UAV program is categorized according to the process of crisis management, and the related work efforts are reviewed along with outstanding research and development issues. This chapter outlines the key development of UAV networks for disaster management applications and addresses open research issues and challenges based on UAV for disaster management. The primary purpose of this work is to deliver technical outcomes that can help enhance people's well-being and advance the state of art in developing a robust disaster management system.

Advanced Computer Science Applications: Recent Trends in AI, Machine Learning, and Network Security. Karan Singh, PhD, Latha Banda, PhD & Manisha Manjul, PhD (Eds.)

4.1 INTRODUCTION

The incidence level of man made and natural disasters is a big concern in both developed and emerging parts of the world. The size and scale of the disaster make it very difficult for people to respond and address the crisis immediately. In some situations, it is almost impossible for people to react timely to the catastrophe. At present, attempts are being made to anticipate and foresee the potential occurrence of a disaster to respond effectively to a crisis amid a tragedy, rapidly and accurately assess the impact, remediate, and re-establish normal conditions. This chapter emphasizes the need to enhance disaster readiness, outlines a vision to exploit recent developments in Unmanned Aerial Vehicles (UAVs), including Wireless Sensor Networks (WSNs) technology, to boost the capability for network-assisted disaster mitigation, prediction, preparedness, and response.

UAVs or drones are a growing technology with an exponentially increasing ability that provides a vast range of effective solutions in the real world in various smart applications. Owing to the exponential growth in demand for civilian UAVs, the use of civil drones needs to integrate into our daily lives. UAVs are an example of new technology that meets the growing demand for the application because of its diverse implementation potential.[1] The main applications of civil UAVs include inaccessible areas of connectivity, medical services, distributing goods, quarantine areas, emergency communication services, healthcare, data gathering, disaster prediction and recovery, public safety, etc.[2–4]

To efficiently carry out complex tasks in real-time, UAV coordination is required in many applications such as monitoring activity, wireless communication across broad areas, disaster recovery, etc.[5] Smart UAVs are expected to change the communication field and incorporate innovations in disaster management applications to minimize risks and provide cost-effective solutions.[3,5] In this article, we discuss the numerous applications based on UAV for disaster management systems and address open issues and research challenges.

Over the last few years, several researchers have contributed enormously to UAVs due to their various benefits in the field of disaster management[6,7] but fail to address open issues and research challenges in the field of disaster management. UAVs contribute a lot to civilian and other critical applications, but still, it needs more concentration toward some areas like disaster management. Some of the applications, like disaster management, are very susceptible to delays and require quick emergency services to save the lives of people. The main motivation behind this article is to address some of the open issues and research challenges need in the field of disaster management for providing better connectivity and coverage and also to timely respond to the victims. In this chapter, the authors try to focus on some of the UAV applications in disaster management and also highlight the most critical open issues and challenges.

The remainder of the chapter is laid out as follows: Section 4.2 discusses the disaster management stage. Section 4.3 focuses on various UAV-based disaster management applications. Section 4.4 highlights the open issues for disaster management; Section 4.5 presents the recent challenges and gives future directions; and finally, in Section 4.6 conclusion is drawn.

4.2 DISASTER MANAGEMENT STAGES

Natural and man made disasters occur globally every day and signify an important feature that affects the growth of development as well as human life. We must learn about the nature of the disaster, its phases, and its components in a position to react to different kinds of natural and man-made catastrophes and develop feasible methods and techniques for managing disasters.

The most critical thing to be addressed when a disaster happens is the safety of human life. In this respect, the Search and Rescue (SAR) operations must be performed effectively and quickly within the first 72 hours after a disaster occurs. In addition to an international SAR methodology and protocol, the International SAR Advisory Group sets out set instructions that specify that the teams have to perform their SAR operation.[8] A team manager will be responsible for the task assignment and local decisions, and all team operations shall be coordinated by a supervisor of the situation. A typical SAR task is performed in four main steps: (i) the commander sets up the search region (communication problems between rescue personnel will be minimum if the search area is smaller), (ii) establishment of a command station in the search region, (iii) first responders were split into rescue personnel and scouts, and (iv) scout teams report to the command station their observations and the rescue officers collect information from the command station to see where to take action. Figure 4.1 presents the role of UAV in different disaster management stages.

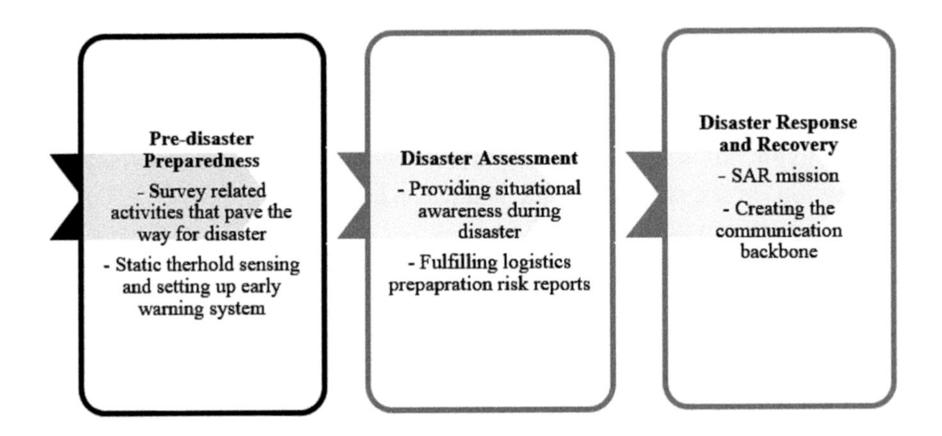

FIGURE 4.1 Role of UAV in disaster management stages.

4.3 UAV-BASED DISASTER MANAGEMENT APPLICATIONS

The technology of UAVs has the potential to reform the accuracy and efficiency of handling disasters. This section addresses many UAV-aided applications and demonstrates UAV's ability to anticipate, forecast, recover, and handle disaster events, including prediction, tracking, early warning systems, disaster response, surveillance, rescue operations, information collection, and logistics.

4.3.1 FORECASTING, MONITORING, AND EARLY WARNING SYSTEMS

Early warning systems, with the advent of telecommunications and sensors, tend to be an exciting choice for the management of disaster to prevent casualties and reduce the economic consequences of the catastrophe event. In general, early warning systems consist of a series of sensors to detect the ecological change and a chain of information communication networks for data transfer to control centers. The functions of an efficient early warning system take care of monitoring, data fusion, risk analysis, and subsequent responses. UAVs, in combination with WSN, are used to offer a detailed and high-efficiency warning for predicting disaster. There are different ways in which UAVs can be used in disaster warnings with WSNs. UAVs can be used to provide connectivity between WSN nodes and decision centers as the data mules or mobile base stations (BSs) in the air. In this case, information can be transferred to airborne UAVs and retrieved from decision centers, avoiding the transmission of multiple intermediate nodes, thus increasing the reliability and accuracy of data transmission.[9,10] The battery-powered node of the WSN network makes energy a key problem in the design, management, and deployment of an early warning system. In this situation, UAV can be used as an aerial charger to recharge the battery into the draining node.

4.3.2 EMERGENCY COMMUNICATION

An essential factor of catastrophe response and recovery is emergency communication. The purpose of emergency communication is to offer communication for the decision-makers, injured people, and disaster rescuers. The effectiveness of disaster management systems, including disaster assessment, SAR, and information gathering, is focused on a robust and easy-going communication network. To make wireless communication more efficient, the interoperability of current technologies of wireless communication systems, like Wireless Fidelity (Wi-Fi), ad-hoc networks, cellular networks, and UAV-based aerial BSs, must be improved while the conventional telecommunications network can provide wireless cellular service and high-speed broadband service. They might be destroyed by large-scale man made or natural disasters, such as BS and cable disruption triggered by an earthquake event, or shutting down power grids in floods. Consequently, the normal means of communication, such as handheld devices, cannot function or may be missing during the disaster. UAV-based approaches were proposed to meet the immediate need for infrastructure-less communication.

Wu et al.[11] suggested a wireless communication network system with a multi-UAV, where every UAV is mounted by an aerial BS, and it offers the ground users wireless communication. The power consumption and trajectory of UAV are taken into account to accomplish justice among terrestrial users, which results in maximization-minimization problems and solves through successive convex optimization techniques. Similar work is also done in paper[12] for robust communication and trajectory design in the presence of jammers based on multi-UAV-enabled wireless networks.

4.3.3 SAR OPERATION

UAVs are known to have tremendous benefits, particularly in SAR operations, public safety, and disaster management. Critical infrastructure, like water and power necessities, telecommunication systems, and transport systems, may be completely or partially affected by a natural or man-made disaster such as tsunamis, floods, earthquakes, etc. To support communication services to assist rescue operations requires fast solutions.[13] Disaster response is a time-critical sprint to find and treat the sufferers as soon as possible. In this situation, the main purpose is to save the lives of people. UAVs can deliver catastrophe warnings in time and support to speed up rescue and recovery activities when public network services are affected. UAVs might help disaster responders to accomplish this aim by rapidly scanning large disaster zones employing navigation sensors and built-in cameras to locate possible victims. A variety of key features should be considered to develop effective SAR based on UAV systems, including the various environmental hazards in service for the UAV network, energy limitations of UAV systems, and the Quality of Service (QoS) of UAV data transmission.

4.3.4 INFORMATION GATHERING BY REMOTE SENSING

The assessment of disaster damage is critical for fast relief measures. Different information is gathered and shared for decision making; thus, for disaster management, information collection or data fusion is necessary. UAV components can be used to spot more details of the catastrophe zones, but with a restricted coverage range. Thus, how to integrate the videos or images captured from multiple sources to create a cohesive model for the disaster scenarios are discussed in articles.[14,15]

4.3.5 LOGISTICS

In disaster management, logistics is one of the most challenging problems. The roads or streets may be obstructed or destroyed during or after a major disaster, and the rescue workers have difficulty reaching the victims and delivering necessary medical treatment. In this situation, UAVs can be used to deliver relief services such as medicine, food, and mobile devices, to the victims of disaster-affected areas. The present civil UAVs could carry a payload up to a few kilograms and could not be a logistically viable solution for repeated demands for emergencies.

In large-scale disaster emergencies, one of the problems is the implementation of a facility site, which is mainly based on the selection of the best site for the emergency centers. For that, the authors of the research article[16] develop a stochastic Mixed-Integer Nonlinear Programming based on the principle that individuals are evenly spread on the edges of a network. Its goal is to determine the best place of relief delivery centers among a range of applicant locations. The authors aim to build a mathematical model that reduces the overall time of people as well as UAVs over several possible scenarios. John et al.[17] proposed coordinated logistics consisting of a moving truck and swarm of UAVs. The moving truck carries the packages and is automatically operated. The UAVs pick up the packages from the truck to provide services to the needy. After the successful delivery of the packages, the UAV will return to the truck to collect more packages until all the packages have been supplied. This study examined how the organized system improves the value of logistics facilities.

4.4 OPEN ISSUES

The involvement of UAVs in disaster management system has numerous research challenges linked to networking as explained below:

4.4.1 UAV DEPLOYMENT

Concerning open issues for UAV deployment, new approaches are required to enhance the 3D deployment of UAVs while taking into consideration their specific features. It is necessary to examine how UAVs could be implemented in cooperation with cellular networks, thereby taking into account mutual interference among terrestrial and aerial networks.[18] One more challenge is to mutually optimize bandwidth distribution and 3D positioning for UAV-BSs to reduce the overall transmission delay of the clients that are aided by UAV-BSs in case of disaster. Also, the deployment of the UAVs should be such that it will provide maximum coverage and connectivity.[19]

4.4.2 MAINTAINING AND CREATING UAV-RELAY NETWORKS

The relay network created by the UAVs is aerial and requires a high level of resistance to communication failures and interference due to changes in motion or changes in energy levels between the UAVs.[20] A two-round process is required, the first round of centralized identification of optimum relay points (called anchors) that link the disaster areas to the closest radio access network and is headed by a second round of decentralized correction for the period of deployment. The issue of UAV backup assignment at each anchor point is similar to reserve backup channels in the cellular situations, with some significant differences: the capacity of the UAV back-up to act as relay anchor that varies over time based on the changes in energy level; in terms of movement near the anchor site. The handoff procedure itself takes resources; a functional failure will occur by changing the UAV's role, among others, from relaying to surveying.

4.4.3 UAV LOCALIZATION

It may be inadequate for UAVs to explore the disaster area autonomously and choose their locations using repeated testing with self-learning techniques given their short flight times and the need for time-bound actions. The authors suggest the usage of partial external inputs to direct the UAVs to create the last hop connection to users and to build a relay network. The last reported information from the cellular database network location and mobile device signals can be leveraged to approximate the number of these impacted individuals and their geographical spread, which calls for a new exchange of signal protocols between networks. The UAV can be guided to high-density locations by strategies such as ant-foraging algorithms, which strengthen the pathways based on the availability or the number of mobile ping requests.[21]

4.4.4 DATA FUSION ISSUES

The UAVs capture images/videos that provide an outline of the circumstances. However, individuals affected can use different social media to exchange images and text messages through the relay UAV network. This provides fine-grained details on the ground that can be fused with high-definition UAV feeds at the control center.[22] The fusion of in-network data with energy constraints within a mobile UAV network has not been fully studied. In addition, the need for fusion of data can affect the UAV network in exciting ways: (i) a more holistic view of the circumstances can take UAVs to different areas; (ii) it can minimize UAV data transfer requirements and thereby save more flight energy. Current channel/source coding from the multimedia sensor network domain is inadequate when the static network topology with separate channel conditions is taken into consideration. Here, both the channel and the topology vary with time.

4.4.5 HANDOVER ISSUES

The handover procedure between UAVs will start early when the UAV approaches the indicated location, but this includes the greater effect of a 3D propagation environment and greater transmission power. Furthermore, the simultaneous moving behavior in air and radio frequency transmissions can also contribute to signal variations due to the Doppler Effect on the incoming UAV. On the other hand, UAVs will position themselves in the air after each other and then start the handover. Nevertheless, there is a trade-off between the benefit of aerial stability with low transmitting power during handover-related communications and the corresponding long time to complete the whole handover procedure.[23,24]

4.4.6 COVERAGE AND CONNECTIVITY

In case of disaster coverage and connectivity plays an important role as most of the terrestrial BS gets destroyed and UAV-BSs provide coverage and connectivity to the ground users.[25] Obstacles influence the performance of coverage of UAV-BS for ground users. One main issue is to optimize the maximum UAV-BS coverage by positioning UAV-BSs optimally dependent on the user's positions and obstacles. Generally, the 3-D placement of the UAV-BSs should be defined so that the maximum coverage of the users is possible, given the obstacles and the positions of the ground users in the area.[26] It is particularly helpful as UAVs fly at high-frequency ranges (e.g., at frequencies of millimeters).

4.4.7 ENSURING ROBUSTNESS, NETWORK SECURITY, AND PRIVACY

Security is one of the major concerns that need to ensure secure data transfer between UAVs and ground stations.[27,28] The emphasis should be made on the security of communication to ensure a robust UAV control network and acquisition of information. Malicious attacks are nearly associated with the operation of the UAV network; robust communication protocols thus play a key role. UAVs are used to collect multimedia information regarding the individuals affected by natural or man-made disasters and pose critical questions on the protection of information and trust issues.[29] In reality, UAV-recorded video footage during the disaster contains a recording of sensitive footage such as dead or injured persons, which should be censored automatically, particularly when the media use the footage.

4.5 RESEARCH CHALLENGES

In this section, some of the major design challenges and considerations of UAV-based disaster management are discussed. Current problems for channel modeling, trajectory optimization, resource management, and network performance are the particular requirements of aerial implementations, taking into account transmission range, delay tolerance, topology changes, performance scalability, mobility issue, and energy constraints. Table 4.1 gives research directions, open issues, and challenges for UAV-based wireless networks. Some of the research challenges are as follows:

4.5.1 CHANNEL MODELING

Different types of channels must be assisted due to the 3D existence of a UAV network. In the case of an aerial system, these links can be either ground-to-air, air-to-ground (A2G), or air-to-air (A2A). There are some open issues for A2G channel modeling. First of all, more accurate channel models are required, derived from measurements in the real world.[30] UAVs are being widely used as A2A channel modeling as a flying BS, drone-user equipment (UE), and also to support the backhaul. There is a prerequisite for a specific UAV-to-UAV model, which can acquire channel and Doppler Effect time-variation due to the UAV movement. In addition, it is important to define multipath fading in A2A communications, thus taking into account the altitude of the UAV as well as the movement of the antenna. UAV's can be designed differently depending on the characteristics of their network, which influence the network-related QoS and thus the supportable traffic.

4.5.2 TRAJECTORY OPTIMIZATION IN UAV

Although the possible mobility of UAVs offers good prospects, it presents new technical problems and challenges. The UAV trajectory must be improved about important performance metrics like delay, spectral efficiency, energy, and throughput. Moreover, the dynamic features and type of UAVs must be taken into consideration in trajectory optimization problems. Whereas a lot of important researches have been conducted on UAV trajectory optimization, some open challenges are still there, which include: (i) UAV trajectory optimization focused on ground user's movement pattern to maximize coverage efficiency, (ii) optimization of the trajectory for reliability maximization and latency minimization in wireless UAV-based networks, (iii) mutual communication, control, trajectory optimization to minimize the fight time of UAVs, and (iv) obstacle sensitive trajectory optimization of UAVs taking into account energy consumption of UAV and delay constraint of the user. Finally, it is another open question for cellular-connected UAV-UE to optimize the trajectory while reducing user's ground interference and being aware of the down-tilt of the antennas in the BSs.[11,31]

4.5.3 RESOURCE MANAGEMENT

One of the important study concerns in UAV-enabled communication systems is resource management. In general, a framework is required, which handles various resources in a dynamic manner such as energy, bandwidth, transmit power, number of UAVs, and flight time of UAV. For example, how to flexibly change the direction of a flying UAV and the transmission power that serves the users on the ground. The main problem, in this case, is to provide optimal mechanisms for the assignment of bandwidths, which can record the impact of UAV's locations, Line-of-Sight interference, mobility, and the distribution of ground user's traffic. Furthermore, efficient scheduling technologies have to be designed to minimize interference in the UAV-supported cellular network between aerial and ground BSs. Moreover, in a heterogeneous flying network and terrestrial BSs, the dynamic spectrum sharing must also be analyzed. Lastly, it is important to design problems for UAV operation to follow acceptable frequency bands (e.g., LTE, Wi-Fi bands).[32,33] Table 4.1 presents research directions, open issues, and challenges for UAV-based wireless networks.

TABLE 4.1 Research Directions, Open Issues, and Challenges for UAV-Based Wireless Network.

References	Research directions	Open issues and challenges
[18–19]	UAV Deployment	• Deployment of UAVs in coexistence with ground networks. • Energy-efficient deployment. • Mutual 3-D deployment and bandwidth allocation.

TABLE 4.1 *(Continued)*

References	Research directions	Open issues and challenges
[30]	Channel Modeling	• Small-scale fading. • A2G path loss models. • A2A channel modeling.
[11, 31]	Trajectory Optimization	• Mutual delay and trajectory optimization. • Path planning with reliable communication. • Energy-aware trajectory optimization.
[32–33]	Resource Management	• Flight time and bandwidth optimization. • Spectrum sharing with terrestrial networks. • Multi-dimensional resource management. • Mutual transmit power and trajectory optimization.
[34–35]	Cellular Network Planning	• UAV optimization. • Cell allocation based on traffic. • Analysis of overheads and signaling. • Cell planning considering backhaul.
[36–37]	Performance Analysis	• Performance analysis considering mobility. • Capturing temporal and spatial correlation. • Analysis of heterogeneous aerial-ground network.

4.5.4 CELLULAR NETWORK PLANNING WITH UAVS

Several important problems must be concentrated on planning an effective UAV system. For instance, what is the minimum number of UAVs needed for providing maximum coverage for a known geographic zone that is moderately covered by terrestrial BSs? Resolving such issues is a special concern where there is no normal geometrical form (i.e., square or disk) in the geographical field of interest. The backhaul-conscious implementation of UAVs as the aerial BSs is another design issue. In this situation, the backhaul compatibility of UAVs, as well as the QoS of their user, should be kept in mind when implementing UAV-BS.[34,35]

4.5.5 PERFORMANCE ANALYSIS

There are several questions yet to be addressed for performance analysis. For example, it is important to describe the performance completely in terms of coverage and capacity of UAV-based wireless networks, comprising both air and land users and terrestrial BSs.[36] Manageable reactions for the possibility of coverage and spectral efficiency are particularly required in aerial-ground heterogeneous networks. Therefore, basic performance assessments must be carried out to determine the intrinsic trade-offs in UAV networks between energy consumption and spectral efficacy. Another issue is determining the UAV performance in wireless networks while taking into account the UAV mobility. A key analysis of mobile wireless networks includes the identification of temporal and spatial variations in different network performance metrics. For example, the trajectory of UAVs needs to be analyzed in terms of power consumption, latency, and throughput. Finally, it is possible to estimate the effect of complex scheduling on the UAV communication systems performance.[37]

4.5.6 REGULATIONS FOR DEVELOPMENT AND DESIGN

Although UAVs acquire their share of national airspace in the country, their deployment must be regulated so that the gains are maximized and the possible damages are minimized. In certain civilian applications, regulation can be a major barrier to UAV deployment. Rules and regulations that are properly designed are realistic. In case of a catastrophe outage, emergency response and rescue measures obey unique acts and guidelines such as disaster mitigation, disaster relief act, etc.[38] to ensure that available services are used efficiently. However, present procedures and regulations are not designed to tackle the use of UAVs in the case of a crisis, and many of them have no guidelines as to how the UAVs are successfully used in disaster management. For example, there are no provisions for necessary characteristics for the dynamic use of UAVs, in particular as part of disaster reduction initiatives such as carrying out infrastructure evaluations and/or study and rescue missions.

4.6 CONCLUSION AND FUTURE DIRECTION

In this article, the first main contribution is given to a comprehensive study on the UAV-enabled disaster management application. It presents the role of UAV in different disaster management phases like forecasting, an early warning system, monitoring, emergency communication, disaster response, surveillance, rescue operation, information collection, and logistics. Moreover, it focuses on open issues in terms of disaster management and gives future directions. In addition, it also presents the recent research challenges faced by UAV-based disaster management. In the future, research can be carried out on how to efficiently deploy a UAV-to-UAV collaboration-assisted network over a large area for the disaster management system.

KEYWORDS

- **disaster management**
- **natural and man-made disaster**
- **early warning system**
- **research and rescue**
- **unmanned aerial vehicles (UAV)**

REFERENCES

1. Mualla, Y.; Najjar, A.; Daoud, A.; Galland, S.; Nicolle, C.; Yasar, A. U. H.; Shakshuki, E. Agent-Based Simulation of Unmanned Aerial Vehicles in Civilian Applications: A Systematic Literature Review and Research Directions. *Futur. Gener. Comput. Syst.* **2019,** *100*, 344–364. https://doi.org/10.1016/j.future.2019.04.051.
2. Shakhatreh, H.; Sawalmeh, A. H.; Al-Fuqaha, A.; Dou, Z.; Almaita, E.; Khalil, I.; Othman, N. S.; Khreishah, A.; Guizani, M. Unmanned Aerial Vehicles (UAVs): A Survey on Civil Applications and Key Research Challenges. *IEEE Access* **2019,** *7*, 48572–48634. https://doi.org/10.1109/ACCESS.2019.2909530.
3. Zhao, N.; Lu, W.; Sheng, M.; Chen, Y.; Tang, J.; Yu, F. R.; Wong, K. K. UAV-Assisted Emergency Networks in Disasters. *IEEE Wirel. Commun.* **2019,** *26* (1), 45–51. https://doi.org/10.1109/MWC.2018.1800160.
4. Ali, K.; Nguyen, H. X.; Vien, Q.-T.; Shah, P.; Raza, M. Deployment of Drone Based Small Cells for Public Safety Communication System. *IEEE Syst. J.* **2020,** *14* (2), 1–10.
5. Khan, A.; Gupta, S.; Gupta, S. K. Multi-Hazard Disaster Studies: Monitoring, Detection, Recovery, and Management, Based on Emerging Technologies and Optimal Techniques. *Int. J. Disaster Risk Reduct.* **2020,** *47* (August 2019), 101642. https://doi.org/10.1016/j.ijdrr.2020.101642.
6. Mozaffari, M.; Saad, W.; Bennis, M.; Nam, Y. H.; Debbah, M. A Tutorial on UAVs for Wireless Networks: Applications, Challenges, and Open Problems. *IEEE Commun. Surv. Tutorials* **2019,** *21* (3), 2334–2360. https://doi.org/10.1109/COMST.2019.2902862.
7. Erdelj, M.; Natalizio, E. UAV-Assisted Disaster Management: Applications and Open Issues. *2016 Int. Conf. Comput. Netw. Commun. ICNC 2016* **2016**.
8. Okita, Y.; Sugita, M.; Katsube, T.; Minato, Y. Capacity Building of International Search and Rescue Teams Through the Classification System: Example From Japan Disaster Relief Team and Insarag External Classification and Reclassification. *J. Japan Assoc. Earthq. Eng.* **2018,** *18* (6), 6_23–6_40. https://doi.org/10.5610/jaee.18.6_23.
9. Erdelj, M.; Król, M.; Natalizio, E. Wireless Sensor Networks and Multi-UAV Systems for Natural Disaster Management. *Comput. Networks* **2017,** *124*, 72–86. https://doi.org/10.1016/j.comnet.2017.05.021.

10. Popescu, D.; Stoican, F.; Stamatescu, G.; Chenaru, O.; Ichim, L. A Survey of Collaborative UAV-WSN Systems for Efficient Monitoring. *Sensors (Switzerland)* **2019,** *19* (21), 1–40. https://doi.org/10.3390/s19214690.
11. Wu, Q.; Zeng, Y.; Zhang, R. Joint Trajectory and Communication Design for Multi-UAV Enabled Wireless Networks. *IEEE Trans. Wirel. Commun.* **2018,** *17* (3), 2109–2121. https://doi.org/10.1109/TWC.2017.2789293.
12. Wu, Y.; Fan, W.; Yang, W.; Sun, X.; Guan, X. Robust Trajectory and Communication Design for Multi-UAV Enabled Wireless Networks in the Presence of Jammers. *IEEE Access* **2020,** *8*, 2893–2905. https://doi.org/10.1109/ACCESS.2019.2962534.
13. Silvagni, M.; Tonoli, A.; Zenerino, E.; Chiaberge, M. Multipurpose UAV for Search and Rescue Operations in Mountain Avalanche Events. *Geomatics, Nat. Hazards Risk* **2017,** *8* (1), 18–33. https://doi.org/10.1080/19475705.2016.1238852.
14. Baker, C. A.; Rapp, R. R.; Elwakil, E.; Zhang, J. Infrastructure Assessment Post-Disaster: Remotely Sensing Bridge Structural Damage by Unmanned Aerial Vehicle in Low-Light Conditions. *J. Emerg. Manag.* **2020,** *8* (1), 27–41. https://doi.org/10.5055/JEM.2020.0448.
15. Chen, Q. J.; He, Y. R.; He, T. T.; Fu, W. J. The Typhoon Disaster Analysis Emergency Response System Based on UAV Remote Sensing Technology. *Int. Arch. Photogramm. Remote Sens. Spat. Inf. Sci.—ISPRS Arch.* **2020,** *42* (3/W10), 959–965. https://doi.org/10.5194/isprs-archives-XLII-3-W10-959-2020.https://doi.org/10.1109/ICCNC.2016.7440563.
16. Golabi, M.; Shavarani, S. M.; Izbirak, G. An Edge-Based Stochastic Facility Location Problem in UAV-Supported Humanitarian Relief Logistics: A Case Study of Tehran Earthquake. *Nat. Hazards* **2017,** *87* (3), 1545–1565. https://doi.org/10.1007/s11069-017-2832-4.
17. Carlsson, J. G.; Song, S. Coordinated Logistics with a Truck and a Drone. *Manage. Sci.* **2018,** *64* (9), 4052–4069. https://doi.org/10.1287/mnsc.2017.2824.
18. Khuwaja, A. A.; Zheng, G.; Chen, Y.; Feng, W. Optimum Deployment of Multiple UAVs for Coverage Area Maximization in the Presence of Co-Channel Interference. *IEEE Access* **2019,** *7*, 85203–85212. https://doi.org/10.1109/ACCESS.2019.2924720.
19. Cetinkaya, O.; Merrett, G. V. Efficient Deployment of UAV-Powered Sensors for Optimal Coverage and Connectivity. *IEEE Wirel. Commun. Netw. Conf. WCNC* **2020,** *2020 May*. https://doi.org/10.1109/WCNC45663.2020.9120738.
20. Zhong, X.; Guo, Y.; Li, N.; Chen, Y.; Li, S. Deployment Optimization of UAV Relay for Malfunctioning Base Station: Model-Free Approaches. *IEEE Trans. Veh. Technol.* **2019,** *68* (12), 11971–11984. https://doi.org/10.1109/TVT.2019.2947078.
21. Yao, R.; Pakzad, S. N.; Venkitasubramaniam, P. Compressive Sensing Based Structural Damage Detection and Localization Using Theoretical and Metaheuristic Statistics. *Struct. Control Heal. Monit.* **2016**. https://doi.org/10.1002/stc.
22. Ji, H.; Luo, X. 3D Scene Reconstruction of Landslide Topography Based on Data Fusion between Laser Point Cloud and UAV Image. *Environ. Earth Sci.* **2019,** *78* (17), 1–12. https://doi.org/10.1007/s12665-019-8516-5
23. Fakhreddine, A.; Bettstetter, C.; Hayat, S.; Muzaffar, R.; Emini, D. Handover Challenges for Cellular-Connected Drones. *DroNet 2019—Proc. 5th Work. Micro Aer. Veh. Networks, Syst. Appl. co-located with MobiSys 2019* **2019,** 9–14. https://doi.org/10.1145/3325421.3329770.

24. Amer, R.; Saad, W.; Marchetti, N. Mobility in the Sky: Performance and Mobility Analysis for Cellular-Connected UAVs. *arXiv* **2019,** *68* (5), 3229–3246.
25. Gupta, A.; Sundhan, S.; Alsamhi, S. H.; Gupta, S. K. *Review for Capacity and Coverage Improvement in Aerially Controlled Heterogeneous Network*, Vol. 546; Springer: Singapore, 2020. https://doi.org/10.1007/978-981-13-6159-3_39.
26. Lyu, J.; Zhang, R. Network-Connected UAV: 3-D System Modeling and Coverage Performance Analysis. *IEEE Internet Things J.* **2019,** *6* (4), 7048–7060. https://doi.org/10.1109/JIOT.2019.2913887.
27. Fotouhi, A.; Qiang, H.; Ding, M.; Hassan, M.; Giordano, L. G.; Garcia-Rodriguez, A.; Yuan, J. Survey on UAV Cellular Communications: Practical Aspects, Standardization Advancements, Regulation, and Security Challenges. *IEEE Commun. Surv. Tutorials* **2019,** *21* (4), 3417–3442. https://doi.org/10.1109/COMST.2019.2906228.
28. Rashid, A.; Sharma, D.; Lone, T. A.; Gupta, S.; Gupta, S. K. *Secure Communication in UAV Assisted HetNets: A Proposed Model*, Vol. 11611; Springer International Publishing, 2019, LNCS. https://doi.org/10.1007/978-3-030-24907-6_32.
29. Sharma, D.; Rashid, A.; Gupta, S.; Gupta, S. K. A Functional Encryption Technique in UAV Integrated HetNet: A Proposed Model. *Int. J. Simul. Syst. Sci. Technol.* **2019,** No. March. https://doi.org/10.5013/ijssst.a.20.s1.07.
30. Khawaja, W.; Guvenc, I.; Matolak, D. W.; Fiebig, U. C.; Schneckenburger, N. A Survey of Air-to-Ground Propagation Channel Modeling for Unmanned Aerial Vehicles. *IEEE Commun. Surv. Tutorials* **2019,** *21* (3), 2361–2391. https://doi.org/10.1109/COMST.2019.2915069.
31. Wang, H.; Ren, G.; Chen, J.; Ding, G.; Yang, Y. Unmanned Aerial Vehicle-Aided Communications: Joint Transmit Power and Trajectory Optimization. *IEEE Wirel. Commun. Lett.* **2018,** *7* (4), 522–525. https://doi.org/10.1109/LWC.2018.2792435.
32. Ceran, E. T.; Erkilic, T.; Uysal-Biyikoglu, E.; Girici, T.; Leblebicioglu, K. Optimal Energy Allocation Policies for a High Altitude Flying Wireless Access Point. *Trans. Emerg. Telecommun. Technol.* **2017,** *28* (4). https://doi.org/10.1002/ett.3034.
33. Anton, S. R.; Inman, D. J. Performance Modeling of Unmanned Aerial Vehicles with On-Board Energy Harvesting. *Act. Passiv. Smart Struct. Integr. Syst.* **2011,** *7977* (540), 79771H. https://doi.org/10.1117/12.880473.
34. Van Der Bergh, B.; Chiumento, A.; Pollin, S. LTE in the Sky: Trading off Propagation Benefits with Interference Costs for Aerial Nodes. *IEEE Commun. Mag.* **2016,** *54* (5), 44–50. https://doi.org/10.1109/MCOM.2016.7470934.
35. Azari, M. M.; Rosas, F.; Chiumento, A.; Pollin, S. Coexistence of Terrestrial and Aerial Users in Cellular Networks. *2017 IEEE Globecom Work. GC Wkshps 2017—Proc.* **2018,** *2018 January*, 1–6. https://doi.org/10.1109/GLOCOMW.2017.8269068.
36. Alsamhi, S. H.; Ma, O.; Samar Ansari, M.; Gupta, S. K. Collaboration of Drone and Internet of Public Safety Things in Smart Cities: An Overview of Qos and Network Performance Optimization. *Drones* **2019,** *3* (1), 1–18. https://doi.org/10.3390/drones3010013.
37. Hayajneh, A. M.; Zaidi, S. A. R.; McLernon, D. C.; Di Renzo, M.; Ghogho, M. Performance Analysis of UAV Enabled Disaster Recovery Networks: A Stochastic Geometric Framework Based on Cluster Processes. *IEEE Access* **2018,** *6*, 26215–26230. https://doi.org/10.1109/ACCESS.2018.2835638.
38. Robert, T. Stafford Disaster Relief and Emergency Assistance Act, Public Law 93-288, as Amended, 42 U.S.C. 5121 et Seq., and Related Authorities. *FEMA* **2019,** No. May. https://doi.org/10.4135/9781452275956.n311.

CHAPTER 5

CROSS-SITE SCRIPTING ATTACK PREVENTION (ON APPLICATION LAYER)

PRABHPREET SINGH SANDHU[1] and LATHA BANDA[2]

[1]*Sharda University, Greater Noida, Uttar Pradesh, India*

[2]*ABES Engineering College, Ghaziabad, Uttar Pradesh, India*

ABSTRACT

In modern web apps, cross-site scripting (XSS) attacks are probably the most prevalent protection concerns. The attacks use web apps programming bugs that have significant repercussions leads to stealing of passwords, web cookies, and other sensitive credentials. XSS happens in intermediate trustworthy websites while viewing content. The client side approach serves as a website proxy for preventing attacks by manually constructing guidelines to avoid attempts at XSS. Customer side strategy successfully defends consumer experience from knowledge leakage. XSS attacks are not complex to conduct though hard to recognize and stop. This essay offers a solution on the customer side for minimizing XSS. Current technologies on the consumer side degrade consumer efficiency contributing to a bad web surf experience. This initiative presents a consumer-side solution to secure XSS, without significantly worsening the experience of consumers. This method safeguards XSS.

Advanced Computer Science Applications: Recent Trends in AI, Machine Learning, and Network Security. Karan Singh, PhD, Latha Banda, PhD & Manisha Manjul, PhD (Eds.)

5.1 INTRODUCTION

Cross-site scripting, is often referred to as XSS, is a sort of attack which collects malicious user credentials: usually in the form of a special hyperlink that saves user credentials. XSS is a flaw in the network protection where a malicious client-side script is inserted into the web page by the intruder. The script code is applied and run transparently in the web server while a user views a web page. The malicious script inherits the privileges of the owner, authentication, etc. The popularity of web-based XSS attack is that software developers frequently have little to no protection history. XSS is an essential weakness for XSS. The consequence is that bad code is installed and made available across the Internet, jammed with security faults. The weakness of the server side, typically due to insufficient input validation procedures, is now being discussed in XSS assaults. For multiple types of request and response data—URL query parameters—URL encoded input ("POST data")—HTTP headers—cookies defense mechanism to protect from XSS can be configured.

The risk of modifying or controlling the activity of the application itself in the browser's HTML documents is risky if manipulated. The capacity for abuse relates explicitly to a malicious programmer's features. JavaScript gives maximum use of the document entity model to view HTML documents. Thus, a script can at least change the document in which it resides: the document may even be erased entirely, and a whole new document can be produced. There are two items of great concern from an attacker's point of view: document-related cookies and access codes. JavaScript gives links to this knowledge as well. You can access database cookies through the call document. Cookie method and access credentials at the application level also are obtained via a form-based login. In this scenario, the user details are placed into input fields that exist in a form. As the form is an integral part of the text, a script is able to view all information in all places or to easily change the form goal Address. The credentials are then redirected to the new goal that is under the attacker's influence. JavaScript is a popular web application creation tool. With the complex design of web applications such as Google Maps, Try Ruby, without customer-facing code inserted into HTML and XHTML sites! And it would not be feasible to Zoho Location. However, as you introduce functionality to a method, the risk for security problems is increased—it is no different to add JavaScript to a web page.

5.1.1 AIMS AND OBJECTIVES

The purpose of this chapter is to carry out suggested rules of operation for the protection of control systems. This chapter in particular:

- describes and define XSS;
- contrasts how XSS impacts protection and stability in XSS-related information systems;
- gives an explanation of how XSS may be used for a control device assault; and
- defines methods for mitigating to protect XSS control systems.

5.1.2 KEY TERMINOLOGY

1. Control Systems

The word control system is used in the chapter as a general concept to include all process management, SCADA, industrial automation, and the associated protection and monitoring systems.

2. Recommended Practice

In this chapter, the suggested method is identified as best practices in industry such as techniques, activities, or methods that have proved to be successful in study, distribution, and assessment. The suggested strategies summarized in this chapter can be treated as recommendations only. With such control device contexts, it might not be feasible to enforce all these values. In this situation, resources owners are advised to collaborate with control device suppliers, ISACS and user groups to define and introduce alternate but successful protections.

5.2 CROSS-SITE SCRIPTING OVERVIEW

XSS is a vulnerability program intrusion leveraging network third party tools to execute a script in the web server or scriptable framework of the user. This happens anytime a browser views or selects a suspicious link to a malicious website. The results of an XSS attack begin with the access between the user and the site server to the Cookie. Which makes it easier for an intruder to hijack the user to the website. Livshits and Erlingsson[15] when XSS is used to trigger more bugs, the most harmful effects exist. The vulnerabilities can enable attackers to stole not only cookies, but also keystrokes, record screen shots, discover and collect data from the network, and access and monitor the computer remotely. Any website or programutilizing web pages user feedback can be susceptible to XSS. These weaknesses get more extreme when an intruder may get the help of an insider (with or without knowledge). Attackers using XSS may also obtain details on possible victims before an assault and use e-mail to specifically target them.

5.3 CROSS-SITE SCRIPTING ATTACK

XSS attacks are certain attack on web applications through which an intruder acquire hold of the user's web, browser to execute a malicious script (commonly HTML or JavaScript code) in the sense of the internet application's confidence. As a consequence, if the applied code is successfully implemented, then intruder will be able to control, actively as well as passively, some critical online-based application browser resource (e.g., session IDs, cookies, etc.). In this portion, we review two major categories of XSS attacks: persistent (stored) and nonpersistent (mirrored) XSS attacks .

5.3.1 PERSISTENT XSS ATACK

Before proceeding, let us first add the former method of assault using the model scenario seen in Fig. 5.2. In this case, we can see the following elements: intruder (A), victim's browser collection (V), weak web application (VWA), malicious web application, trustworthy domain (TD), and malicious domain (MD). We broke the assault in two key phases. In the first stage (refer Fig. 5.2, stages 1–4), user A (attacker) registers in VWA's application and publishes the following HTML/JavaScript code as MA message:

Entire HTML/JavaScript code of the MA message is then stored on TD (trustworthy domain), and is ready to be shown by any other VWA user in VWA repository (see Fig. 5.1, step)

```
<HTML>
<title>Welcome!</title>
Hi everybody! See that picture below, that's my city, well where I come from ...<BR>
<img src="city.jpg">
<script>
document.images[0].src="http://www.malicious.domain/city.jpg?stolencookies="+document.cookie;
</script>
</HTML>
```

FIGURE 5.1 Content of message M_A.

Then the corresponding vi id cookie saved in the browser's depository of each victim vi is driven to an external repository of stolen cookies discovered in MD (malicious domain) at the second level (see Fig. 5.2 stages 5i–12i), and for any victim vi–to–V who shows the message MA. The details contained in this stolen cookie archive can eventually be used to enable an intruder to enter VWA using the identities of other users.[14]

As we observed in the previous case, the harmful JavaScript code inserted in web server by intruder is continually retained in the data repository of the server. The user's browser enables a script to enter its cookies registry while the user of an application loads malicious code to its own browser and because the code is submitted from the sensitive background of the program.[13] Therefore, the script is entitled to steal the sensitive information of the victim of the attacker's malicious background and hence bypass any JavaScript engine's specific protection policy which limits data access to only scripts which are of the same origin when the information has been build up.

In addition to stealing browser data tools, the usage of the previous strategy. In the message inserted by the intruder that simulates, for example, a user logout from the website of the company and provides a fake authentication type that will store victim credentials (like authentication, password, hidden questions/answers, etc.) into the malicious background, we imagine an extended JavaScript file. If the information has been retrieved, the script will redirect the application's flow back to the previous state or allow use of the compromised details to log in to the application's website.

Persistent XSS assaults are typically correlated with web apps with poor feedback authentication systems for message boards. There are several well-known instances of recurrent XSS attacks correlated with such applications. A constant XSS assault on Hotmail, for example, was noticed in October 2001. In an assault so that the remote intruder was able to rob Hotmail users' .NET Passport IDs by gathering their browser cookies with the aid of a technique close to that seen in Figure 5.2. Similarly, on October 2005, the Samy worm was used to disseminate itself over MySpace user accounts via a well-known recurrent XSS assault on the MySpace online social network. More recently, a different Google social network, Orkut, was also targeted by a related persistent XSS assault on November 2006. According to sources, Orkut was susceptible to cookies by merely inserting the robbery script in the attacker's profile. Any other person accessing the profile of the attacker was then identified and the identities were moved to the account of the attacker.

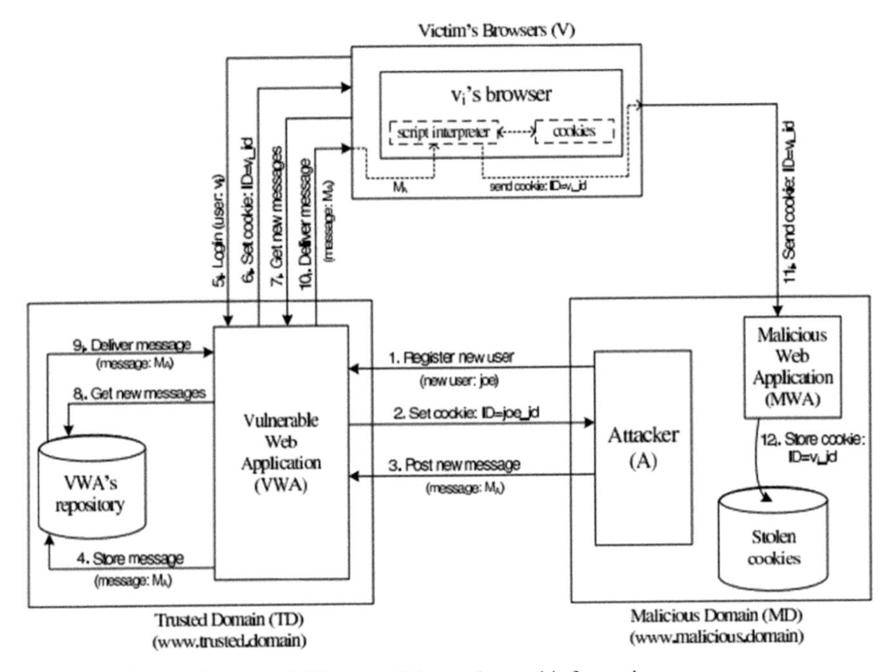

FIGURE 5.2 Analyzing and filtering of the exchanged information.

5.3.2 NONPERSISTENT ATTACK

In this segment, we discuss a difference in the fundamental XSS attack mentioned above. This second type, defined in this chapter as a Non-Persistent XSS attacks, is a sensitive web application to the use of information given by the user to create a page for that user that is mirrored in the literature. In this case, the malicious code itself is explicitly reflected to the recipient by a third party process instead of being retained as a malicious code contained in the attacker's letter. The striker could trick the victim to click on a connection with dangerous code by, for example, using a spoofed email. If so, the code shall eventually be restored to the customer albeit from the reliable sense of the website of the program. The victim's browsers then execute their code inside the trust realm of the application similar to the attack scenario seen in Figure 5.2 which will enable it to transmit related details (e.g., cookies which session IDs) without breaching the same browser's policy of origin.[11]

- XSS nonpersistent attacks are by far the most prevalent kinds of XSS attacks on existing apps, which are sometimes paired to achieve their targets (e.g., capture confidential details for customers, such as card numbers) by other tactics, including phishers which social engineering. Due to the existence of such a version, that is, because it is not indefinitely held on the application platform and the need for techniques from third parties, professional attackers execute nonpersistent XSS assaults and spam attacks. It is very interesting to know the harm incurred by these assaults.

5.4 PREVENTION TECHNIQUES

Although the architecture of web apps has improved efficiently after XSS attacks were recorded first, such attacks continue every day. Since the late 1990s, attackers have been willing, while secured by standard network protection strategies including firewall and cryptographic mechanisms, the continued implementation of XSS attacks through Internet appliances. Using some safe technologies will help to minimize the issue. But they are not always necessary. For example, the usage of safe coding practices (i.e., the model recommended to identify anomalous executions) and/or protected programming frameworks is frequently relegated to standard implementations and might not be helpful in addressing the network paradigm. In comparison, general methods for input validation are mostly concentrated on numerical details or binding authentication, although XSS attack prevention can often discuss input string validation.[10]

This scenario demonstrates the lack of usage of simple protection advice as specific steps to ensure the protection of web apps, which contributes to the need for external safety protocols where XSS assaults have been prevented. In this portion, we present detailed approaches to identify and avoid XSS assaults. The presentation of these methods has been organized into two main categories: evaluate and philter knowledge exchanged; and apply web browsers in the runtime.

In order to overcome both nonpersistent and persistent threats, most, if not all, existing web framework, which permits the usage of rich material while knowledge is shared between browser and web server, introduces simple content filtering schemes. This simple filtering will simply be enforced by establishing an agreed list of characters and/or special marks, such that all features not included in this list are clearly denied. Alternatively, encoding may often be used to render certain characters or tags less dangerous in order to enhance the filtering process. We do think, though, that these key techniques are too restricted for trained attackers to escape.

The literature also reports the use of policy-based strategies. Scott and Sharp[1], for example, suggest a proxy server for the database of the application to handle both the input and output data streams. Their filtering mechanism implements a variety of policy guidelines defined by developers of web applications. Although their technique has made significant improvements compared with the above-mentioned basic mechanisms, this approach remains very limited. We suspect that professional attackers may use their lack of study of syntactical constructs to avoid detection mechanisms and to hit harmful questions. It is obviously possible to prevent such filters simply utilizing standard expressions. Secondly, the specifics of the policy terminology suggested in its work have not been explicitly recorded, and we know that it is not trivial and potentially error-prone to use it for specifying general filtering regulations for every potential application/browser pair. Third, the filtering proxy may be put on the server side to easily enforce restrictions on the scalability of the program.

More recent server-based filtering agents were also mentioned by Su and Wasserman[2] for similar purposes. A filtering proxy is intended by Pietraszeck and Vanden-Berghe[3] to be put on the web application server side such that trustworthy and untrusted traffic may be segregated into different channels. The author proposes to carry out the partitioning process by means of a thin-grained taint analysis. In addition, they demonstrate how to achieve their proposal by changing manually a server-side PHP interpreter for the monitoring of previously defaced details for each string entries. The key drawback of this strategy is that every web application that is introduced in a foreign language does not defend itself or is susceptible to usage of resources, such as code wrappers, by third parties. The strategy suggested relies on the operating environment, which greatly affects its portability. The administration of this idea is already nontrivial and theoretically error-prone of every conceivable pair of software/browsers. Su and Wasserman[2] also plan to screen out inappropriate data sources a syntactic criterion. They evaluate questions easily by covering the false argument to prevent the final stage of an assault and identify misuse. In addition, the writers conducted and tested tests with five real-life situations that prevented deceptive material and generated no false positive. However, the aim of their method seems to encourage programmers to prevent bugs on the server side as early as possible rather than on the customer's side.

The incorporation of a certain filtration and/or review procedures on the customer side, such as described by Ismail et al.[4], are often presented in related solutions. A consumer side filtering procedure is suggested by Kirda et al.[5] in concern to avoid XSS attacks via the avoidance of inappropriate Link contacts from the victim's browsers. The writers adopt this method by blacklisting the links contained on web pages, distinguish good and poor URLs. In this way, the client-side proxy refuses the redirection to URLs connected with such blacklisted connexions. We may not think this is adequate to identify or deter complicated XSS assaults. Only the blacklisting approaches will identify specific XSS attacks based on the same root breach. In order to bypass such a mitigation function, alternate XSS methods as suggested by Alcorna[6] can be used or some other limitation that is not attributable to input validation. The authors introduce a modern client-based proxy that analyses the data shared between the device and the server of the web application. Their method of analysis is aimed at detecting harmful requests from the intruder to the victim (e.g., the nonpersistent XSS scenario showed in Section 5.2). If a harmful request is observed, the proxy re-encodes the characters of such a program and attempts to prevent the attack's progress.

Clearly, the key drawback with this strategy is that nonpersistent XSS attacks can only be prevented; compared to the prior solution, the strategy only tackles attacks focused on HTML/JavaScript technology. In summary, while proposals based on analysis and filtering are the security method as well as strategies that have been most commonly implemented until now, they have major shortcomings on the identification and precaution from compound XSS attacks in existing web based applications. While technically, we accept that such methods for filtering and analyzing can be suggested as an simple job, we still think that its execution is quite difficult in practice (particularly with applications with high client-side processing, such as applications centered in Ajax). First of all, utilizing filtering and review proxies, in particular on the server side, creates major efficiency and scalability limits for a particular web application. In comparison, deceptive scripts may be inserted inside the documents shared in a somewhat blurred manner (e.g., hexadecimal encoding or later encoding) such that certain filters/analysts became less suspect. In the end, many of the well-documented XSS attacks in JavaScript are embossed and incorporated in HTML papers; it is possible to employ other techniques such as Java, Light, ActiveX, etc. For this purpose, the conception of a general filtering or review mechanism that can tackle potential misuse of such languages appears to us to be quite complicated.

5.4.1 RUNTIME ENFORCEMENT FOR WEB BROWSERS

Alternatives for evaluating and filtering site material on any server or client proxies attempt, by suggesting techniques for applying the runtime meaning of a stage, that is, a web viewer, to remove the need for intermediate items. For starters, the authors suggest an auditing framework for the Mozilla web browser JavaScript interpreter. They use an intrusion prevention system that identifies misuses of JavaScript and takes proper countermeasures for the safety of the browser (e.g., XSS attack) during execution of JavaScript operations. The key thrust behind this strategy is to identify instances where executing a script written in JavaScript requires abusing browser tools, such as the redistribution of website cookies to untrusted parties—who breach the same web browser origin policy. The authors present this approach in their work and assess the total expense of the interpreter of the browser. Such an overhead tends to grow tremendously as do the amount of script operations. Therefore, during the study of nontrivial JavaScript-based routines, we note scalability drawbacks of this method. In comparison, the only way to avoid JavaScript-based XSS attacks is by their approach. In order to handle the auditing of various interpreters, such as Java, Flash, and other, we have no further progress to our knowledge. The use of tiny tests is a different way to conduct the code audit to ensure that tools in the browser are not exploited. You will find an expanded version of the Mozilla web browser JavaScript interpreter which uses taint power. Their approach to testing is the same as those audit procedures suggested on the server side for script analyses (e.g., at the website of the app or an intermediate proxy) in the previous section. But as Hallaraker and Vigna's work in 2005[7] and Jovanovic et al.'s[8] proposal use a Dynamic JavaScript Review Code to determine where application objects (e.g., user identity, cookies) is transferred by the server JavaScript Parser, without using intrusion mitigating technologies. If this is done, the user is told to accept or reject the transfer. Even if this final draught is based on a solid fundamental principle, there may be major drawbacks. First, under the final user decision, the browser defence adds another security layer. Sadly, most of the web app users are not really aware of the hazards and will immediately support the browser's appeal. A second limitation of this strategy is that the complex auditing of all information transfers cannot be assured. The authors have to add a static analysis to their dynamic approach to solve this situation, any time they see that dynamic analysis is not adequate. This limitation potentially contributes to scalability disadvantages in medium and large-scale scripts. It is also fair to assume how good and how working they are in our handling of our motivative problem, which we deem excessively expensive, is their static study. Furthermore, this proposal continues to address the single case of XSS-based attacks on JavaScript, comparable to many other literary proposals, despite a variety of other languages like Java, Flash, AtivaX, etc., being discussed.

Jim et al.[9] provide writers with a protocol approach to handling the web browser against an XSS attack that includes a list of activities, for example, to accept or deny a specific script in the text exchange mechanism between the server and its client. By using these actions and like the Mozilla Firefox plugin extension noscript, a browser can then select whether to execute or deny a browser interpreter script, for example, whether a browser resource may or may not be manipulation by another script. As Jim et al.[9] point out, there are some analogies to host-based intrusion detection approaches in the plan, not only to execute a local surveil to detect program-based maluse but more importantly, to define the required behavior, using white listing scripts and boxes. But we consider their approach is too restricting, particularly by using their recommendation of different browser tools by using sandboxes—which, we think, may explicitly or indirectly impact various sections of the same text, which would obviously affect the correct usability. We also conceive of a lack of semanticize in the policy vocabulary Jim et al.[9] proposed in the context for exchanging policies.

5.4.2 DATA INPUT VALIDATION

Input validation is achieved so as to guarantee that only correctly formulated data is inserted into a workingflow in a system of knowledge, that malformed data is prevented in a database and that multiple components are malfunctioning. Validation of information can take place into the data flow as quickly as possible, ideally after the data is collected from the third party. Input validation should be subject to details from all possibly un-trustful channels including not just web-facing clients but also backend feeds from aliens, vendors, collaborators or regulators, who could all be corrupted on their own and begin to relay malformed knowledge. Input validation is not the only way to avoid XSS, SQL injection and other attacks covered in respective cheat sheets but also may greatly mitigate their effect if properly applied. Syntactic and semantic feedback confirmation should be implemented. The proper syntax of structural fields (e.g., SSN, date, currency symbol) can be used for syntactic validating. In a particular business setting, textual validation shall ensure consistency of its principles (e.g., the start date is pre-end date, the price inside the anticipated range). In coping with the user's (aggressor) order, it is often advised to stop attacks as soon as possible. The authentication of data should be used until a submission is evaluated to identify unwanted information.

5.4.3 DATA SANITIZATION

Data Sanitization ensures that the data contained on a memory unit is intentionally deleted, indefinitely and irreversibly, and therefore unrecovered. A sanitized system contains no residual evidence, and the evidence can never be retrieved except with the aid of sophisticated forensic instruments. Data sanitization is accomplished by three methods: physical removal, cryptographic eradication, and data eradication.

5.4.4 BY APPLYING THIRD PARTY FIREWALL

Installing a more efficient Firewall along with system and server security mechanism on both sender and receiver end, to keep a tap and track on the data and data activity before sending the data from sender and receiving the data at receiver end. Moreover, other permissions and requests can also be managed by taking more security measures.

5.5 SUMMARY AND COMMENTS ON CURRENT PREVENTION TECHNIQUES

In summary, we conclude that the ideas evaluated are not adequately developed and therefore need to develop to handle our issue area better. In addition, in order to effectively overcome the XSS risks for present web applications, we assume that an arrangement between server and browser-based solutions is required. While, in contrast to any server or client-based proxy solutions, we agree that there are obvious advantages to compliance of web browsers (e.g., bottleneck situations and scalability situations in which the processing and retrieval of knowledge shared is performed through an indirect proxy in server or client), the collection of acts that should eventually by the client side. There are several other compliance practices to remember, such as authenticating all sides before policy meetings and a range of asset safety measures at the consumer level. Indeed, we are working in this way to develop and incorporate a policy-dependent web browser compliance with the XACML approach on the server side and to switch the X.509 certificate and Secured Shell (SSL) protocol between client and server.

5.6 FUTURE ASPECTS

A safer feedback processing is the most critical strategy for alleviating XSS assaults. Each when user feedback is implemented in a web page, encoding methods should be included. Encoding processes may be supplemented by frameworks for authentication of inputs in many other situations. The healthy handling of the feedback supplied by the user may take care of which webpage perspective injects the information provided by the user. Therefore, secure management of user-supplied information must be included in the client and server-side origins of web apps to avoid all forms of XSS assaults. Taking into account all the investigation weaknesses in the latest defensive XSS attack strategies, a new protective XSS technology is required to satisfy all of the requirements as follows:

- Automatic mechanism to differentiate from malicious inserted code amongst valid JavaScript code.
- Speed testing of XSS attack detection should be carried out easily and rapidly.
- Context Specific sanitation routines have to be incorporated in the web server source code.

5.7 CONCLUSIONS

The growing usage of the network model for the creation of ubiquitous apps is giving window to new protection risks to the background foundation of those apps. web-based application makers must acknowledge the use of support resources to promise a placement free of risks, like safe coding methods, stable programming prototype and, particularly, design mechanisms for the implementation of applications that are based on web. However, new techniques are being invented by attacker in order to hack the application. The importance of those attacks can be shown from the ubiquitous existence of such software apps in, for instance, essential vital structures in sectors such as government bodies, finance, sectors in health care, and so on. In this article, we have analyzed a particular instance of assault against online apps. This shows that the presence of XSS vulnerabilities on a web server may entail a massive danger for the framework along with the users. In a survey conducted by us, we have given current methods for the protection of XSS attacks on compromised systems, addressing their pros and cons. While handling nonpersistent or persistent XSS attacks, there are currently quite gripping solutions which include interesting methods to to give best solutions for the issues. Certain failures, those do not have adequate protection and can be easily dodged, some of them are so complicated that they are impossible in actual scenarios. We conclude that an effective approach to deter XSS attacks should be implemented for the protection policies established over the end-point and at the side of server as well. The behavior of the browsers resources must be specified and implemented by their creators and administrators. Working in this direction we added an extension for the Mozilla's Firefox framework that extends its framework's same origin policies in order to carry out policies of XACML defined at the side of server, and shared between server and client via certificates of X.509 over the protocol of SSL. Our goal is to subsist with not only JS-based XSS assaults, but also any other scripting language deployed across current web browsers and potentially dangerous for the security of certain browser resources associated to specific web-based apps. We have revised our strategy and dealt with some of the main problems.

KEYWORDS

- **code injection attacks**
- **security rules**
- **software protection**
- **cross-site scripting**
- **web proxy**

REFERENCES

1. Scott, D.; Sharp, R. Abstracting Application-Level Web Security. In *11th International Conference on the World Wide Web*, 2002; pp 396–407.
2. Su, Z.; Wasserman, G. The Essence of Command Injections Attacks in Web Applications. In *33rd ACM Symposium on Principles of Programming Languages*, 2006; pp 372–382.

3. Pietraszeck, T.; Vanden-Berghe, C. Defending Against Injection Attacks Through Context Sensitive String Evaluation. In *Recent Advances in Intrusion Detection (RAID 2005)*, 2005; pp 124–145.
4. Ismail, O.; Etoh, M.; Kadobayashi, Y.; Yamaguchi, S. A Proposal and Implementation of Automatic Detection/Collection System for Cross-Site Scripting Vulnerability. In *18th International Conference on Advanced Information Networking and Applications (AINA 2004)*, 2004.
5. Kirda, E.; Kruegel, C.; Vigna, G.; Jovanovic, N. Noxes: A Client-Side Solution for Mitigating Cross-Site Scripting Attacks. In *21st ACM Symposium on Applied Computing*, 2006.
6. Alcorna, W. Cross-Site Scripting Viruses and Worms—A New Attack Vector. *J. Netw. Security* July **2006,** *2006* (7), 7–8.
7. Hallaraker, O.; Vigna, G. Detecting Malicious JavaScript Code in Mozilla. In *10th IEEE International Conference on Engineering of Complex Computer Systems (ICECCS'05)*, 2005; pp 85–94.
8. Jovanovic, N.; Kruegel, C.; Kirda, E. Precise Alias Analysis for Static Detection of Web Application Vulnerabilities. In *2006 Workshop on Programming Languages and Analysis for Security*, 2006; pp 27–36.
9. Jim, T.; Swamy, N.; Hicks M. Defeating Script Injection Attacks with Browser-Enforced Embedded Policies. In *International World Wide Web Conference, WWW 2007*, May 2007.
10. Anderson, A.; Lockhart, H. SAML 2.0 Profile of XACML v2.0. Standard, OASIS, Feb 2005.
11. Amit, Y. XSS Vulnerabilities in Google.com, Nov 2005.
12. Hansen, R. XSS Cheat Sheet for Filter Evasion. http://ha.ckers.org/xss.html
13. Howard, M.; LeBlanc, D. *Writing Secure Code*, 2nd ed.; Microsoft Press: Redmond, 2003
14. Larson, E.; Austin, T. High Coverage Detection of Input-Related Security Faults. *USENIX Security Simposium*, 2003; pp 121–136.
15. Livshits, B.; Erlingsson, U. Using Web Application Construction Frameworks to Protect Against Code Injection Attacks. In *2007 Workshop on Programming Languages and Analysis for Security*, 2007; pp 95–104.

CHAPTER 6

TBM-BASED CHARGER DEPLOYMENT TECHNIQUE IN THE INTERNET OF THINGS

ANKITA JAISWAL, SUSHIL KUMAR, and PANKAJ KASHYAP

School of Computer and System Sciences, Jawaharlal Nehru University, New Delhi, India

ABSTRACT

Wireless Power Transfer technology is a convenient and reliable method for supplying power to wireless devices. In recent years, this technique has been practically used in Internet of Things for power supply to the wireless devices. This technique efficiently solves the finite battery power problem of the devices. Previously proposed models on optimal charger deployment have not considered the effect of data traffic distribution in the network on energy consumption of devices. Traffic distribution causes the nonuniform energy consumption of devices. Because of it, the network performance deteriorates. In this context, we propose a method to find optimal location for charger deployment based on transferable belief model. Further, we determine the maximum energy supplied by charger in the network for calculation of combined belief to select the optimal location. An algorithm to find the optimal number of chargers to be deployed in the network has also been proposed. This algorithm minimizes the number of chargers, resulting to cost-effective network. It also optimizes the network performance by increasing the network lifetime. The results obtained from simulation show that the proposed technique for charger deployment performs better with

Advanced Computer Science Applications: Recent Trends in AI, Machine Learning, and Network Security. Karan Singh, PhD, Latha Banda, PhD & Manisha Manjul, PhD (Eds.)

less chargers than random deployment approach. The performance of the network has also been determined based on variable parameters of the proposed algorithm.

6.1 INTRODUCTION

In the last few years, considerable attention has been given on Internet of Things (IoT) market and technologies.[1,2] It is considered to be the technological revolution for the upcoming generation.[3,18] Generally, batteries are used as a power source for energy-limited IoT devices (sensors), but they face problem of restricted lifespan because of the finite battery size. Numerous researches have been done on effectively dealing with the problem of finite battery power, comprising the efficient power utilization,[4] effectual routing protocol,[5] and so on. Nevertheless, the above solution did not solve the fundamental aspect of the problem. Even though they can successfully increase the lifespan of the network, despite that the network cannot be utilized in the end, when the battery of the sensors totally drained out.

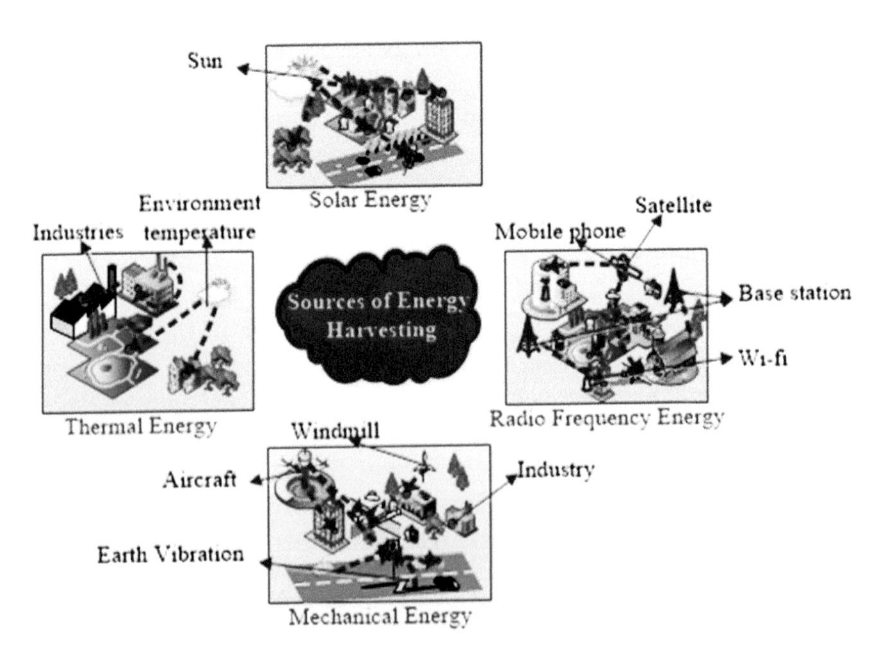

FIGURE 6.1 Sources of energy harvesting.

Generally, sensors can be recharged in two different ways. One way is to allow sensors to recharge their batteries with the energy harvested from the surroundings. However, energy harvesting using ambient sources, for example, wind, solar, vibration, etc. (see Fig 6.1) is not as effectual as anticipated because of its uneven and unpredictable nature. Hence, the contemporary approach in wireless energy transfer using radio frequency signal can be embraced to enhance the lifespan of IoT-enabled system. Energy harvesting from a source that emits RF signal is more well-founded and controllable to assure a definite amount of needed energy transfer as compared with natural sources. Another approach is to deploy chargers in the locality of sensing devices and recharge their batteries by exploiting the wireless energy transfer technique. This new technology has allured attentiveness of industrial professional and academic researchers, because of its high stability nature and convenient use. Since chargers are expensive, there should be minimum number of chargers deployed in the region. Contrarily, the performance of the sensor is significantly affected by the amount of the energy provided by the charger. Thus, efficaciously minimizing the quantity of chargers while also improving the efficiency of the entire wireless network becomes an important problem in IoT system. In spite of the fact that, a lot of research works relating to the placement of chargers have been published in the last few years, these researches did not consider some critical aspects. As an illustration, former research has solved the charger's deployment problem without considering the effect of traffic flow in the network.[8,9] Former studies assume an even power consumption of the devices, yet it is not the case in a practical scenario. In this context, this chapter proposes the novel data traffic-conscious charger deployment method based on transferable belief model (TBM) for an IoT system. The target is to enhance network lifespan with optimal number of chargers. The key contributions of the chapter can be summarized as follows:

(1) Firstly, we develop a model for optimal charger deployment based on transferable belief model.
(2) Secondly, we propose an algorithm to find optimal number of chargers to be deployed in the network.
(3) Finally, the proposed work is simulated against state-of-the-art algorithms.

The rest of the chapter is organized in following sections. Section 6.2 presents the related works. In Section 6.3, we present the system model. Section 6.4 presents the optimal TBM model for charger deployment. Section 6.5 presents the proposed algorithm. Section 6.6 presents the simulation results and their analysis. Finally, the conclusion of the chapter is given in Section 6.7.

6.2 RELATED WORK

In this section, a qualitative review on optimal charger deployment in wireless sensor networks to enhance the network lifetime is presented.[6,10] Two heuristic algorithms, namely; the greedy cone covering and the adaptive cone covering are proposed for optimum placement of directional chargers in the network.[11] Deployment region has been discretized into grids and chargers' charging range has been modeled as a cone. Since charging space is considered as a cone, accurate value of sensor's charged energy may not be calculated. The problem of optimal placement of mobile chargers with the knowledge of event happening has been considered and is solved by utilizing the integer linear programming method.[12] Profit minimizes with growth in the number of missions and sleep scheduling of sensor has not been considered, which may diminish the performance of the network. Optimization algorithm based on particle swarm optimization method has been proposed to optimally place omnidirectional chargers in the monitoring region to enhance the lifetime of the network.[13] Optimization problem has been formulated as a function of space between nodes and charger, and angle between antenna of charger and vector going from charger to sensor. Limitation of the solution is that, it is practically applicable only on indoor wireless rechargeable sensor network. The problem of optimal number of charger deployment has been considered as a minimum dominating set problem.[14] The notion of virtual grid has been utilized to represent the monitoring region. An efficient algorithm has been designed for optimal charger deployment by using the knowledge of sensors' mobility path.[15] Moreover, for large number of grid points, there is an increase in computational complexity of the algorithm with decrease in distance between the grids. A scheduling problem for the placement of a charging vehicle with multiple wireless chargers has been considered to provide energy to a large-scale wireless network.[16] However, under the novel charging model, the traveling path of the vehicle is complicated because the vehicle needs to do both the task of deployment and collection of chargers when charging task is done.

The above studies did not consider some main factors. As an illustration, former researches have solved the charger deployment problem without considering the traffic flow of the network and assumed an even power consumption of the devices which is not the case in a practical scenario.

6.3 PROPOSED SYSTEM MODEL

In this section, the network model for an IoT-enabled wireless rechargeable sensor network is presented. We assume that the monitoring area is circular in shape with radius R. Let S number of sensors is randomly deployed in the region; C be the number of chargers and a sink is placed at the center of the circular region. According to the assumption of the network, it can be observed that sensed information in the whole network eventually reach the center of the monitoring region. Sensors near the sink exhaust their battery faster in this type of sensor deployment approach. Hence, for even distribution of sensors, we divide the monitoring area in concentric circles with approximately equal gap. Data traffic follows geometric distribution when transmitted from outer annulus to inner annulus toward the sink by the sensors.[17] The objective of the proposed work is to deploy optimal number of directional chargers. Each charger can be placed anywhere in the monitoring area. It is considered that each node can be charged by multiple chargers and each charger can provide energy to multiple sensors. The wireless chargers can be placed in an unlimited number of locations in the real monitoring environment and sensors can get affected by the chargers differently. The selected region should be divided into limited subregions, so that the problem can be solved and charger deployment location limited to a finite number of positions. We assume that the circular monitoring area is first divided into many concentric circles of equal gap and after that it is divided into sectors, which means each sector in partitioned into a number of area and each sector is given a number to identify it. The different sector partitions are to be taken as charger deployment locations as shown in the Figure 6.2.

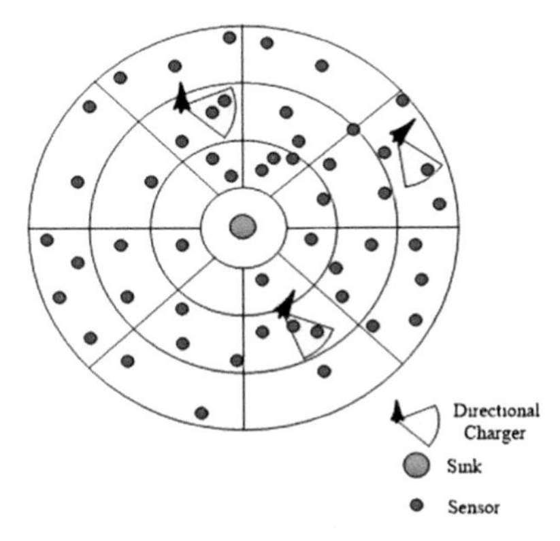

FIGURE 6.2 Network architecture with chargers' deployment.

6.3.1 RECEIVED SIGNAL STRENGTH INDICATOR MODEL FOR SENSOR'S LOCATION CALCULATION

In this section, from each location for charger deployment, distance of every sensor node is calculated. Each node determines its distance from the charger deployment locations using received signal strength indicator. Equation 6.1 computes the received signal strength.

$$I_r = I_o + 10\ \upsilon log\left(\frac{d_r}{d_o}\right) + \delta \tag{6.1}$$

$$D_r = d_r 10^{\frac{RSSI-\alpha}{10\beta}} \tag{6.2}$$

Where I_o (in dbm) is the reference signal strength at distance d_o ($d_o = 1\ m$), υ is path loss exponent ($2 \leq \upsilon \leq 4$), and d_r is the actual distance. δ represents Gaussian random variable, which has mean zero and variance $\sigma^2(4 \leq \sigma \leq 12)$. By eq 6.2, the distance between charger deployment location and node is calculated. Where α represents the strength of received signal in 1 m distance from the charger deployment location without any obstacle.

6.3.2 ENERGY MODEL

In this section, we present the energy model of the proposed network architecture. We assume that the sensor nodes are operating uninterrupted for T_{act} seconds, for harvesting duration of T seconds. The duty cycle is defined as

$$m = \frac{T_{act}}{T} \tag{6.3}$$

The network functionality can be activated through many external activities or through regular data collection with rate m throughout the application period, $0 < n < 1$. We presume that the data transmission rate is not uniform therefore, it is represented by a random variable. It is assumed that the wireless chargers are directional charger and its coverage area is assumed to be a sector with radius w and charging angle be θ. However, energy received by the sensors that are outside of the charging region is too small. Let the maximum transmit power of cth charger be P_c and maximum battery capacity of each sensor be B_{max}. Let Hr_k be the harvesting rate of sensor node k. The energy harvested at a sensor node k from charger c is given as

$$EH_k = \eta_k P_c (d_{kc})^{-\upsilon} \varepsilon_k X_k T \tag{6.4}$$

where $(0 < \eta_k < 1)$ represents the energy harvesting coefficient for sensor node k from charger c, $0 < \varepsilon_k < 1$ represents the sensor k energy conversion efficiency, d_{kc} is the distance between the node k and charger c, and υ is the path-loss exponent. X_k represents the composite fading channel. Because of the random nature of X_k, the harvested energy EH_k is also a random variable. The mean value of harvested energy can be written as

$$\overline{EH_k} = \eta_k P_c (d_{kc})^{-\upsilon} \varepsilon_k \overline{X_k} T = Hr_k T \tag{6.5}$$

EH_k is deterministic but varies for different nodes because the distance between the charger and the different sensor nodes is not same. Next, we compute the total energy utilized by sensor k in time T.

When the network is activated, it uses distributed or centralized mechanisms for steady-state, even, collision free function and we presume that to achieve these each sensor node consumes j Joule of energy. Each sensor node consumes h Joule-per-bit on an average for processing and transmission. The rate of data transmission (in bps) is represented by a random variable Ψ with PDF $P(\Psi)$. The application layer of each receiver sensor node receives data with the rate of z bps. At the same annulus, data is transmitted by n homogeneous sensors, for each receiver sensor node in each annulus the ratio $\frac{z}{n}$ is defined as coupling point. Because of the variability in transmission rate of data per node, we experience the following two cases: (i) underloading of receiver sensor node, where $\Psi < \frac{z}{n}$ and it consumes "idle" energy with rate i Joule-per-bit (J/b); (ii) overloading of receiver sensor node, where $\Psi > \frac{z}{n}$ and it consume energy with rate p J/b for buffering the data before transmission. We give components for the energy consumed by sensor node in the duration of mT seconds during which the sensor is in active mode, that is, transmitting (h J/b), receiving (g J/b), buffering (p J/b), beaconing (i J/b), sensing and runtime operations. Each sensor node consumes energy given by $m\mathrm{Th}\int_0^{\infty}\Psi \mathrm{P}_{\mathrm{y}+1}(\Psi)\, d\Psi = m\mathrm{ThE}_{\mathrm{y}+1}[\Psi]$ J in order to transfer its own data along with the data transferred to it from y other sensor nodes. $E_{y+1}[\Psi] \equiv (\mathrm{y}+1)\mathrm{E}[\Psi]$ and for each sensor node that is not a relay $\mathrm{E}[\Psi]$ is the expected transmission rate. The energy consumed by each sensor node for receiving and buffering data from y other nodes before transferring it is, $m\mathrm{Tg}\int_0^{\infty}\Psi \mathrm{P}_{\mathrm{y}}(\Psi)\, d\Psi = m\mathrm{TgE}_{\mathrm{y}}[\Psi]$ J. The idle energy consumed by each sensor node is, $mTi\int_{0}^{\frac{z}{n}}(\frac{z}{n}-\Psi)P_{y+1}(\Psi)d\Psi$ J. The penalty energy consumed by each sensor node for data buffering is, $mTp\int_{\frac{z}{n}}^{\infty}(\Psi-\frac{z}{n})P_{y+1}(\Psi)\,d\Psi$ J. Thus, the total consumed energy of sensor node k is given as

$$E_p = j + mT\times\left[E_{y+1}[\Psi]\left(h+\frac{gy}{y+1}\right)+i\int_{0}^{\frac{z}{n}}\left(\frac{z}{n}-\Psi\right)P_{y+1}(\Psi)d\Psi+p\int_{\frac{z}{n}}^{\infty}\left(\Psi-\frac{z}{n}\right)P_{y+1}(\Psi)\,d\Psi\right] \quad (6.6)$$

We are using the relationship $\forall y>0\ :\ E_{y+1}[\Psi]=\frac{y+1}{y}E_y[\Psi]$ in eq 6.6, because with respect to y the expected transmission rate linearly increases in a homogeneous WSN. Adding and subtracting $mTp\int_{0}^{\frac{z}{n}}\left(\Psi-\frac{z}{n}\right)P_{y+1}(\Psi)\,d\Psi$ in E_n, we get:

$$E_p = j + mT\times\left[E_{y+1}[\Psi]\left(h+\frac{gy}{y+1}+p\right)-\frac{zp}{n}+(i+p)\int_{0}^{\frac{z}{n}}\left(\frac{z}{n}-\Psi\right)P_{y+1}(\Psi)\,d\Psi\right] \quad (6.7)$$

Clearly, the energy consumption is influenced by the coupling point, $\frac{z}{n}$, and by the data transmission rate PDF per sensor node, $P_{y+1}(\Psi)$. Since we have mentioned earlier that, data traffic flow follow geometric distribution, $P_{y+1}(\Psi)$ with probability of success q = 1/(y + 1)r is given as

$$P_{y+1}(\Psi)=(1-q)^{\Psi-1}q \quad (6.8)$$

where *r* is the network data rate. For $\Psi > 0$ in this case, the expected value of Ψ is $E_{y+1}[\Psi] = (y + 1)$ r bps. By using eq 6.8 in 6.7, we obtain:

$$E_p = j + mT\times\left[(y+1)r\left(h+\frac{gy}{y+1}+p\right)-\frac{zp}{n}+(i+p)\left(\frac{z}{n}\left(1-(1-q)^{\frac{z}{n}}\right)-\left(\frac{1-(1-q)^{\frac{z}{n}}}{q}\right)-q\frac{z}{n}(1-q)^{\frac{z}{n}}\right)\right] \quad (6.9)$$

6.3.3 *OPTIMAL CHARGER DEPLOYMENT WITH TBM*

In this section, we consider a TBM[18]-based approach to deploy optimal number of chargers to maintain the network functional, so that network lifetime would be enhanced. Also, this approach is used to find optimal location in the network, where charger can be placed. The TBM works in 2 level, namely, credal and pignistic. In the credal phase, a charger C_v is placed at each partition of sector in random fashion, thereafter it is rotated in 360° directions to find the direction in which charger can supply maximum amount of energy in the network. To calculate the belief ($b_{kv}(x_v y_v)$) of each sensor depends upon two factors, first, the sensor must be in the range of charger and second the amount of energy harvested by the sensor must satisfy the constraint, E^{thr} of minimum energy harvest. In the pignistic phase, the collective belief $Bet\ P(x_v y_v)$ is calculated as given in eq 6.15 by using the belief computed at credal phase and is used to find optimal location for charger deployment. The calculation of beliefs and collective beliefs in the reference of TBM are described in the following section.

6.3.4 FRAME OF DISCERNMENT (FOD)

In TBM, a set μ that contains every possible states of the network system is called FoD. In the reference of optimal charger deployment, the set of locations in the monitoring area is considered as FoD, and is given as

$$\mu = L_{x,y}(C_v), \qquad \forall C_v \in C \tag{6.10}$$

Where $L_{x,y}(C_v)$ represents the location information of the candidate location for charger deployment and C denote set of chargers in terms of coordinates.

6.3.5 BELIEF CALCULATION

Let C_v denotes a charger at location $(x_v y_v)$, w be its charging radius, θ be its charging angle and R_v be a distance such as $0 \leq R_v \leq w$. The certain charging zone of sensor k is when it lies within the sector of angle θ and radius $(w - R_v)$ $(w - R_v)$. In this range, sensor is charged maximum. The uncertain charging zone of sensor is complement of certain zone and is divided into two sub-parts: partial ignorance zone and total ignorance zone. Partial ignorance zone is defined as sector of angle θ with radius between $(w - R_v)$ and w. And total ignorance zone is defined as complement of partial ignorance zone. Sensor k provides information on energy charged by charger Cv placed at location $(x_v y_v)$ with a belief b_{kv} and is given as

$$b_{kv}(x_v y_v) = \frac{\mathrm{Hr_k T}}{B_{max}}, \qquad 0 \leq b_{kv}(x_v y_v) \leq 1 \tag{6.11}$$

Altogether, depending on the distance between a sensor and a charger, the sensor provides the following belief functions:

- Within certain charging zone, the sensor produces belief and nonbelief function as follow

$$b_{kv}(x_v y_v) = 1$$

$$b_{kv}(\mu) = 0$$

- Within partial ignorance zone, the sensor produces belief function as follow

$$b_{kv}(x_v y_v) = \frac{\text{Hr}_k \text{T}}{B_{max}}, \qquad 0 \le b_{kv}(x_v y_v) \le 1 \tag{6.12}$$

$$b_{kv}(\mu) = \left(1 - \frac{\text{Hr}_k \text{T}}{B_{max}}\right) \tag{6.13}$$

- Within total ignorance zone, the sensor produces belief function as follow

$$b_{kv}(\mu) = 1$$

6.3.6 BELIEF COMBINATION

Applying a conjunctive combination, we construct a new belief representing the consensus of the belief obtained from sensors for optimal location of charger C_v. We combine the belief of all sensors that are within the range of charger and whose charged energy from charger C_v at location $(x_v y_v)$ is greater than some threshold, E^{thr}. Similarly, combined belief is calculated for all locations μ.

$$b_{kv}^{cl}(x_v y_v) = \prod_{k=1}^{S} b_{kv}(x_v y_v) + \frac{b_{av}(x_v y_v) b_{bv}(x_v y_v) b_{lv}(x_v y_v) \ldots b_{kv}(x_v y_v)}{1 : S - 1 terms, a, b \ldots l = 1 \ldots S, a \ne b \ne l \ne k} \sum_{k=1}^{S} (1 - b_{kv}(x_v y_v)) \tag{6.14}$$

6.3.7 DECISION MAKING

The pignistic transformation that allows the construction of probabilities that is used for selection of optimal location is based on selecting the location with maximum pignistic probability.

$$Bet\,P(x_v, y_v) =$$

$$\prod_{k=1}^{S} b_{kv}(x_v y_v) + \frac{b_{av}(x_v y_v) b_{bv}(x_v y_v) b_{lv}(x_v y_v) \ldots b_{kv}(x_v y_v)}{1 : S - 1 terms, a, b \ldots l = 1 \ldots S, a \ne b \ne l \ne k} \sum_{k=1}^{S} (1 - b_{kv}(x_v y_v)) + \frac{\sum_{k=1}^{S} (1 - b_{kv}(x_v y_v))}{S} \tag{6.15}$$

Thus, optimal location for charger deployment is selected whose pignistic probability is max(*Bet* $P(x_v y_v)$).

After obtaining the optimal location of charger, next objective of the proposed work is to find the optimal number of chargers such that each sensor has minimum residual energy E_{res}, which can keep the network alive.

The problem can be formulated as

$$C_{opt} = \min\left(C, such\,that\left(\overline{\mathrm{EH_k}} - \mathrm{E_p}\right) = E_{res}\right) \tag{6.16}$$

for all sensors k, where C_{opt} represents the optimal number of chargers. To solve the problem 6.16, an algorithm is given in Algorithm 1.

6.4 PROPOSED ALGORITHM

Algorithm 6.1 *Belief-Based Optimal Charger Deployment (S,* L^c*, sec,* N_{sec}*, C,* C_{opt}*)*

//S=Total number of sensor nodes with its location, L^c*=Number of partitions in each sector,* N_{sec} *=Total number of //sectors, C=Number of directional chargers,* C_{opt} *= Minimum number of chargers,* L^c_{set}*= Set of belief of each sector, //*Bet^c *= Belief at each sector at location* L^c*,* L^{opt} *= Set of optimal location of maximum belief*

Input:

N_{sec}, *sec, C,* L_{opt}*, a*

Output:

C_{opt}

1. ***Initialize*** N_{sec}*=8, sec=0, C=0,* L_{opt} *=* Φ*, a=1*
2. ***while*** *(a=1) do*
3. ***for*** *(sec= 1+2sec to* N_{sec}*-1)*
4. ***for*** *(*L^c *=1 to 3)*
5. *Find belief Bet* $P(x_v, y_v)$ *for each* L^c *by using eq 6.15 and store into set* L^c_{set} *=* $\{Bet^1, Bet^2, Bet^3\}$
6. Bet^c_{max} *= max* $\{L^c_{set}\}$
7. $L_{opt} = L_{opt}$ *U* $\{Bet^c_{max}\}$
8. ***end*** *for*
9. *sec=sec++*
10. ***end*** *for*
11. L_Bet^c_{max} = max $\{L_{opt}\}$
12. *Place charger at location* $P(x_v, y_v)$ *where* Bet^c_{max}

13. *C=C+1*

14. *Calculate* $\left(\overline{\mathrm{EH}_\mathrm{k}} - \mathrm{E_p}\right)$, $\forall\, k \in S$

15. ***if*** $\left(\overline{\mathrm{EH}_\mathrm{k}} - \mathrm{E_p}\right) \geq E_{res}$, $\forall\, k \in S$ *then*

16. *{a=0; return* C_{opt}*=C}*
17. ***else***
18. ***for*** *(sec= 2(sec) +2 to* N_{sec}*)*
19. ***repeat*** *step 4 to13*

20. *Calculate* $\left(\overline{\mathrm{EH}_\mathrm{k}} - \mathrm{E_p}\right)$, $\forall\, k \in S$

21. ***if*** $\left(\overline{\mathrm{EH}_\mathrm{k}} - \mathrm{E_p}\right) \geq E_{res}$, $\forall\, k \in S$ *then*

22. *{a=0;return* C_{opt}*= C}*
23. ***else***
24. *Go to step 3*
25. ***end*** *while*

In this section, an algorithm is proposed to find optimal number of chargers to be deployed in the network to keep it functional. The input parameters in the proposed Algorithm 6.1 are N_{sec}, sec, C, L_{opt}, a and output parameter is C_{opt}. Here, we start by considering that the monitoring region is first divided into L^c number of concentric circles and then partitioned into N_{sec}, number of sectors, which can be identified by an identifier as even or odd sector. In order to find minimum number of chargers, first we choose a random position in the first partition of sector-1 (odd sector) and place a charger. After that we rotate the charger continuously in 360° direction to find the direction in which maximum energy is harvested by each sensor that are covered by the charger and has minimum harvesting energy greater than E^{thr}. We use the energy of sensors to calculate belief of the sensors. Similarly, we calculate the belief at all locations in all odd sectors and choose the location (x_v, y_v) with maximum belief using eq 6.15, where we place first charger (Step 3–11) and increase charger by one. Then we check $\left(\overline{\mathrm{EH}_\mathrm{k}} - \mathrm{E_p}\right) \geq E_{res}$ for all sensors. If this condition is true for all sensors, then algorithm terminate (Step 14–16) and we get optimal number of chargers C_{opt}. Otherwise, same above steps are repeated for even sectors (18–22). We reapply the strategy until all sensors meet its energy requirement $\left(\overline{\mathrm{EH}_\mathrm{k}} - \mathrm{E_p}\right) \geq E_{res}$. So, finally we get optimal number of chargers and their locations required for the network.

6.4.1 PERFORMANCE EVALUATION

In the given section, the performance of the proposed TBM-based approach for optimal charger deployment is evaluated. We have showed the simulation results using different parameters which impact the performance of the proposed algorithm. To evaluate the performance of the proposed algorithm, the parameters which are used for better understanding of the scenario are charger's transmission range, that is, charger's charging radius w and angle θ, number of sensor nodes S and total monitoring area with dimension, R. Next, the proposed algorithm is compared with the random deployment approach. The corresponding simulation parameters are provided in Table 6.1.

TABLE 6.1 Simulation Parameters.

Parameter	Value	Parameter	Value
η_k	0.8	m	0.112
υ	2	Θ	45°, 90°, 135°
ε_k	1	H	2.292×10^{-5} J/b
T	21600 *sec*	P_c	46 *dBm*
z	144 *kbps*	j	165.6Mj
p	3.893×10^{-5} J/b	B_{max}	0.0005 J
i	2.173×10^{-5} J/b	g	2.923×10^{-4} J/b J/b

6.4.2 EFFECT OF THE SENSOR QUANTITY ON AVERAGE ENERGY SUPPLIED BY THE CHARGERS IN THE NETWORK

This section analyzes the effect of the sensor quantity on average energy supplied by the chargers in the network. Here, different charging angle of the charger, $\theta = \{45°, 90°, 135°\}$ is considered. Sensors are deployed randomly in the monitoring area of radius 100 *m* and it varies from 20 to 180. Number of chargers are fixed, that is, $C = 6$ with radius 20 *m*. It can be analyzed from Figure 6.3(a) that the average energy supplied in the network increases with increase in the number of sensors. However, for $S > 80$ the change in average energy supplied in the network is not enough to be noticed. The reason behind it is that each sensor can be charged through multiple chargers and at some point, no extra energy can be supplied in the network when the network reach at its stable state. And it can also be observed that energy supplied in the network increases with increase in the charging angle of chargers.

6.4.3 IMPACT OF CHARGING ANGLE θ

Here, the effect of charging angle θ on the quantity of chargers required to charge the network efficiently is analyzed.

From Figure 6.3(b) it can be observed that the quantity of chargers needed to charge the sensors minimizes with increase in the charging angle of chargers. The reason behind it is that as the charging angle increases, the charging region of charger also increases, which can cover a greater

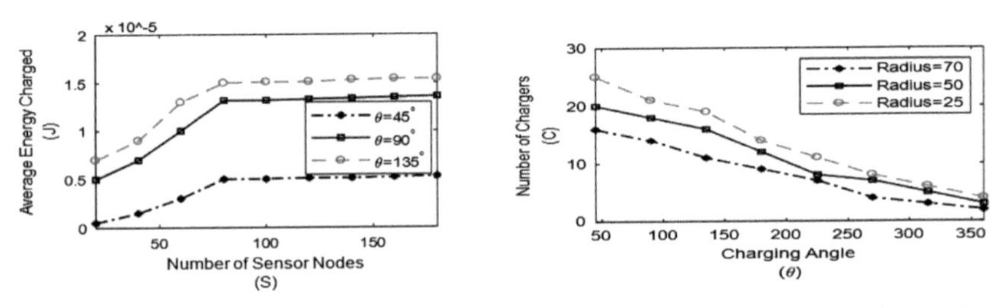

FIGURE 6.3 (a) Number of sensor nodes vs average energy charged, (b) Charging angle vs number of chargers.

number of chargers at a time. It can also be observed that charger with larger charging radius can supply more energy in the network, which in turn decreases the number of chargers required for the network to be functional. Here we consider radius w = {25 m, 50 m, 70 m] for the evaluation work.

6.4.4 NUMBER OF CHARGERS REQUIRED BY VARYING NUMBER OF SENSORS NODES AND TRANSMISSION RADIUS WITH DIFFERENT DIMENSION OF MONITORING AREA

From Figure 6.4, we can observe that for different size of monitoring area and number of sensor nodes, when the radius or transmission range of the charger increases, the quantity of chargers' requirement for supplying energy in the network decreases. The reason behind it is that when the transmission range increases more number of sensors can be in the range of charger that can recharge their battery. Accordingly, the number of chargers required for the network minimizes. Here, it can be observed that chargers' requirement is minimum for larger value of charger radius, that is, 70 and charger requirement is more for small value of radius. It can also be analyzed that with increase in the number of sensor nodes, minimum charger requirement to keep network alive also increases. The reason behind it is that, as the sensor nodes in the network increases, the energy requirement of the network also increases accordingly to perform the required operation. Another criterion that can be analyzed for performance evaluation of the network is its dimension, R. In a large monitoring area, sensors are deployed randomly and are scattered in all direction. For such a network, more chargers are required for supplying energy to all the sensing nodes. In Figure 6.4(a)–(c), to analyze the network performance, different dimension of monitoring area is considered, such as area with radius 50, 100, and 200 m. With charger transmission radius 50 and number of sensor node 100, the number of chargers required are five for area with radius 50 m Whereas, for the same parameter and area with radius 100 and 200 m, the chargers' requirement is seven and nine, respectively.

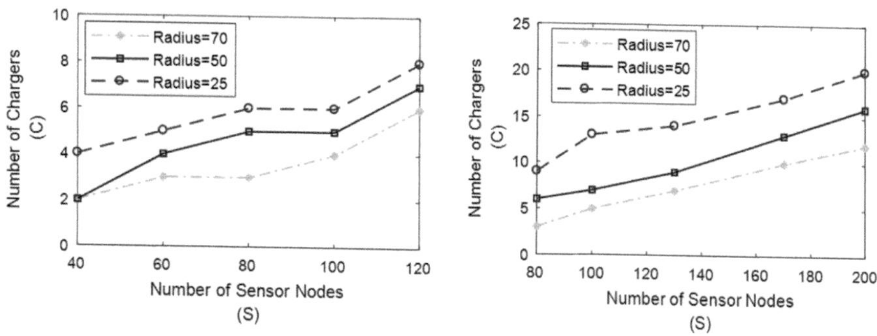

FIGURE 6.4 Number chargers required by varying number of nodes and transmission range when (a) R = 50m, (b) R = 100 m, (c) R = 200 m.

6.4.5 *PERFORMANCE COMPARISON OF TBM-BASED APPROACH WITH RANDOM DEPLOYMENT APPROACH FOR OPTIMAL CHARGER DEPLOYMENT*

Here, the comparison of the proposed charger deployment algorithm with the random deployment in terms of number of sensors and number of chargers is presented. For evaluation, we consider the directional charger with radius 70 and charging angle 135°. The radius of circular area for charger deployment is considered of 100 *m*. It is shown from Figure 6.5 that the proposed algorithm requires a smaller number of chargers than random deployment approach to be deployed in the network for supplying continuous energy to keep the network functional. This enhances the network performance in terms of cost and network lifetime.

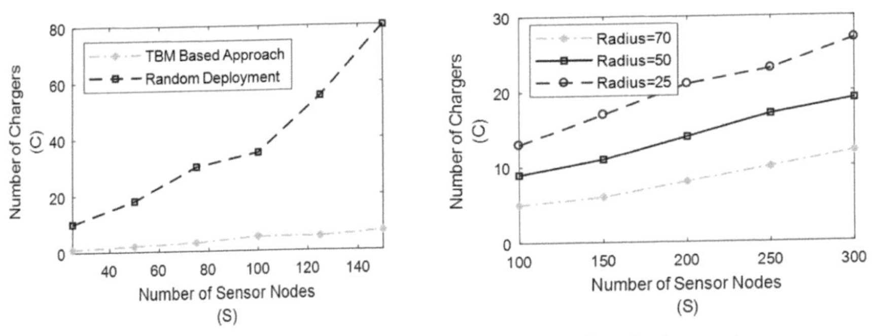

FIGURE 6.5 Comparison of TBM-based approach with random deployment.

6.5 CONCLUSIONS

In this chapter, we have proposed a TBM-based solution for optimal charger deployment problem in the network. Here, we present an algorithm based on belief function to find out an optimal charger deployment strategy. Belief is calculated by using the amount of energy harvested by each sensor in the network from the charger. We have also considered the effect of traffic flow in computation of average energy consumption of the sensors, which has been neglected by many researches. The main goal of the proposed work is to find optimal locations for the deployment of minimum number of chargers. This enhances the network performance in terms of cost and network lifetime. Here, we consider directional chargers for the network because of its low cost and stable power coverage. The simulation of our proposed work is performed on Matlab 2017a platform. From the simulation result, it has been clearly shown that our approach outperforms the random deployment approach for optimal charger placement. We have also checked the performance of the proposed algorithm based on variable parameters. Still there are many factors that need to be considered, including charger recharging, omnidirectional charger, etc. In future direction, different parameters can be considered to optimize the proposed algorithm. Later, we can consider different approaches to develop more innovative algorithm to solve the complex problem, with better network lifetime.

ACKNOWLEDGMENT

The work is supported by Jawaharlal Nehru University under the grant UPE-II.

KEYWORDS

- **optimal charger deployment**
- **Internet of Things**
- **radio frequency energy transfer**
- **transferable belief model**

REFERENCES

1. Farhan, L.; Kharel, R.; Kaiwartya, O. et al. Twards Green Computing for Internet of Things: Energy Oriented Path and Message Scheduling Approach. *Sustain. Cities Soc.* **2018,** *38*, 195–204.
2. Kaiwartya, O.; Abdullah, A. H.; Cao, Y.; Lloret, J.; Kumar, S.; Shah, R. R.; Prakash, S. Virtualization in Wireless Sensor Networks: Fault Tolerant Embedding for Internet of Things. *IEEE Internet Things J.* **2018,** *5* (2), 571–580.
3. Kumar, K.; Kumar, S.; Kaiwarty, O.; Cao, Y.; Lloret, J.; Aslam, N. Cross-Layer Energy Optimization for IoT Environments: Technical Advances and Opportunities. *Energies* **2017,** *10* (12), 2073–2113.
4. Aanchal, K. S.; Kaiwartya, O. et al. Green Computing for Wireless Sensor Networks: Optimization and Huffman Coding Approach. *Peer-to-Peer Netw. Appl.* **2017,** *10* (3), 592–609.
5. Kumar, V.; Kumar, S. Energy Balanced Position-Based Routing for Lifetime Maximization of Wireless Sensor Networks. *Ad Hoc Netw.* **2016,** *52*, 117–129.
6. Sun, G.; Liu, Y.; Yang, M.; Wang, A.; Zhan, Y. Charging Nodes Deployment Optimization in Wireless Rechargeable Sensor Network. *GLOBECOM IEEE Global Commun. Conf.* **2017,** 1–6.
7. Fu, L.; Cheng, P.; Gu, Y.; Chen J.; He, T. Optimal Charging in Wireless Rechargeable Sensor Networks. *IEEE Trans. Vehicular Technol.* **2016,** *65* (1), 278–291.
8. He, T.; Chin, K. W.; Soh, S. On Maximizing Min Flow Rates in Rechargeable Wireless Sensor Networks. *IEEE Trans. Ind. Inform.* **2017**, *14* (7), 2962–2972.
9. Arivudainambi, D.; Balaj, S. Optimal Placement of Wireless Chargers in Rechargeable Sensor Networks. *IEEE Sens. J.* **2018,** *18* (10), 4212–4222.
10. Li, Y.; Chen, Y.; Chen, C. S.; Wang, Z.; Zhu, Y. H. Charging While Moving: Deploying Wireless Chargers for Powering Wearable Devices. IEEE *Trans. Vehicular Technol.* **2018,** *67* (12), 11575–11586.
11. Jiang, J. R.; Liao, J. H. Efficient Wireless Charger Deployment for Wireless Rechargeable Sensor Networks. *Energies* **2016,** *9* (9), 696.

12. Erol-Kantarci, M.; Mouftah, H. T. Mission-Aware Placement of RF-Based Power Transmitters in Wireless Sensor Networks. In *IEEE Symposium on Computers Communications (ISCC)* 2012; pp 000012–000017.
13. Chen, Y. C.; Jiang, J. R. Particle Swarm Optimization for Charger Deployment in Wireless Rechargeable Sensor Networks. In *26th International Telecommunication Networks and Applications Conference (ITNAC)* 2016; pp 231–236.
14. Lin, T. L.; Li, S. L.; Chang, H. Y.; Chang, W. K.; Lin, Y. Y. An Effective Wireless Charger Deployment Method for Complete Coverage in Wireless Chargeable Sensor Networks. In *International Conference on Networking and Network Applications (NaNA)* 2016; pp 379–382.
15. Chiu, T. C.; Shih, Y. Y.; Pang, A. C.; Jeng, J. Y.; Hsiu, P. C. Mobility-Aware Charger Deployment for Wireless Rechargeable Sensor Networks. In *14th Asia-Pacific Network Operations and Management Symposium (APNOMS)* 2012; pp 1–7.
16. Zou, T.; Xu, W.; Liang, W.; Peng, J.; Cai, Y.; Wang, T. Improving Charging Capacity for Wireless Sensor Networks by Deploying One Mobile Vehicle with Multiple Removable Chargers. *Ad Hoc Netw.* **2017,** *63*, 79–90.
17. Dohare, U.; Lobiyal, D. K.; Kumar, S. Energy Balanced Model for Lifetime Maximization in Randomly Distributed Wireless Sensor Networks. *Wireless Pers Commun.* **2014,** 407–428.
18. Kashyap, P. K.; Kumar, S.; Dohare, U.; Kumar, V.; Kharel, R. Green Computing in Sensors-Enabled Internet of Things: Neuro Fuzzy Logic-Based Load Balancing. *MDPI Electron.* **2019,** *8* (4), 384–405.

CHAPTER 7

DATA MINING FOR THE INTERNET OF THINGS: A SURVEY

MUKESH KUMAR[1], SUSHIL KUMAR[1], and SANEH LATA YADAV[2]

[1]*School of Computer & Systems Sciences, Jawaharlal Nehru University, New Delhi, India*

[2]*School of Engineering and Technology, K. R. Mangalam University, Gurugram, Haryana, India*

ABSTRACT

It was like a dream, to connect everything on this earth for computation and communication, but the Internet of things (IoT) makes this "impossible" possible. With the introduction of the Internet of Things, devices with varying size and computational capabilities can be connected for communication. These communication devices produce data which is converted into knowledge for making decisions. The sensor devices deployed in IoT environment generate huge amount of data and all data are not useful. It occupies storage space that is not cost effective and energy efficient. Therefore, useful information needs to be extracted to make appropriate decisions. Extraction of knowledgeable data from raw data is possible with the help of data mining. Data mining for IoT is used to formulate an intelligent environment. Therefore, this chapter elaborates various data mining techniques used for IoT, their applications, challenges in developing IoT environment, and few open research issues.

Advanced Computer Science Applications: Recent Trends in AI, Machine Learning, and Network Security. Karan Singh, PhD, Latha Banda, PhD & Manisha Manjul, PhD (Eds.)

7.1 INTRODUCTION

Internet of things (IoT) is a popular research topic in technology, where different kinds of devices connect with each other through the Internet. The devices are termed smart objects/things. These smart devices have the ability to sense the environmental conditions to make a decision as per the predefined constraints. It is a global platform to create a smart environment where things communicate, compute, coordinate, and make decisions. Such an environment minimizes man power globally. Therefore, things are identified uniquely and automatically.[1–3] After a vast study on IoT, it came into view that IoT has many applications and standards followed with various challenges. Different surveys overview about five layered architecture of IoT to describe overall working design. The five layers are edge technology, access gateway, Internet, middleware, and application. To study IoT, there are three different angles such as the Internet, things, and semantics. The sensor nodes deployed to form an intelligent environment produce huge amount of data which occupies storage space and consumes lots of energy. All the data produced by sensor devices is not useful. Therefore, it wastes the memory space and degrades the energy. As the devices are battery powered, it is wise to utilize the available energy in an effective and efficient manner. The data stream in IoT systems is increasing continuously that is used to develop business models, customized products enabled with personalized services.[4–6]

The infrastructure is well understood, but the question is how the produced signals or data can be formed into knowledgeable data. The answer can be data mining (DM). Data mining is helpful in finding a solution for this issue of extracting important data from raw data. Data mining is the technology which extracts hidden information from raw data. This is known as knowledge discovery in databases. The integration of KDD and DM facilitates to generate highly intelligent and operational systems. Data mining has various technologies to extract useful information. All the available technologies are application dependent. A vast research has been done to develop data mining technologies in IoT to strengthen the performance of smart environments. It makes IoT smarter with intelligent services. This article elaborates a detailed study of data mining techniques for IoT.[10,11]

This chapter is organized as: Section 7.2 explores the related work to describe applications of IoT-based system and various data mining techniques for the same. Section 7.3 elaborates rules for selecting DM techniques in different IoT environments. A comprehensive comparison of the DM algorithms is shown in a tabular form, followed with challenges in developing such an environment with DM techniques in Section 7.4. Section 7.5 discusses about few research issues in the discussed problem. Section 7.6 draws a brief conclusion of the chapter.

7.2 RELATED WORK

This section addressed the applications of smart environment followed with the introduction of DM techniques in the IoT system.

7.2.1 APPLICATIONS OF IOT-BASED SYSTEM

Healthcare applications: Nowadays, the medical equipment in hospitals are trained and intelligent equipment carries health data of thousands of patients which provides essential additional information regarding the disease. Such information provides further treatment interventions and potential preventive measures in different cases.[7]

Monitoring patient remotely: Health care professionals, family members, and other professionals care takers involved in treatment can monitor and optimize real-time changes in patient's health when they are not able to reach at location. This application reduces the need of medical professionals and other medical equipment significantly. This application benefits in taking decisions for any critical situation with different standard opinions. Such monitoring techniques benefit the patients whose routine health monitoring is mandatory. This is a preventative and early diagnose technique.[8]

Monitoring remote locations: This application includes similar features of remote patient monitoring. Remote locations or out of human reach locations benefitted with this application to monitor various events remotely.[8]

Monitoring assistive equipment: Assistive equipment are deployed in smart homes or hospitals especially for disabled and elderly peoples to enhance the quality of life. Such applications are life-changing advancements in technology. Examples for such assistive equipment are smart wheelchairs, wheelchair management systems etc. to monitor the status and location of users. Smart homes are equipped with such devices to control the accessibility and enhance the quality of life. Such devices are capable of collecting potential information that defines the basic routine of user.[9]

Traffic monitoring: Vehicular traffic or live traffic can be monitored before the worst situation and action can be taken accordingly. The status of path in near future during a drive can be monitored and an alternate way can be generated or shown.[9]

The smart objects used in IoT environments generate massive information, which is used in various applications. The information from smart devices is extracted to convert it into useful information. Extraction process eliminates garbage information from raw data and extracts hidden information. This can be done through data mining algorithms. The IoT platform is increasing at a very fast rate; hence, the smart devices produce large amounts of data. The large data is termed big data that needs to be analyzed to make it useful at its maximum capacity. Data mining discovers novel, useful, and interesting information patterns from a set of large data. Then it applies a data mining algorithm on it to extract hidden information from it. Knowledge extraction, knowledge discovery databases (KDD), data archeology, data pattern analysis, information harvesting etc. are few terms used in data mining (DM). The DM process works in developing an effective and efficient model, capable of generalizing new data, discovers specific information from a large set of databases, data warehouse, and from other data repositories. The DM process includes few steps such as data preparation, data mining, and data presentation. Data preparation steps make data prepared which consist of three substeps such as integrating data from various sources, cleaning of noise from data, and making it ready for preprocessing. To evaluate or find useful patterns from the collected data to classify data for knowledge discovery, DM steps are followed. Data presentation step makes the extracted data presentable for the viewer.[10]

7.2.2 DATA MINING TECHNIQUES

The data around us is useless until it is processed under data mining techniques. With the utilization of DM techniques, the IoT environment becomes intellectual. For automatic data analysis, data mining techniques are divided into supervised, unsupervised, and reinforcement learning. The analysis of data under DM techniques provides more precise results as it goes through multiple layers. Supervised and unsupervised learning together for automatic data extraction is also considered machine learning techniques. The raw data is collected from various IoT devices which is further forwarded for the knowledge discovery process. In knowledge discovery, data is preprocessed to mold raw data into a relevant format for analysis. Under preprocessing, various actions such as feature selection, extraction (eliminating garbage information), noise abstraction, normalization dimension reduction are performed. After preprocessing the data goes under a data mining process where pattern discovery, recognition, abstraction, filtering, and event sequence detection are performed. Data preprocessing and data mining are together known as deep learning.[24,25] After DM techniques, the data is utilized for decision-making, automation, and optimization purposes by the IoT infrastructure. Data mining techniques are divided into four broad categories such as classification, clustering, association rules, and frequent pattern discovery method.[11,20]

7.2.2.1 CLASSIFICATION

Categorizing the available data with respect to some predefined targets is known as classification of data. It is main goal is to forecast the target class for accuracy. Classification is a type of supervised learning process because target labels are supposed to be known before the preprocessing. The classifier or prediction function requires training, so as to classify unlabeled data. The labeled data is used to train the classifier. In the initial stage, the classifier is built from a set of rules by previously available data. The data can be labeled or unlabeled. The labeled data is also known as the training set of data and unlabeled data is also known as the testing set of data. The classifier is first constructed by training data; then validation is done through testing data followed with analysis of data to classify the data in an appropriate class. Classification algorithm computes the probability of relevance of an item to a particular class, and then compares the cutoff value. The performance of classification is computed by evaluating accuracy level and error rate. To classify the data, decision tree induction, neural networks, Bayesian network, support vector machine, rule-based classification, classification by backpropagation, deep neural network, frame-based and ensemble methods are used.[40] For large-scale complex applications, fusion of different classification techniques is adopted.[21] The most suitable classification methods for today's IoT environment are rule-based, support vector machine, and association-based analysis. An intellectual model can be developed using the hidden Markov model of data mining. In biomedical and smart city applications naïve Bayes, Gaussian naïve Bayes, Bayesian belief network, artificial neural network, and ensemble method are most suitable.[12,13,51]

7.2.2.2 CLUSTERING

Dividing the data into meaningful groups is known as clustering. The data in a group constitute similar features. Clustering is an unsupervised learning process because it does not require prior knowledge to group the data into clusters. Various clustering techniques are hierarchical clustering, partitioning algorithm, co-occurrence, scalable high dimensional clustering, K-means, K-nearest neighbor, K-medoids, grid-based clustering, etc. In a smart IoT-based environment, cloud-based distributed clustering is more suitable than centralized clustering. Data in cloud-based distributed clustering are accessible by everyone. This feature has its own pros and cons. This faces privacy issues.[14–19]

7.2.2.3 ASSOCIATION RULE OR FREQUENT PATTERN MINING

A set of objects that appear repeatedly are known as frequent patterns. In the felicitous environment, frequent pattern mining provides better analytical understanding. Association rule helps in predicting an accurate pattern of an event. Mining the relevant frequent pattern is also known as sequential pattern mining. Sequential pattern mining is more attractive than frequent pattern mining that can analyze a sequence of events in a particular time frame. It is used for event discovery and event recognition. The occurrence of an event can be measured through frequent pattern mining. For example, in the medical field, such mining is used to diagnose the early occurrence of a disease. It observes the gradual internal changes in the body and extracts some useful pattern that might be the initial symptoms of a big disease. It observes the deviation in patients' health. Support and confidence are the two terms used to predict the early signs of a disease.[35–39] Various association rules for frequent pattern mining are Boolean association rule (a priori knowledge-based algorithm), class sequential rule mining, and clustBigFIM.[48–50]

7.2.2.4 OTHER MINING TECHNIQUES

The unforeseen useful information from raw data is called anomalous objects or outliers. Outliers are different from regular data objects and provide interesting inherent features. Outlier deviations are very useful in IoT applications such as smarthome, smart traffic, smart agriculture, and packing systems. There are four attractive outlier approaches named statistical distribution-based outlier detection, distance-based outlier detection, density-based local outlier detection, and deviation-based outlier detection.[44] In a study provided by Bishwas and Mishra in 2015, an IoT-based environment is developed to monitor health where a biometric sensor and Arduino UNO-based setup is developed to monitor health parameters.[22,23] After this outlier detection mining is applied to extract anomalous information for an emergency like situation. This is a cluster-based analysis framework with recursive principal component analysis to enhance the effectiveness of the system. The outlier approach achieves fast convergence.[32,33]

This is concluded after studying various DM techniques that an algorithm must include time scan, multilevel, multi-dimensional parallel real-time stream processing, and analysis capability to improve the effectiveness and convergence rate. All the techniques are application dependent; therefore, applicability of various techniques varies. The comprehensive comparison of various DM algorithms is addressed in Table 7.1.

7.3 DATA MINING TECHNIQUES FOR IOT

This section elaborates about the most applicable data mining technology for IoT which is suitable for the development of high-performance systems. The data mining algorithms such as classification, clustering, association rules, and frequent pattern methods are already discussed earlier in detail. To describe IoT, two simple phrases are used, "data about things" and "data generated by things," that refer to data to define things and data captured from sensor devices also refer to "big data." The amount of big data is around zettabyte which cannot be handled with the traditional data analysis tools. Single-storage systems cannot store zettabytes of data. Therefore, the traditional tools are not able to process and analyze big data. There are few traditional data mining methods such as random sampling, data condensation, divide and conquer, and incremental learning that can handle and analyze big data produced from IoT devices.[45,47] These methods capture interesting patterns from sensor devices to reduce the complexity of input data. Few traditional data mining methods are able to reduce patterns as well as number of dimensions to improve the convergence rate. Such methods are helpful in developing applications like smart homes or smart cities. As discussed in the previous section, KDD is successfully applied to extract hidden information from raw data with the following steps: selection, preprocessing, transformation, DM, and interpretation. Data preprocessing includes selection, preprocessing, and transformation; decision-making includes interpretation and evaluation steps. Data preprocessing steps are taken before DM steps and decision-making steps are taken after DM steps. Data mining steps are responsible for data extraction from the output of data processing and forward then for decision-making steps, where transformation is done. It is understood that all the attributes are not useful for mining; therefore, the selection step selects key attributes.

Various data mining algorithms are used to enhance the intelligence of the IoT system which is used to foresee the actions of occupants in a smart environment. The data mining technologies are not restricted up to smart environments but prove their efficiency in other domains well. Various studies for the relevant topics provide the successful use of data mining technologies in smart or self-intelligent environments. It enhances the smartness of provided IoT infrastructure. The examples for smart infrastructure are event detection spots, smart supermarkets, traffic management, and various transportation management systems. Such infrastructure improves the overall performance of activities. Data sources for such infrastructure can be deployed as sensor devices. Data mining technologies extract the interesting patterns of information from sensor devices. To extract important information, metaheuristic algorithms are also used which provides optimized solutions.

There are few rules on the basis of which these DM algorithms are adopted in different IoT environments. The rules can be adopted as per the rules discussed below:

Rule 1. Divide the DM technologies into two classes depending upon the characteristics of the problem. Classification and clustering into one class and association and frequent pattern in another class.

Rule 2. Classification works more suitably with labeled data as well as unlabeled data while clustering works for unlabeled data only. Therefore, divide the problem further accordingly.

Rule 3. Frequent pattern method works when the data is in a particular sequence while association deals with a set of relevance of data. There is no particular order of data in association rule-based events. Therefore, decide the problem further accordingly.

All the above-mentioned DM technologies have their own pros and cons. The need of IoT environment is changing with respect to modernization of living style.[30,31] However, single DM technique will not work effectively in large-scale smart environment. Therefore, a combination of DM technologies at different levels is adopted for better results. For example: (a) clustering and classification are combined to work as an unsupervised learning system, where an automatic set of classifiers is generated through clustering without any prior knowledge of input patterns, then incoming patterns are classified through classification methods. (b) Classification and clustering are combined to make another integrated system where classifiers are generated through classification methods from known dataset and new classifiers are added to existing classifiers through clustering. This combination acts as semi-supervised learning methods. These methods are capable of handling data from IoT dynamically. (c) Clustering, classification, and frequent pattern methods form another combination to make a single system for analysis of information. The three DM technologies can be arranged in different orders to make a new system depending upon the requirement of the problem statement. In the same way clustering, classification, and association rules are also arranged in different two orders to make a single system for data analysis. These systems can perform their tasks repeatedly in a loop to create better solutions. Such systems can be viewed in smart health analysis tools, smart homes, smart cities, etc.

TABLE 7.1 Different Data Mining Algorithms for IoT Systems.

Data Mining Algorithms	Techniques	Objective	Source of Raw Data
Classification	KNN, Naïve Bayes, Logistic regression, Support Vector Machine[34]	Event detection, traffic management, parking management, action discovery, recognition, identification, and prediction	Text data, sensor data signals, video camera, microphone, smart meters, smart energy devices, smart phone, wearable sensor devices, smart health care devices, and machineries
Clustering	K-means,[42] K-anonymity, micro aggregation Extended finite automation	Performance measurement, enhancement of quality of life, energy preservation, security, and privacy	
Association Rule	Residual method, unsupervised and probabilistic IPCL data fusion technique	Relevant action prediction	
Frequent Pattern	Fp-growth, episode discovery sequential pattern mining,[43] unsupervised discontinuous varied-order sequential miner	Tag management for RFID,[29,43] event behavior analysis, pattern recognition	

7.4 CHALLENGES WITH DATA MINING FOR IOT

This is understood that without data mining technologies, the dream of a smart environment is just a dream. But data mining technologies, along with cloud computing techniques, make this dream more applicable. The demand of society from technology is increasing day by day, so with traditional DM techniques, it is not possible to connect everything and compute information effectively. After all the mentioned applications there are few open issues with DM techniques for IoT systems. These issues are related to scalability of big datasets.

1. **Issues with infrastructure:** The decentralized and heterogeneous nature of IoT system affects DM techniques. Smart environments must support decentralized data storage and computing capabilities. Existing smart environments have centralized systems for computing and storage systems which need to be decentralized for better performance. Centralized system consumes more energy. It is observed that decentralization is not required in all the situations of IoT, but a completely centralized system increases the energy consumption level. Sometimes, the performance of an IoT system is degraded because of not having accessibility to all the data. The existing DM technologies are designed for small-scale smart applications; therefore, they provide low computation and low throughout when applied to large-scale infrastructure. For large-scale infrastructure, some cloud-based systems should be developed. But cloud-based systems face different challenges in terms of cost of computing.
2. **Issues with data:** Data preprocessing, information extraction, and retrieval are the three ways to deal with large-scale data produced from IoT devices. The sensor devices have limited size of memory; therefore, redundant data and unimportant data need to be eliminated from the storage space of sensor devices to upgrade overall execution of the system. The solutions such as dimension reduction, data compression, and data sampling are adopted. Acquisition, deposition, analysis, and integration employed various issues that affect the performance of the system. However, there are several standard protocols that define the connectivity parameters of different sensor devices, so as to make the produced information useful. But the meaning of input produced from heterogeneous sensor devices is not the same in all the applications. This issue can be tackled through few technologies such as ontology, semantic web, and extensible markup language (XML).[41] But these technologies are not appropriate to produce final solutions.

3. **Issues with algorithms:** For a complex smart application environment, it is very difficult to select DM technologies for integration. The order and selection of DM modules to design an optimal solution for the extraction of useful information is also challenging. Overfitting is another issue under an algorithm issue. This can be explained as, the labeled data are used to train the classifier and labeling of data is very expensive. The labeled data is also known as training patterns. The more the training patterns, the higher the accuracy rate. The balance between cost and accuracy of a model is challenging for the selection of an algorithm.
4. **Issues with privacy and security:** A promising paradigm, IoT with numerous applications in different domains, faces security and privacy issues. The applications like smart meters, monitoring patient remotely, smart cities, waste management, and industrial controllers demand security for their data, which is violating in the current scenario. The security of data is an essential requirement of personal data such as living patterns, habits, preferences, and social requirements.[25–28] Massey, Anton et al explain a framework to observe the privacy policies of the system. It explains the concept for using the suitable positions to apply privacy policies and examines about the input and output security concerns. It is a five-stage policy framework for security and privacy of smart environment. Evan and Eyers et al.[55] examined about the usage of tagging techniques for the sake of privacy which also helps the system under the flow of information. But this methodology is expensive in terms of processing, storage, and communication. A trust-based model for privacy preservation is developed by Appavoo, Chan et al.[56] with an objective to improve privacy. This methodology ensures and restricts the unauthorized user's accessibility. An anonymization algorithm is produced by Otgonbayar, Pervez et al., which supports the k-anonymity privacy model.[57] It examines the similar input and groups them into clusters and utilizes a time-based sliding window technique for anonymizing the input data. It supports rapid cluster formation. The model is validated on real-time dataset and proves its effectiveness with minimum information loss and higher convergence rate. Cryptographic techniques also show a great privacy concern; the model is proposed by Alelaiwi et al.[58] The model is expensive but very effective in real-time medical domains. It is a multi-party framework that hides the data from attackers. In this type of model, every sensor node contains

a private secret key to ensure the privacy of the information shared between the devices.[46] A sensitive and privacy-based environment framework is proposed by Perez et al. for health care and automation system applications.[59] This approach is also based on cryptographic techniques. This model ensures the secure exchange of data between devices and handles protected data. A modern privacy framework is addressed by Ge, Hong et al.[60] To tackle new security issues that incorporate different phases such as data processing, security model generation, visualization, analysis, and model updates, this is applied on IoT generator nodes, security model generator, and security evaluator. This model is a potential security defender. The computerized numerical control information-based privacy mechanism is addressed by Li and Li which is a lightweight authentication method for security of information based on organizational characteristics in the IoT environment.[61] The protocol includes five parts like system setup phase, sensor node registration phase, user node registration phase, login phase, authentication and session key agreement phase. A series of analysis is done to prove the efficiency of secure environment. This protocol employs a double privacy protection strategy.[52,53]

5. **Networking issues:** When devices are connected to share data, there must be appropriate signal quality to protect the data integrity. The devices in the IoT system are different in terms of scalability and computing capability; they might generate data in different ways and therefore connectivity between these devices may cause an issue. Connecting between different devices depends on various factors such as availability, interoperability, cost, scalability, reliability, coverage, data rate, and power consumption. The sensor devices are connected through low-range wireless communication network and gathered data is forwarded through large-scale wireless communication network. Robust and scalable connection is a mandatory requirement for the IoT network. The type of network selection is based upon type of application to maintain a proper network connection. This issue should be taken care of on a priority basis.[54]

7.5 OPEN RESEARCH ISSUES

1. As the sensor devices produce a huge amount of data and sensor devices are battery powered, therefore an energy-efficient mechanism

should be developed with some improvement scopes with the passage of time. Such an environment will always be in the scope of improvements. That is why energy issue in IoT environments will always be a hot topic of research.

2. Along with energy, the huge amount of produced data may suffer from congestion issues. Therefore, congestion management or congestion control could be the sensitivity research topic in the IoT environment.
3. The sensor devices have a limited storage capacity and this is an era of technology; therefore, cloud-based or fog-based mechanism should be developed with some advanced features to incorporate storage concept.
4. The new equipment for smart environment should match the interoperability with old-generation equipment.
5. Security and privacy is always a prime concern for any technology to prevent data from cyber-attacks or unauthorized access.

7.6 CONCLUSIONS

This article elaborates about different data mining techniques that can be used in Internet if things to improve the overall performance of the system and environment. Different data mining algorithms such as classification, clustering, association rule, and frequent pattern mining work with various technologies that provide a large scale of applications. Applications of Internet of things are also discussed with example. This article concludes that with the increase in demand of users for a smart environment, integration of data mining technologies is becoming mandatory. This chapter also draws attention on challenges for developing Internet of things environment with data mining technologies.

KEYWORDS

- **data mining**
- **sensors**
- **Internet of Things**
- **smart environment**

REFERENCES

1. Atzori, L.; Iera, A.; Morabito, G. The Internet of Things: A Survey. *Comput. Netw.* **2010,** *54* (15), 2787–2805.
2. Li, S.; Da Xu, L.; Zhao, S. The Internet of Things: A Survey. *Inf. Syst. Front.* **2015,** *17* (2), 243–259.
3. Miorandi, D.; Sicari, S.; De Pellegrini, F.; Chlamtac, I. Internet of Things: Vision, Applications and Research Challenges. *Ad hoc Netw.* **2012,** *10* (7), 1497–1516.
4. Bandyopadhyay, D.; Sen, J. Internet of Things: Applications and Challenges in Technology and Standardization. *Wireless Personal Commun.* **2011,** *58* (1), 49–69.
5. Domingo, M. C. An Overview of the Internet of Things for People with Disabilities. *J. Netw. Comput. App.* **2012,** *35* (2), 584–596.
6. Kulkarni, R. V.; Förster, A.; Venayagamoorthy, G. K. Computational Intelligence in Wireless Sensor Networks: A Survey. *IEEE Commun. Surveys Tutorials* **2010,** *13* (1), 68–96.
7. Kortuem, G.; Kawsar, F.; Sundramoorthy, V.; Fitton, D. Smart Objects as Building Blocks for the Internet of Things. *IEEE Internet Comput.* **2009,** *14* (1), 44–51.
8. Alam, K. M.; Saini, M.; El Saddik, A. Toward Social Internet of Vehicles: Concept, Architecture, and Applications. *IEEE Access* **2015,** *3*, 343–357.
9. Alam, M. A. U.; Roy, N.; Misra, A.; Taylor, J. CACE: Exploiting Behavioral Interactions for Improved Activity Recognition in Multi-Inhabitant Smart Homes. In *2016 IEEE 36th International Conference on Distributed Computing Systems (ICDCS)*; IEEE, June 2016; pp 539–548.
10. Bin, S.; Yuan, L.; Xiaoyi, W. Research on Data Mining Models for the Internet of Things. In *2010 International Conference on Image Analysis and Signal Processing*; IEEE, April 2010; pp 127–132.
11. Rashidi, P.; Cook, D. J.; Holder, L. B.; Schmitter-Edgecombe, M. Discovering Activities to Recognize and Track in a Smart Environment. *IEEE Trans. Knowl. Data Eng.* **2010,** *23* (4), 527–539.
12. Phyu, T. N. Survey of Classification Techniques in Data Mining. In *Proceedings of the International Multi Conference of Engineers and Computer Scientists* Mar **2009,** *1* (5).
13. Roddick, J. F.; Spiliopoulou, M. A Survey of Temporal Knowledge Discovery Paradigms and Methods. *IEEE Trans. Knowl. Data Eng.* **2002,** *14* (4), 750–767.
14. Chen, F.; Deng, P.; Wan, J.; Zhang, D.; Vasilakos, A. V.; Rong, X. Data Mining for the Internet of Things: Literature Review and Challenges. *Int. J. Distrib. Sens. Netw.* **2015,** *11* (8), 431047.
15. Jain, A. K.; Murty, M. N.; Flynn, P. J. Data Clustering: A Review. *ACM Comput. Surveys (CSUR)* **1999,** *31* (3), 264–323.
16. Xu, R.; Wunsch, D. Survey of Clustering Algorithms. *IEEE Trans. Neural Netw.* **2005,** *16* (3), 645–678.
17. Ng, R. T.; Han, J. CLARANS: A Method for Clustering Objects for Spatial Data Mining. *IEEE Trans. Knowl. Data Eng.* **2002,** *14* (5), 1003–1016.
18. Guha, S.; Meyerson, A.; Mishra, N.; Motwani, R.; O'Callaghan, L. Clustering Data Streams: Theory and Practice. *IEEE Trans. Knowl. Data Eng.* **2003,** *15* (3), 515–528.
19. Chiang, M. C.; Tsai, C. W.; Yang, C. S. A Time-Efficient Pattern Reduction Algorithm for K-Means Clustering. *Inf. Sci.* **2011,** *181* (4), 716–731.

20. Anvari-Moghaddam, A.; Monsef, H.; Rahimi-Kian, A. Optimal Smart Home Energy Management Considering Energy Saving and a Comfortable Lifestyle. *IEEE Trans. Smart Grid* **2014,** *6* (1), 324–332.
21. Bijarbooneh, F. H.; Du, W.; Ngai, E. C. H.; Fu, X.; Liu, J. Cloud-Assisted Data Fusion and Sensor Selection for Internet of Things. *IEEE Internet of Things J.* **2015,** *3* (3), 257–268.
22. Biswas, S.; Misra, S. Designing of a Prototype of E-health Monitoring System. In *2015 IEEE International Conference on Research in Computational Intelligence and Communication Networks (ICRCICN)*; IEEE, 2015; pp 267–272.
23. Brdiczka, O.; Reignier, P.; Crowley, J. L. Detecting Individual Activities from Video in a Smart Home. In *International Conference on Knowledge-Based and Intelligent Information and Engineering Systems*; Springer: Berlin, Heidelberg, 2007; pp 363–370.
24. Chen, Q.; Wang, W.; Wu, F.; De, S.; Wang, R.; Zhang, B.; Huang, X. A Survey on an Emerging Area: Deep Learning for Smart City Data. *IEEE Trans. Emerg. Topics Computat. Intell.* **2019,** *3* (5), 392–410.
25. Choi, W.; Shah, P.; Das, S. K. A Framework for Energy-Saving Data Gathering Using Two-Phase Clustering in Wireless Sensor Networks. In *The First Annual International Conference on Mobile and Ubiquitous Systems: Networking and Services, 2004. MOBIQUITOUS 2004*; IEEE, 2004; pp. 203–212.
26. Chowdhary, R. R.; Chattopadhyay, M. K.; Kamal, R. IoT-Based State of Charge and Temperature Monitoring System for Mobile Robots. In *Innovations in Electronics and Communication Engineering*; Springer: Singapore, 2020; pp 401–413.
27. European Commission (EC). Europe 2020: A Strategy for Smart, Sustainable and Inclusive Growth. *Working Paper {COM (2010) 2020}*, 2010.
28. Da Xu, L.; He, W.; Li, S. Internet of Things in Industries: A Survey. *IEEE Trans. Ind. Inform.* **2014,** *10* (4), 2233–2243.
29. Keller, T. *Mining the Internet of Things-Detection of False-Positive RFID Tag Reads Using Low-Level Reader Data* (Doctoral dissertation), 2011.
30. MacQueen, J. Some methods for Classification and Analysis of Multivariate Observations. In *Proceedings of the Fifth Berkeley Symposium on Mathematical Statistics and Probability* June **1967,** *1* (14), 281–297.
31. Safavian, S. R.; Landgrebe, D. A Survey of Decision Tree Classifier Methodology. *IEEE Trans. Syst. Man Cybern.* **1991,** *21* (3), 660–674.
32. Masciari, E. (**2007,** September). A Framework for Outlier Mining in RFID data. In *11th International Database Engineering and Applications Symposium (IDEAS 2007)*; IEEE, 2007; pp. 263–267.
33. Knorr, E. M.; Ng, R. T. A Unified Notion of Outliers: Properties and Computation. *KDD* **1997,** *97*, 219–222.
34. Fleury, A.; Vacher, M.; Noury, N. SVM-Based Multimodal Classification of Activities of Daily Living in Health Smart Homes: Sensors, Algorithms, and First Experimental Results. *IEEE Trans. Inf. Technol. Biomed* **2009,** *14* (2), 274–283.
35. Koperski, K.; Han, J. Discovery of Spatial Association Rules in Geographic Information Databases. In *International Symposium on Spatial Databases*; Springer: Berlin, Heidelberg, 1995; pp 47–66.
36. Han, J.; Cheng, H.; Xin, D.; Yan, X. Frequent Pattern Mining: Current Status and Future Directions. *Data Mining Knowl. Disc.* **2007,** *15* (1), 55–86.

37. Zhao, Q.; Bhowmick, S. S. Sequential Pattern Mining: A Survey. *ITech. Rep. CAIS Nayang Technol. Univ. Singapore* **2003,** *1* (26), 135.
38. Agrawal, R.; Imieliński, T.; Swami, A. Mining Association Rules Between Sets of Items in Large Databases. In *Proceedings of the 1993 ACM SIGMOD International Conference on Management of Data*; 1993; pp 207–216).
39. Bekri, F. E.; Govardhan, A. Association of Data Mining and Healthcare Domain: Issues and Current State of the Art. *Global J. Comput. Sci. Technol.* **2011**.
40. Ranka, S.; Singh, V. CLOUDS: A Decision Tree Classifier for Large Datasets. In *Proceedings of the 4th Knowledge Discovery and Data Mining Conference* **1998,** *2* (8).
41. Brunner, S.; Kucera, M.; Waas, T. (**2017,** June). Ontologies Used in Robotics: A Survey with an Outlook for Automated Driving. In *2017 IEEE International Conference on Vehicular Electronics and Safety (ICVES)*; IEEE, 2017; pp 81–84.
42. Dhillon, I. S.; Guan, Y.; Kulis, B. Kernel k-Means: Spectral Clustering and Normalized Cuts. In *Proceedings of the Tenth ACM SIGKDD International Conference on Knowledge Discovery and Data Mining*, 2004; pp 551–556.
43. Ding, K.; Jiang, P. RFID-Based Production Data Analysis in an IoT-Enabled Smart Job-Shop. *IEEE/CAA J. Automatica Sinica* **2017,** *5* (1), 128–138.
44. Ester, M.; Kriegel, H. P.; Sander, J.; Xu, X. A Density-Based Algorithm for Discovering Clusters in Large Spatial Databases with Noise. *Kdd* **1996,** *96* (34), 226–231.
45. Gole, S.; Tidke, B. Frequent Itemset Mining for Big Data in Social Media Using ClustBigFIM Algorithm. In *2015 International Conference on Pervasive Computing (ICPC)*. IEEE, 2015; pp 1–6.
46. Gu, T.; Wang, L.; Chen, H.; Tao, X.; Lu, J. Recognizing Multiuser Activities Using Wireless Body Sensor Networks. *IEEE Trans. Mob. Comput.* **2011,** *10* (11), 1618–1631.
47. Guo, H.; Liu, J.; Zhao, L. Big Data Acquisition Under Failures in FiWi Enhanced Smart Grid. *IEEE Trans. Emerg. Topics Comput.* **2017**.
48. Han, J.; Cheng, H.; Xin, D.; Yan, X. Frequent Pattern Mining: Current Status and Future Directions. *Data Mining Knowl. Disc.* **2007,** *15* (1), 55–86.
49. Huang, K. Y.; Chang, C. H.; Lin, K. Z. Prowl: An Efficient Frequent Continuity Mining Algorithm on Event Sequences. In *International Conference on Data Warehousing and Knowledge Discovery*; Springer: Berlin, Heidelberg, 2004; pp 351–360.
50. Jensen, S. K.; Pedersen, T. B.; Thomsen, C. Time Series Management Systems: A Survey. *IEEE Trans. Knowl. Data Eng.* **2017,** *29* (11), 2581–2600.
51. van Kasteren, T.; Krose, B. Bayesian Activity Recognition in Residence for Elders, 2007.
52. Lyu, L.; Nandakumar, K.; Rubinstein, B.; Jin, J.; Bedo, J.; Palaniswami, M. PPFA: Privacy Preserving Fog-Enabled Aggregation in Smart Grid. *IEEE Trans. Ind. Inf.* **2018,** *14* (8), 3733–3744.
53. Roman, R.; Zhou, J.; Lopez, J. On the Features and Challenges of Security and Privacy in Distributed Internet of Things. *Comput. Netw.* **2013,** *57* (10), 2266–2279.
54. Chen, L.; Ren, G. The Research of Data Mining Technology of Privacy Preserving in Sharing Platform of Internet of Things. In *Internet of Things*; Springer: Berlin, Heidelberg, 2012; pp 481–485.
55. Evans, D.; Eyers, D. M. Efficient Data Tagging for Managing Privacy in the Internet of Things. In *2012 IEEE International Conference on Green Computing and Communications*; IEEE, 2012; pp 244–248.

56. Appavoo, P.; Chan, M. C.; Bhojan, A.; Chang, E. C. Efficient and Privacy-Preserving Access to Sensor Data for Internet of Things (IoT) Based Services. In *2016 8th International Conference on Communication Systems and Networks (COMSNETS)*; IEEE, 2016; pp 1–8.
57. Otgonbayar, A.; Pervez, Z.; Dahal, K. Toward Anonymizing IoT Data Streams Via Partitioning. In *2016 IEEE 13th International Conference on Mobile Ad Hoc and Sensor Systems (MASS)*; IEEE, 2016; pp 331–336.
58. Tso, R.; Alelaiwi, A.; Rahman, S. M. M.; Wu, M. E.; Hossain, M. S. Privacy-Preserving Data Communication Through Secure Multi-Party Computation in Healthcare Sensor Cloud. *J. Sign. Process. Syst* **2017,** *89* (1), 51–59.
59. Pérez, S.; Rotondi, D.; Pedone, D.; Straniero, L.; Núñez, M. J.; Gigante, F. Towards the CP-ABE Application for Privacy-Preserving Secure Data Sharing in IoT contexts. In *International Conference on Innovative Mobile and Internet Services in Ubiquitous Computing*; Springer: Cham, 2017; pp 917–926.
60. Ge, M.; Hong, J. B.; Guttmann, W.; Kim, D. S. A Framework for Automating Security Analysis of the Internet of Things. *J. Netw. Comput. App.* **2017,** *83*, 12–27.
61. Li, Y.; Li, M. A Privacy Protection Mechanism for Numerical Control Information in Internet of Things. *Int. J. Distrib. Sens. Netw.* **2017,** *13* (8), 1550147717726312.

PART II
AI and Machine Learning

CHAPTER 8

CLASSIFICATION OF WEB USER INTEREST LEVEL USING WEB USAGE MINING

TENDE IVO SAKE[1], LATHA BANDA[2], and DEVENDRA GAUTAM[2]

[1]*Department of Computer Sciences, Sharda University and Engineering, Greater Noida, Uttar Pradesh, India*

[2]*ABES Engineering College, Ghaziabad, Uttar Pradesh, India*

ABSTRACT

The analysis of web user behavior has long been a trending topic in the research field of data mining. This is due to the steady increase in internet users and likewise the enormous amount of data that are found and collected in the or server log in the form of clickstream as users browse the internet. In order to analyze web user behavior, web usage mining is used. But a significant amount of time is consumed during the analysis of the web data; hence the user's behavior can be retrieved by analyzing the clickstream of the users that are classified as highly interested in the website. This is because online user click behavior is practically motivated by their interest. This chapter seeks to classify and cluster users into three different level of interest; highly interested, averagely interested, and less interested following information gotten from NASA dataset. Before the classification, the data is preprocessed to extract useful features and clean all the irrelevant records that might affect the results and also to obtain a structured data within which users and sessions can be easily identified.

Advanced Computer Science Applications: Recent Trends in AI, Machine Learning, and Network Security. Karan Singh, PhD, Latha Banda, PhD & Manisha Manjul, PhD (Eds.)

8.1 INTRODUCTION

The size of information found in World Wide Web (WWW) has been on constant increase since it discovery in 1989.[12] This is due to the rise in the number of users and how in today's era, everyone is dependent on WWW to get information about any query.[1] With the internet being a public space, having a website on the internet enable businesses, enterprises, government, companies, and individuals to put forth their information to the public, yet it is important for user to be interest to consume this information. As users of the internet, we all spend a significant amount of time searching for this information and our behaviors are all different.[3] This behavior can be depicted by analyzing the web data that are recorded as user take perform clicks. Web log files are files that lists the actions of user that have been occurred when browsing website. These log files reside in the web server. Web log files contain information about User name, IP address, Timestamp, Access request, number of bytes transferred, and User agent. Analysis of these log files gives navigation behavior of the user.[4]

The application of data mining techniques to discover patterns from the WWW,[5] is known as web mining. This can provide so much information but usually the result is based on the focused application. Exploring the way users interact with web contents and where their attention focuses onto during navigation, is one of the key issues in this field of web mining.[6] It is very important to understand how a website is being used by users.[7] Beyond the knowledge discovered in the log file, they also provide a clear status and health of the site.[8]

The various ways to look at user interest can be; which user is interested in the site, which content is more interesting to users. For an ideal case, an interest user will be the user the frequent the website every day and spend considerable amount of time on most/ all the pages on the site and a noninterested user will be the one that have visited the site for a single moment within a large time frame and usually visited just 10% of the of all the site pages.

The number of visitors on a website is easy to determine using software tools like Google analytics, at the backend of the website using php scripts, using plugins cases in WordPress websites. Also, there exist log analyzers tools like AWSTATS, DEEP LOG ANALYZER, and WEKA TOOL. These tools can provide useful information about number of visitors, Referral links, popular links, bounce rate, and much more, Still, from the users which clicked and go through the site, it does not necessarily mean they are interested. Hence it is of high importance to mine this data and analyzed it to better understand the interestingness of users. Weblog file can be in the form of CVS file format or ARFF file format and usually stored in server.[9]

8.1.1 WEB USAGE MINING

Web usage mining is the leading research area in Web Mining concerned about the web user's behavior.[10] The process of web usage mining involves the application of data mining techniques to uncover useful patterns in the weblog made by user click streams.[27] The primary task of web usage data on a particular website is to capture web users behavior.[11] Web usage mining can be classified according to kinds of usage data examined. In our context, the usage data is web log data, which contain the information regarding the user navigation. This information is in the form of click streams that are logged into the server during users' session. We deal with the web server logs which maintain the history of page requests. Web log files are the files that contain complete information about the users browse activities on the web server

8.1.2 WEB LOG STRUCTURE

During a user's navigation session, all activity on the website is recorded in a log file by the web server.[16] The structure of the web log is as shown on Figure 8.1.

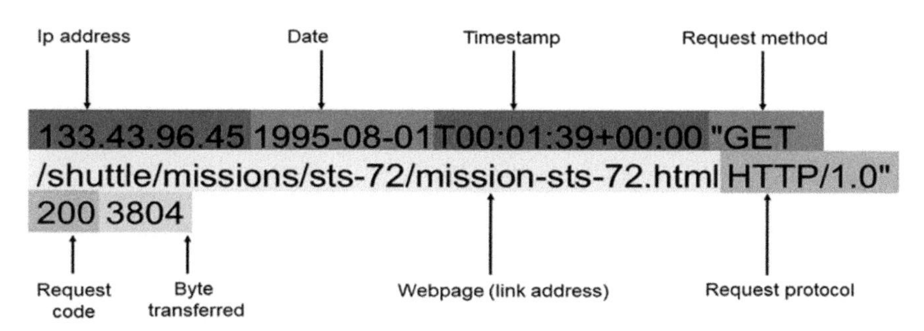

FIGURE 8.1 Structure of web log.

Ip address: This is the unique address of the user PC at the time of accessing the web.

Date: This corresponds to the day in which the log was recorded

Timestamp: this is the specific time in which a user clicks a link on the web site

Request Method: this is the method used during data request. It can either be a POST, HEAD, or GET request.

Webpage (link address): this is the specific link clicked by the user at that specified stamp

Request protocol: this is the means through which client-side and server-side computers communicate when the request is made.

Request Code: this refers to the http request codes. There are various types of request codes; informational responses (100–199), successful responses (200–299), redirects (300–399), client errors (400–499), and server errors (500–599).

Byte Transferred: this is the amount of data downloaded by the user when the request is made or when the page is clicked. It is in the byte unit.

8.2 RELATED STUDIES

August et al. classifies the content of a site as interested or not of interested to users according to its usefulness to the users of the site. This was done using the decision table in which the depth of visit was also a factor of interest.[12] This is an efficient way to measure if a user is interested in a website because if a site uses the hierarchical model and has five levels in its link structure, with a user visits and reaches the fifth level, then, it should be considered as an interested user. But this is not the case for all website as some website models are sequential, matrix of database.

Diep et al. described an unsupervised method for obtaining interest of user from weblog. Users are grouped together with respect to their similarity in interest in page visited. This method is effective for creating clusters comprising of who-visit-what parameter.[13]

Anandhi et al. uses a classification technique to define user behavior as being either frequent, synthetic, or potential and aim to predict the navigational pattern of the potential users. These types of users are defined as those that visit a website but perform no transaction and are likely to segregate from the weblog.[2] Classifications are employed that is used to predict the target class based on web log data and then apply clustering techniques to group the data based on visited pages. These types of users can also be term users with low level interest to the site. The importance of finding the navigational pattern of potential users is to understand the steps needed to convert them into frequent users or high-level interested users.

Suneetha et al.[14] proposed a two-phase model that extract focused group of interested users using the C4.5 algorithm. He considers user to be interested or not interested using; time spend on a session, request method (POST or GET) number of pages visited and depth access from particular pages. But a user can also show high degree or interestingness based on only the amount of time spend on the site irrespective of the pages visited. Also, not all website possesses the POST request capabilities are they are solely designed to provide information to uses without asking for user data. For example, blog web sites.

Kumar et al.[15] aim to design a web browser that is responsive to user's interest. He proposed that a combination of user's scroll position and real-time action can be determine to give the level of interest by using the Neurosky EEG sensor. In this scenario, web pages are included or removed from a website if the level of interest by users is higher or lower than a threshold. This idea is very important when targeting component that are dynamic on the site, for example, Ads and forms. But for the sake of search engine optimization, removing and putting pages might cause serious indexing problem. This is done in order to increase the user engagement or satisfaction while browsing.

8.3 PROPOSED METHODOLOGY

The workflow design for this work involves;

a) Data preprocessing
b) Pattern discovery
c) Pattern analysis (Classification using decision table)

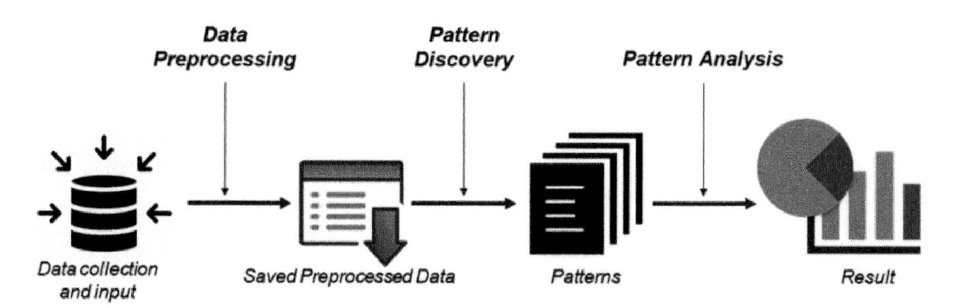

FIGURE 8.2 Modified work flow.[10]

8.3.1 DATASET COLLECTION AND INPUT

The weblog data used for this work NASA server log was collected within a 7 days period from 1/8/1995 to 7/8/1995. The log file is in CSV file format and made up of 268,345 records. The raw file is noisy, unstructured, and contain several irrelevant records and thus required to be preprocessed before analysis.

8.3.2 DATA INPUT

The collected data is then input into an SQL database for analysis. Using the SQL commands, the data can be selected, deleted, and updated within tables. A PHP script is used to write the necessary functions.

8.3.3 DATA PREPROCESSING

Data preprocessing converts the raw data into the data abstractions necessary for pattern discovery.[16] During this stage, the noisy data are removed. These are data that are meaningless, have unnecessarily information and usually increases the amount of storage space required and can also adversely affect the results of any data mining analysis. The processes involved in data preprocessing include; data cleaning, session identification, and user identification.

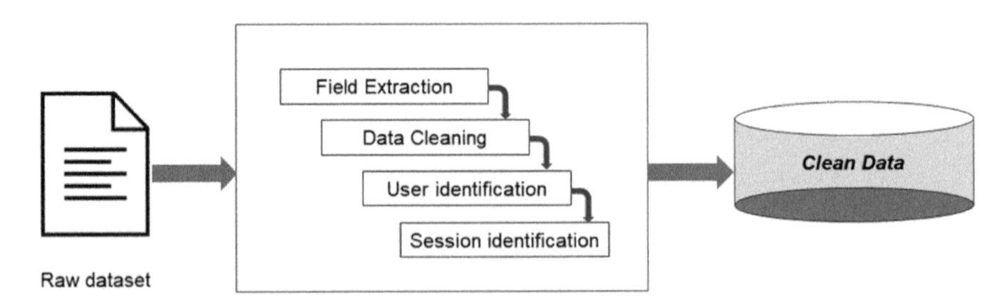

FIGURE 8.3 Data preprocessing.[16]

8.3.3.1 FIELD EXTRACTION

The log section contains different fields that should be discrete out for the handling. The way toward isolating fields from the single line of the log document is known as field extraction. The server utilized various characters which acts as dividers. Some commonly known dividers are "," or space characters. With the NASA web data, space character is used within every single log entry to identify and extract the various fields. The Field Extract algorithm is given below.

8.3.3.2 DATA CLEANING

During this stage, records pointing to file types other than that pointing to web pages are eliminated. Such records are links to files such as; images (png, jpeg, jpg, gif, xbm), documents (pdf, doc, docx), javascript, css, or link not referring a particular file. Also records that are having request codes referring to fail requests or indicating server failure are deleted. A non-numeric ip address is usually an address pointing to a website or a server. In this case, it is not considered as a during user identification.

SQL code for removing irrelevant data

(1) Start
(2) DELETE FROM table WHERE link NOT LIKE htm AND html AND php AND apsx
(3) DELET FROM table WHERE ip NOT NUMERIC

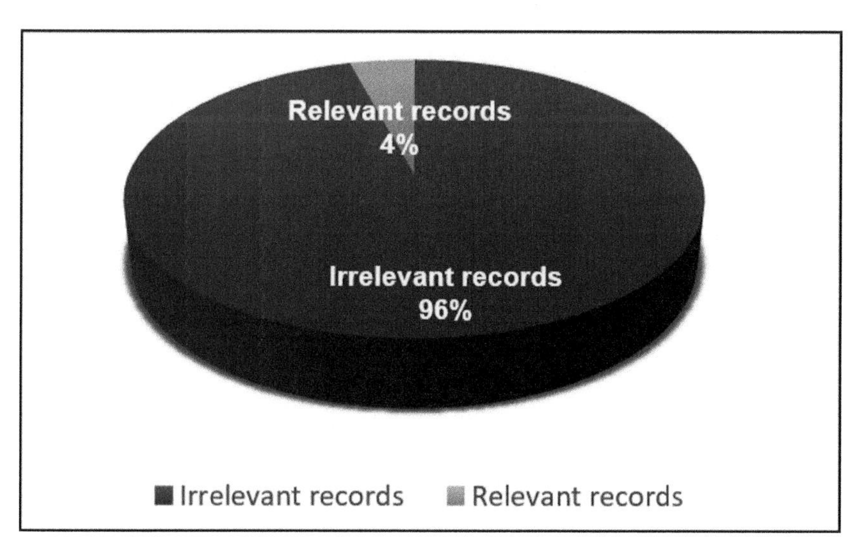

FIGURE 8.4 Showing percentage of relevant and irrelevant records.

8.3.3.3 USER IDENTIFICATION

During user identification, users are identified using their unique ip addresses. During this stage, certain assumptions are considered;

1. If an ip address is found in more than one places within the record, it is considered as the same user.
2. Users are identified by the computer ip, irrespective of the human using it at the moment.

8.3.3.4 SESSION IDENTIFICATION

A session is the total time devoted to an activity. In computer systems, a user session begins when a user logs in to or accesses a particular computer, network, or software service. It ends when the user logs out of the service, or shuts down the computer. In this case, the sequence of URLs navigated by a user for a period of 20 minutes (1200 seconds) is treated as a session. Session represents the navigation pattern of a user.[4] Some assumptions are made during this stage;

(1) If difference between two consecutive timestamps in the log for a user is greater than 20 minutes, it is considered two sessions.
(2) Sessions are taken irrespective of the date; meaning it does not matter if a session cut across two different days

Algorithm for session identification

```
START
Total = SELECT * FROM table
For (x < total)
        result_ip = SELECT DISTINCT ip FROM table;
        While(row_ip = fetching records(result_ip))
        INSERT row_ip into array distinct_ip[ ];
For (y < result_ip)
        user_time = SELECT TIME from table WHERE ip is row_ip[y];
        While(time_ip = fetching records(user_time))
                Insert time_ip into array time[ ];
        While(z < row_ip)
                session = absolute(time[z+1]—time[z])
```

```
                If(session <= 1200)
                        Output session;
                        Return session;
                    endif;
                z++;
                endwhile;
        y++;
        endfor;
STOP
```

8.3.4 PATTERN DISCOVERY

Classification is used to classify user's interest level. The classification of users is based on a decision table classifier. A decision table or decision matrix is a tabular representation of decisions in a system, which is constructed to analyze the patterns found.

8.3.4.1 CONSTRUCTION OF DECISION TABLE

8.3.4.1.1 Attributes

C1: Total Number of visited pages, C2: Number of sessions, C3: Total amount of time spend, C4: Number of pages per session, and C5: Amount of Time spend per session.

TABLE 8.1 Action Table.

Attributes	Conditions		
	LI	**AI**	**HI**
C1	< 5	< 5 & < 10	> 10
C2	= 1	> 1 & < 2	> 3
C3	< 600	> 600 & < 900	> 900

LI: Less Interested, AI: Averagely Interested, HI: Highly Interested

8.3.5 PATTERN ANALYSIS

8.3.5.1 CLASSIFICATION

Only attributes C1, C2, and C3 are considered to create the classifier because C4 and C5 are simply multiples of C1 and C2. The output from Table 8.2 above is a combination of results from classifiers C1, C2, and C3, irrespective of the order in which the decision tree test is carried out. Thus, if the result of a test route gives [LI, AI, HI] from the three classifiers, it implies the user is averagely interested.

TABLE 8.2 Decision Table.

Decision Routes			Outcomes
LI	LI	LI	Less Interested
LI	LI	AI	Less Interested
LI	LI	HI	Averagely Interested
LI	AI	AI	Averagely Interested
AI	AI	AI	Averagely Interested
LI	AI	HI	Averagely Interested
AI	AI	HI	Highly Interested
AI	HI	HI	Highly Interested
HI	HI	HI	Highly Interested
HI	HI	LI	Highly Interested

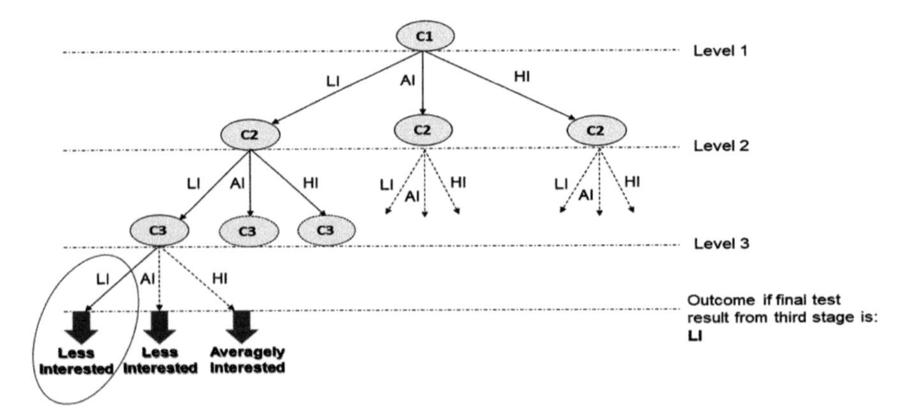

FIGURE 8.5 Decision tree.

8.3.5.2 CLUSTERING

Clustering is a discovery process in data mining, which groups set of data items, in such a way that maximizes the similarity within clusters and minimizes the similarity between two different clusters.[17] Based on the above classifier, the user is grouped into less interested user, averagely interested user, and highly interested user. The user who has only a single page visit are also classified and grouped as not interested users.

8.4 EXPERIMENTAL RESULT

These parameters are then used to produce a decision table, that will be used to evaluate user interest level.

TABLE 8.3 Size of Record.

Processes	Record	Size
Raw data	268345	26.8MB
After removal of nonwebpage links	54553	5.8MB
After removal of non-numeric ip addresses	11893	1.9MB
After removal of fail requests	11405	1.8MB

TABLE 8.4 Unique Records.

Records	Quantity
Total Records	11405
Unique Ip addresses	3191
Unique links	439

It was observed at this stage that from the 11405 records, 1328 records have only one access to a page and the site. This information can be used to obtain the bounce rate of the website. A website's bounce rate is calculated by dividing the number of single-page sessions by the number of total sessions on the site. Hence the bounce rate for this site based on this amount of data collected and for this period of duration is;

$$\frac{1328}{11405} \times 100 = 11.64\%$$

This record indicates that 11.64 % of the visitors that visits the site initially, left without browsing any further. Depending on the type of website you have, the type of advertising you do, and the type of user on your site, the average number of page views per session can range from 1.2 sessions to 10 sessions. For ecommerce sites, 5–10 page views per session is a reasonable figure. Within the table, the 1328 users that have just one-page view are removed from the record, since a single page view clear indicate no interest in the site. These users might have landed on the page due to ads, touch screen error for the cases of users with smart phone and tablet, or just users scouting the internet looking for interesting topics to read. This leaves us with a total of 1863 user records to consider for classification.

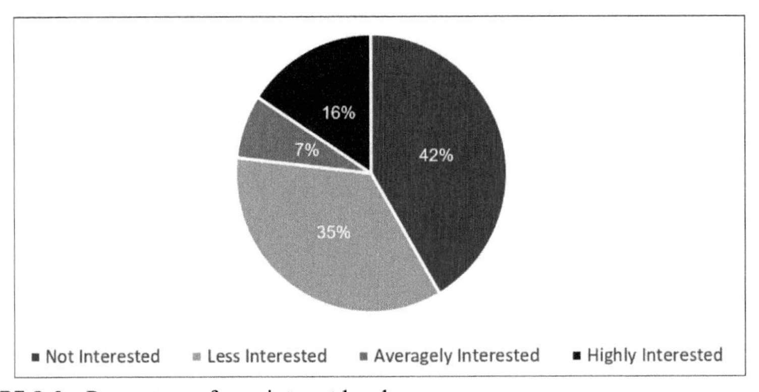

FIGURE 8.6 Percentage of user interest level.

8.4.1 SYSTEM EVALUATION

A windows 10 operating system with an intel Core i5 CPU and a 12GB Ram is use. XAMP software is used to run the database with holds the collected data.

8.5 DISCUSSION

From the 268345 records found, only 11405 were valid records. This accounts for only 4% of the data, the remaining 256940 (96%) of the records being termed noisy and irrelevant. Web administrators need to continuously analyze their website, examine the extent to which their goals are achieved,[1] to gain a deep understanding of users' behavior, develop their future strategies and goals, and ultimately improve user experiences. Importance for working with interested users: (1) to get the behavior of web users from their frequent patterns, (2) to understand the content which is useful to users, and (3) the less interested users which are users that usually have just a single visit, accounts for greater percentage of the bounce rate. In order to make a correct conclusion on whether a user is IN or UIN in a website, the particular website in question should be of consideration. Meaning some websites are having higher number of pages than others and hence the

8.6 CONCLUSION AND FUTURE WORK

This study describes a process of analyzing a weblog to determine users with different level of interest to the website. The dataset used in this study is the web data from NASA collected within a 7 days period. During the process, data processing is carried out to clean off irrelevant and noisy data to obtain a structured tabulated dataset. It is observed that up to 96% of the records are irrelevant. After the preprocessing of the web data, a classification decision table is used to classify user interest level. Clustering is used to group users with similar interest level after the classification process. With the large amount of data involve, finding online user behavior can be tedious. Hence this study seeks to find users that are most interested in the web site, within which the result can be used to find frequent navigational patterns.

In the future study, the number of bytes transferred for each request a user makes, can be added as an attribute. More so, interesting topic that liable to be a focus point for future research in the analysis of user behavior using bounce rate and rate of failed request in a network. This is important because such data are also being recorded and are usually considered as noisy tuples when performing data cleaning.

KEYWORDS

- **user interest**
- **web usage mining**
- **classification**
- **session**
- **data preprocessing**

REFERENCES

1. Tyagi, N.; Gupta, S. K. Web Structure Mining Algorithms: A Survey. *Adv. Intell. Syst. Comput.* **2018,** *654*, 305–317.
2. Anandhi, D.; Ahmed, M. S. I. Prediction of User's Type and Navigation Pattern Using Clustering and Classification Algorithms. *Cluster Comput.* **2019,** *22*, 10481–10490.
3. Anitha, V.; Isakki, P. A Survey on Predicting User Behavior Based on Web Server Log Files in a Web Usage Mining. *2016 Int. Conf. Comput. Technol. Intell. Data Eng. ICCTIDE 2016* 2016.
4. Anupama, D. S.; Gowda, S. D. Clustering of Web User Sessions to Maintain Occurrence of Sequence in Navigation Pattern. *Procedia Comput. Sci.* **2015,** *58*, 558–564.
5. Stakhiyevich, P.; Huang, Z. Building User Profiles Based on User Interests and Preferences for Recommender Systems. *Proc.—2019 IEEE Int. Conf. Ubiquitous Comput. Commun. Data Sci. Comput. Intell. Smart Comput. Netw. Serv. IUCC/DSCI/SmartCNS 2019* **2019,** 450–455.
6. Naderkhah, F.; Moradi, H. A Website Analytics System Considering Users' Category. *2019 IEEE 5th Conf. Knowl. Based Eng. Innov. KBEI 2019* **2019,** 764–769.
7. Oo, H. Z.; Saing, N.; Kham, M. Pattern Discovery Using Association Rule Mining on Clustered Data. *Int. J. New Technol. Res.* **2018,** *4* (2), 7–11.
8. Dubey, S. M.; Shrivastava, D.; Dwivedi, A. K. Analysis of the Event-Log Using Apriori Algorithm. **2015,** *2* (7), 6–11.
9. K.Mishra, S.; Richaria, V.; Sharma, V. Recognition of Interested Web Users Behavior. *Int. J. Comput. Appl.* **2013,** *61* (6), 14–17.
10. Raut, K. S. Research on Web Log Mining to Predicting User Behavior through Session. *Int. J. Res. Appl. Sci. Eng. Technol.* **2018,** *6* (7), 743–745.
11. Suthar, P. Supervised Learning Techniques in Web Usage Mining : Comparison and Analysis. **2017,** *1* (3), 12–18.
12. August, J.-; Mishra, P. S. Classification of Web Users into Interested Users and Not Interested Users by Using Decision Table Available Online at Www.Ijarcs.Info Classification of Web Users into Interested Users and Not Interested Users by Using Decision Table. **2020,** No. June.
13. Diep, N. N.; Van Tien, N.; Anh, N. H.; Phuong, T. M. An Unsupervised Method for Web User Interest Analysis. *Proc.—2019 6th NAFOSTED Conf. Inf. Comput. Sci. NICS 2019* **2019,** 27–
14. Suneetha, K. R.; Krishnamoorthi, R. Classification of Web Log Data to Identify Interested Users Using Naive Bayesian Classification. *Int. J. Comput. Sci. Issues* **2012,** *9* (1), 381–387.
15. Kumar, A.; Bose, J.; Bansal, D. A Web Browser Responsive to the User Interest Level. *12th IEEE Int. Conf. Electron. Energy, Environ. Commun. Comput. Control (E3-C3), INDICON 2015* **2016,** 1–6.
16. Aye, T. T. Web Log Cleaning for Mining of Web Usage Patterns. *ICCRD2011—2011 3rd Int. Conf. Comput. Res. Dev.* **2011,** *2*, 490–494. https://doi.org/10.1109/ICCRD.2011.5764181.
17. Hamed, M.; Elhebir, A.; Abraham, A. Discovering Web Server Logs Patterns Using Clustering and Association Rules Mining. **2015,** *3*, 159–167.

CHAPTER 9

DESIGN AND DEVELOPMENT OF AI-ASSISTED SMART VENTILATORS

ANSHUL BHARDWAJ and CHANDRA SHEKHAR SINGH

Dronacharya College of Engineering, Gurgaon, India

ABSTRACT

In this chapter, we present the design and development of AI-assisted smart ventilators with the aim that this ventilator is going to be smart, low cost, portable, and economical enough to handle the diseases of the patient particularly the problem associated with COVID-19. Here we have used sensors like pressure sensors, element or spo2 sensors, flow sensors, a raspberry pi microcontroller to support and ensure the correct functioning of varied sensors and valves to style the required system. We have conjointly designed a software system model mistreatment MATLAB that uses artificial neural networks and machine learning. The software system model uses numerous parameters like patient's age, heartbeat, temperature, blood pressure, air exchanging capability per minute, etc. associate degrees on the basis of these parameters a health score is generated, and a recommendation for the patient is given. The recommendation therefore generated can facilitate the doctors and employees to higher perceive patient conditions and treat them consequently. The benefit of utilizing AI in ventilators is it will facilitate the early recognition of the difficulty or health problem in an exceedingly patient.

Advanced Computer Science Applications: Recent Trends in AI, Machine Learning, and Network Security. Karan Singh, PhD, Latha Banda, PhD & Manisha Manjul, PhD (Eds.)

9.1 INTRODUCTION

A ventilator is a device that supports or reproduces the way toward breathing by pumping air into the lungs, thus sometimes individuals likewise allude to it as a vent or breathing machine. It is a machine that gives mechanical ventilation by moving breathable air into and out of the lungs, to convey breaths to a patient who is truly unfit to inhale, or breathing deficiently.[3] Modern-day ventilators are mechanized chip-controlled machines.[2] In 21st century, the requirement for the ventilators has gotten more essential uniquely in 2020 in view of Coronavirus outbreak all through the world, which offer ascent to the need of more keen, effective, versatile, and minimal effort ventilators.[4] These issues can be addressed by utilizing artificial intelligence. The AI utilizes different strategies and calculations to gather information, measure this information, compute, and foresee the outcome. This strategy can be artificial neural networks (ANN), Machine Learning (ML) and so forth. ANN is a bionic investigation strategy for Artificial Intelligence. The advantages of utilizing AI in ventilator is it can help in the early recognition of the issue or illness in a patient. It can likewise assist with lessening the dangers related with the use of mechanical ventilators like guaranteeing the legitimate pressing factor and wind stream is kept up to help the patient appropriately breathe. Diminishing the danger of lung harm due to over expansion or imploding of air sacks because of dull opening and shutting.[5]

"A Ventilator Weaning Prediction System (VWPS) utilizing ANN, VWPS is an easy to understand framework dependent on the patients' clinical information. It assists specialists with anticipating dispassionately and adequately whether weaning is suitable for persistent. A ventilator weaning achievement rate forecast framework (VWPS) that can assist clinical experts with anticipating weaning achievement rate dependent on clinical test information. The framework creates its forecasts dependent on ANN procedures. VWPS uses MATLAB codes and interfaces together to show the expectation results on an easy-to-use interface."

An India-Dutch start-up mentored by Hyderabad Security Cluster (a Government of Telangana activity), has created savvy ventilators that are coordinated with man-made consciousness (AI) and AI (ML). LEVEN Medical has concocted three models of ventilators—SMART Ventilator, C5 COVID-19 Ventilator, and ICU Ventilator. These ventilators cannot just auto change oxygen stream of a patient however can begin following of all contacts of a basic COVID-19 suspect, places visited throughout the most recent 14 days, send computerized cautions to specialists at the exact second a patient comes in. They likewise send cautions to specialists/medical caretakers and alarms to relatives in the event of a crisis, permit far off observing while the specialist is away from the emergency clinic too.

In this chapter, we presented the approach to design and development of the design and development of AI-assisted smart ventilators with the aim that this ventilator is going to be smart, low cost, portable, and economical enough to handle the diseases of the patient particularly the problem associated with COVID-19. Section 9.2 discuss the Prototype ventilators model consists of various interfacing sensors with microcontroller to accurate measurement of the human body parameters such as SpO_2 level, Flow level, Pressure sensor, Temperature, etc. and automatic calibration of valves in the ventilators. Section 9.3 discusses the flow chart of the overall designed system and implementation in MATLAB and also its efficiency in the measurement. Section 9.4 concludes the overall work and the future scope of the system.

9.2 SECTION II: PROTOTYPE MODEL OF SYSTEM

Mechanical Ventilators used in ICU basically operates on three values (i) FR (respiratory frequency); (ii) ratio of inspiration-to-expiration (I:E); and (iii) the air volume ratio (Vt). These values are previously setup manually according to various physiological condition of the human body. The air and oxygen are the deciding parameters for the calibration. Figure 9.1 shows how these parameters regulate the pressure of the valves. ANN predictive model implemented in MATLAB is used to predict the calibration values. Figure 9.1 shows how the mechanical prototype model of the ventilators interfaced with electronic sensors which are accurate and low cost.

The model diagram appeared beneath utilizes the accompanying sensors, which are as per the following

The various sensors used are:

Pressure sensors are used to measure the amount of pressure applied during the process of inhaling and exhaling. It was connected in such a way that it sent measurements to an Arduino Uno board that processes them and transmits the processed knowledge to the Raspberry Pi. The Raspberry Pi takes these measurements and feeds the ANN algorithm. Oxygen or spo2 sensors: It is used to measure the oxygen or spo2 level applied. Flow sensors: it is used to detect the airflow based on the differential pressure applied

The various valves used are:

Relief valve "It is controlled to ensure that the pressure of the gas supplied to the patient never rises above a set maximum level. The relief valve acts as a backup safety mechanism and opens if the pressure exceeds a safe level, thereby dumping excess gas to atmosphere" (Julienne LaChance, 2020).

Antisuffocation check valve "It is used to check whether there is a situation of suffocation or not in the system. The check valve is oriented such that air can be pulled into the system in the event of system failure, but that air cannot flow outward through the valve" (Julienne LaChance, 2020).

The microcontroller used is raspberry pi which ensures the proper functioning of each sensor and valve used in the system.

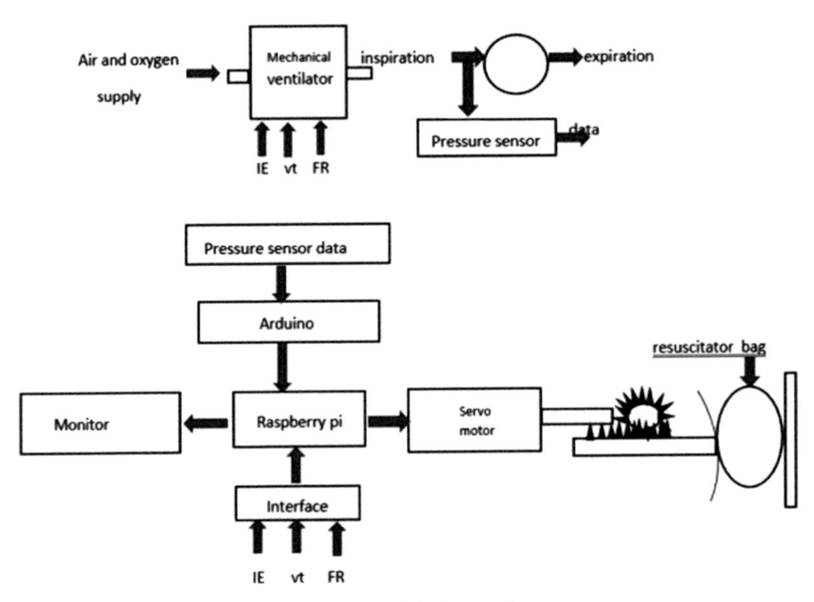

FIGURE 9.1 Prototype model interfaced with electronic sensors.

9.3 SYSTEM IMPLEMENTATION IN MATLAB

In MATLAB R2016b ML toolbox, we implemented ANN-based model to predict the parametric values of the ventilators. First, the data processing is done to extract information and useful data using ML predictive model which is trained on the testing subset data. ANN model is trained from training data using backpropagation model to generalize the model. The weights are adjusted using supervised learning. The repetitive iteration is done until error is minimized. ANN classifier implemented by cycles of forward propagation followed by backpropagation. The backpropagation is used as a optimization problem to obtain optimum results.

TABLE 9.1 Input Parameters for MATLAB ANN Model.

Sr no	Parameters	Characteristics
1	Age	Age is perhaps the most threat factors for individuals over 50 years of age
2	Heartbeat	It ought to be 60 and 120 per minute
3	Systolic Pressure	It ought to be somewhere in the range of 90 and 150 mmHg
4	Body Temperature	It ought to be between 35.5° C and 37.5° C
5	Max Pressure of Inhaling	It uses to be the huge alluding that assessing patient could immediacy relax. It ought to be between 10.1 and 14.7 psi
6	Max Pressure of Expiration	It ought to be in excess of 30 cmH_2O
7	Index Number of Shallowly Quick Breath	It should under 105 per minute
8	Air Exchanging Capacity per minute	It ought to be more modest than 10 1/min

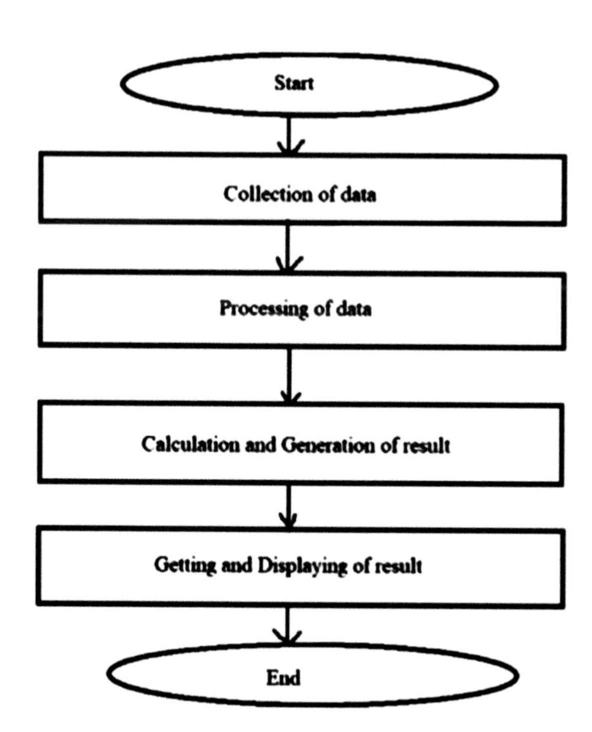

FIGURE 9.2 Flow chart of the process.

The MATLAB model of our work is shown below:

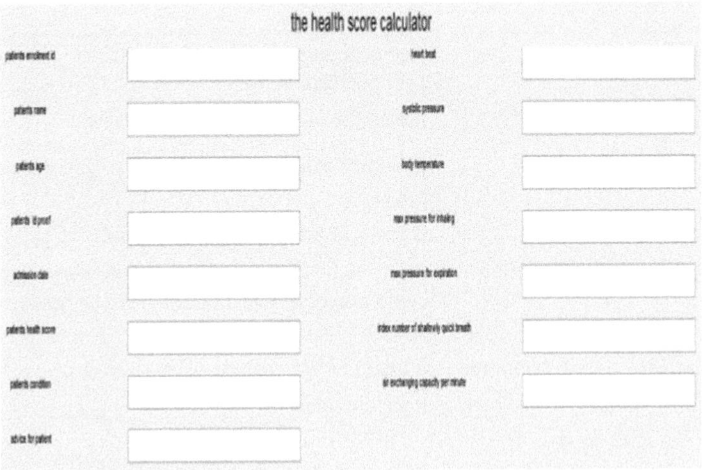

FIGURE 9.3 The MATLAB model.

9.4 RESULT AND DISCUSSIONS

The client enters the fundamental data and the different parameter values that are shown on the model. The model at that point gathers this information and afterward utilizes ANN to ascertain the score and to create the outcome appropriately. The score is determined dependent on the parameters, every parameter is given five-point dependent on its value like very bad = 1 and very good = 5. For instance, in the event that body temp is between 35.5 and 35.9, at that point reading is 5, on the off chance that it is between 37.1 and 37.5, at that point it is 1 and in the event that it is in the middle of 36.3–36.7, at that point it will be 3. Once the outcome is produced it is shown on the system and the guidance is created in like manner. On the off chance that the score is above 70%, at that point just the patient is viewed as healthy and a guidance for its release will be given, else the patient is not solid and a guidance will be produced in like manner.

9.5 CONCLUSION AND FUTURE SCOPE

In this chapter, we have portrayed the plan and advancement of AI-based smart ventilator to the investigation of different boundaries of a ventilator.

Firstly, we have built up the model ANN and ML. When the model was created, we are utilizing it to anticipate the health score and counsel the specialists and staff likewise dependent on the information we get for the calculation and afterward we are combining this system software on the model we have portrayed in our chapter, to make it shrewder and more effective. In future, we can add more parameters like blood sugar level, cough, capacity of urine, calcium, nutrition rate, and so on with the goal that it tends to be utilized for the treatment of different sicknesses also.

KEYWORDS

- **ventilator**
- **sensors**
- **MATLAB**
- **artificial intelligence**
- **artificial neural networks**
- **machine learning**

REFERENCES

1. Degner, M.; Ewald, H. A New Compact and Low-Cost Respirator Concept for One Way Usage. *IFAC-Papers OnLine* **2018,** *51* (27), 367–372.
2. Brossier, D.; El Taani, R.; Sauthier, M.; Roumeliotis, N.; Emeriaud, G.; Jouvet, P. Creating a High-Frequency Electronic Database in the PICU: The Perpetual Patient. *Pediatr. Crit. Care Med.* **2018,** *19* (4), e189–e98.
3. Tehrani, F. T. A Control System for Mechanical Ventilation of Passive and Active Subjects. *Comput. Meth. Prog. Bio.* **2013,** *110*, 511–518.
4. Ranney, M. L.; Griffeth, V.; Jha, A. K. Critical Supply Shortages—The Need for Ventilators and Personal Protective Equipment during the Covid-19 Pandemic. *N. Engl. J. Med.* **2020,** *382*, e41.
5. Gattinoni, L.; Quintel, M.; Marini, J. J. Volutrauma and Atelectrauma: Which Is Worse? *Crit. Care* **2018,** *22*, 1–3.
6. Deshmukh, A. D.; Shinde, U. B. A Low Cost Environment Monitoring System Using Raspberry Pi and Arduino with Zigbee. In *Proceedings of the 2016 International Conference on Inventive Computation Technologies (ICICT)*, Vol. 3; Coimbatore, India, 26–27 Aug 2016; pp 1–6. 38.
7. Brown, B.; Roberts, J. Principles of Artificial Ventilation. *Anaesth. Intensiv. Care Med.* **2019,** *20*, 72–84.

8. Amato, M. B. P.; Marini, J. J. Pressure-Controlled and Inverse-Ratio Ventilation. In Principles and Practice of Mechanical Ventilation; Tobin, M. J., Ed., 3rd ed.; McGraw-Hill: New York, 2013; pp 227–251.
9. Nguyen, H. Q.; Loan, T. T. K.; Mao, B. D.; Huh, E. N. Low Cost Real-Time System Monitoring Using Raspberry Pi. In *Proceedings of the 2015 Seventh International Conference on Ubiquitous and Future Networks*; Sapporo, Japan, 7–10 July 2015; pp 857–859.

CHAPTER 10

DESCRIPTIVE REVIEW ON A NEPALI SIGN LANGUAGE RECOGNITION SYSTEM

KANCHAN SAPKOTA, SAILESH RANA, SANTOSH KHANDAL, YUGARAJ TAMANG, and PANKAJ SHARMA

Department of Computer Science and Engineering, Sharda University, Greater Noida, Uttar Pradesh, India

ABSTRACT

Sign language, being the basic means of communication between specially-abled people, that is, deaf, and normal ones, a massive amount of research works has been carried out in this field. However, since sign language varies from country to country and region to region, and also most of the people do not have enough knowledge about this medium of communication, it is extremely difficult to build a robust and efficient system that can solve sign language recognition (SLR) problems. Some sign languages for which active research has been done are Indian Sign Language, American Sign Language, British Sign Language, etc. However, when we talk about Nepali Sign Language (NSL), the research work is extremely limited. In this chapter, we review different vision-based as well as sensor-based models of NSL recognition using different approaches that were published from 2015 to 2019. Among different techniques, vision-based approaches along with deep learning algorithms were mostly used. Classification algorithms included: Convolutional Neural Network, Random Forest Algorithm, K-Nearest

Advanced Computer Science Applications: Recent Trends in AI, Machine Learning, and Network Security. Karan Singh, PhD, Latha Banda, PhD & Manisha Manjul, PhD (Eds.)

Neighbor, and so on. All the papers have recognized static gestures with most of them achieving an accuracy of more than 90%. With every new proposed model, there is always a remarkable improvement in SLR in terms of accuracy, training time, and computational expense. However, there are still some challenges we may face while developing a system for NSL Recognition.

10.1 INTRODUCTION

Sign language is a visual language used by physically impaired individuals who cannot speak or hear for communication in their day-to-day conversation activities. They interact by expressing signs using their hand shapes, orientation, and movement of hands. Like any other language, its sole purpose is to make such individuals able to interact with normal people. However, most normal persons may not clearly understand the sign language. As a result, there is a huge communication gap between persons having disabilities and the general public. Human translators to some extent contribute in bridging the gap, but it is not enough as they are not available 24 × 7 for interpreting the language. In such a situation, it is vital to exploit the technologies that we have today to support them overcome their disabilities and fill up that communication gap. One such technology is Sign Language Recognition (SLR). Using this technology, we can develop a system that can interpret gestures and signs into texts and vice versa. In the past, different SLR systems have been developed by many developers around the globe, but none of them are flexible or cost-effective for the end-users.

The sign language in use at a particular place depends on the culture and spoken language at that place. Nepali Sign Language (NSL) is the mode of communication for the deaf community in Nepal. It is a standard and well-developed way of communication for hearing impaireds in Nepal. It consists of both word-level and finger-spelling gestures.

There are many research works and studies on popular sign languages like the Indian Sign Language, the British Sign Language, the American Sign Language, and so on. However, when it comes to NSL, there are only a handful of researches on it.

This survey reviews various studies done on NSL and discusses the different methodologies implemented for its recognition in recent years. We have focused on the latest papers as they contain state-of-the-art techniques to solve the problem. We have reviewed various vision as well as sensor-based models used along with different learning algorithms to detect and recognize static gestures. Among the papers we have studied, none have dealt with dynamic gestures. The queries addressed in this survey are: which techniques are used for hand detection and segmentation? Which learning algorithms are used for the recognition of gestures? What type of datasets are used? How accurately the systems recognize sign languages? What is the type of gesture (finger-spelling or isolated or continuous)? A total of four papers were reviewed and analyzed.

The remainder of this chapter is arranged as follows. Section 10.2 includes a deep review and analysis of the papers on the basis of dataset, human–computer interaction (HCI) approach, used methodologies, and so on. In Section 10.3, we have discussed the limitations and drawbacks of each research work. Finally, we have concluded the review, and proposed our system that overcomes the mentioned limitations of other papers in Section 10.4.

10.2 LITERATURE REVIEW

[1] Finger Spelling Recognition for NSL - Vivek Thapa, Jhuma Sunuwar, and Ratika Pradhan

This paper has proposed a system to recognize NSL where the hand gesture is recognized using shape information. Here, Vertex Chain Code (VCC) and Freeman Chain Code are used as the feature extractors to identify distinguishable information from the hand gesture which is later used by the K-Nearest Neighbor (KNN) to classify the gesture.

The dataset consists of five different gestures representing NSL consonants GHA, CHHA, BA, MA, and SA. The image is obtained using a USB camera and is converted into a binary image that is then cropped and resized to 320 × 320. This depletion in the image size helps make the processing faster. Image brightness and contrast are also altered to enhance the feature of the image. After this, the hand gesture is segmented from the image using skin color detection and blob analysis.

After this preprocessing, the center point of the image is determined, and it is separated by drawing a line between the line and the boundary of the image at the central angle and the intersection point. After that, points are linked, and using the Freeman chain code process, the chain code is obtained. From this chain code, we get the first difference code, and finally, the shape number that acts as a unique feature for each gesture.

During the training phase, a set of images is fed into the system to extract the shape number/information of the input gesture and a feature set is created and stored in the database. This will be used for comparison while testing.

The shape number of the performed gesture is acquired during the recognition process using the same chain code system. Now, this shape number is compared with the one that is saved in the database during the learning phase. Since the training and testing gesture are not the same, the KNN classifier matches all possible occurrences of a related sequence in the shape number. Two gestures with shape numbers close to each other are considered to be a similar gesture and are said to be recognized.

The accuracy is determined by dividing the total number of correctly predicted gestures by the total number of tested gestures. The accuracy of recognition using the radial approach for sampling along with the Freeman chain code was 56%. Similarly, when the grid line method was applied for sampling along with Freeman chain code, the accuracy was 48% and for the radial approach with the VCC method, the accuracy was 44%. Hence, the radial approach and Freeman chain code used for sampling and extracting feature, respectively, were found to be comparatively better than the other two methodologies.

[2] Hand Gesture Vocalizer for Dumb and Deaf People - Sanish Manandhar, Sushana Bajracharya, Sanjeev Karki, Ashish Kumar Jha

This chapter has proposed the system to recognize NSL through the device named Glove Controller. This Glove Controller consists of two parts: one is Flex Sensor that detects finger movement and the other is Accelerometer that detects hand movement. Flex Sensor is a hardware device that measures the amount of deflection and provides us electrical resistance value. If the resistance value given by the flex sensor is high, the flex sensor is bent or deflected more and vice versa. In this way, we can know the finger movement of the user and this resistance value is used for feature extraction in the next stage.

According to this article, both regression and classification use a supervised learning algorithm named as Random Forest Algorithm (RFA). During the training method, the RFA operates by constructing a decision tree. After the training process is done, the votes from decision trees are collected and then aggregate to give the output class or predicted class. This classifier generates the set of decision trees from the randomly collected training sets and these sets are used to collect the votes and the mode of these votes results in the final class.

In the chapter, it is highlighted that the RFA is very effective as it applies the weight concept and a high error rate signifies the low weight value. There are mainly two stages in this algorithm. The first one is the Random Forest creation stage and the second is the Random Forest prediction stage. In the first stage, decision trees are created during the training session of the system and the sets of decision trees are made by Random Forest Classifier. In the second stage, the prediction is done to identify the class from the trained data sets available for the system. Prediction is done only after the vote is collected from the sets of decision trees made by Random Forest Classifier. To prevent the system from overfitting and to train the model with less variance, the datasets for all the alphabets are merged and rearranged and the same is done for other words. Image data sets of alphabets and commonly used words are used to train the machine. In that collected dataset, the machine was able to give an accuracy of 96.8%.

[3] Two Dimensional (2D) Convolutional Neural Network (CNN) for NSL Recognition - Drish Mali, Rubash Mali, Sanjeeb Prasad Panday (PhD), Sushila Sipai

This chapter has proposed a system that recognizes static NSL gestures using 2D CNN. In this chapter, the authors have trained three CNN models for recognizing 5, 7, and 9 gestures. Here, the manual collection of image dataset is done using the webcam of a laptop.

Both single- and double-handed static gestures are collected for the dataset. For each gesture, 400 red masked frames were shot using a 1-mega-pixel camera, i.e., 100 and 300 for test and training purposes respectively. The red mask of images was obtained by HSV. The images were resized to 64×64 size to make it computationally efficient. Then, re-scale, horizontal flip, shear, and zoom operations were performed for data augmentation to increase the size of the dataset. Now, we have 1500 masked images for each gesture in training sets.

These preprocessed images are fed into the CNN model one by one. The first layer containing 32 kernels of size 3×3 performs the convolutional operation on the input image to produce 32 feature maps. Now, these feature maps are sent to the Max pool layer having 2×2 pool matrix to downsample it by 75%, while still preserving the key features. In both layers, the Rectified Linear Unit is used as an activation function to get rid of less than 0 pixel values from the feature vector. The feature maps are flattened to a 1D vector and then fed into a fully connected layer with

128 neurons. Finally, it passes into the output layer with nodes equal to the number of classes.

The accuracy for the five gestures model was 95.4% on the test sets of 500 images. Similarly, 94.57% and 94.11% accuracy were obtained for seven and nine gestures models, respectively. The accuracy of the model starts to decrease as the number of gestures increases. Red gloves for hand segmentation and CNN as classifiers were effective to produce high accuracy for the model.

[4] Hand Gesture Recognition for NSL Using Shape Information—Jhuma Sunuwar and Ratika Pradhan

This chapter approaches the task of SLR using the digital image processing technique of freeman chain code along with Run Length Encoding to get the unique shape numbers of different static NSL gestures. Here the authors justify their work saying if a sign language gesture is given then the gesture is classified based on the closest relation to the pre-existing database of sign gestures.

The dataset used in this chapter consists of 35 static consonants sign gestures out of a total of 49 gestures of NSL. The dataset was created manually using a four-megapixel camera and then adjusting the contrast and brightness of the images to improve the quality of the dataset. The images were captured in the plain background with no skin-colored objects nearby so that it may not pose as noise during the preprocessing phase.

In the preprocessing phase to get the area of relevance, that is, the hand region, the color cue model was used. In this approach, a predefined color range of skin is used to segment the hand region from the other parts of the image that does not come under the skin color range. After the segmentation, the images are then binarized and then certain morphological operations like dilation, erosion, and filling of holes are applied to remove unwanted noise from the images. Following the noise removal, blob analysis is done in the image. The largest blob in the image is then extracted considering it as the hand region. The extracted blob, that is, the hand region is then resized to a standard size to remove any size constraints.

After the preprocessing phase, the hand sign images are segmented, noise is reduced and resized to the standard size. The images are then overlapped with a grid of size 25 × 25. To get the sample boundary point, the edges or contour of the images nearest to the grid line were taken. It was done to reduce the points that represent the boundary and smooth out the variations in the contour. Since each gesture is classified using the unique shape number, it was derived with the help of freeman chain coding. In this method, the boundary points are used to get a compact representation of the image in the form of chain codes which are translation invariant. After the chain code for the image is obtained, it is further compressed using Run Length Coding, and the first difference is calculated from which the final shape number for the image is then generated. The final shape number is then stored in the database using a proper data structure so that it can be used for a new classification in real time.

For the classification of a new sign gesture, all the previous applied steps are used to get the final shape of the number of the test sign image. The shape number is then matched across the database created using the previous methods and the maximum count of the matching sequence between the database and test gesture is taken, which is then used to classify the gesture into its corresponding NSL alphabet.

10.3 FINDINGS AND DISCUSSION

Despite being the prominent mode of communication for the deaf community in Nepal, its research and development are still in the rudimentary stage in our country. In this survey, we reviewed four articles on Nepalese SLR systems using different learning algorithms that were published in the last 5 years. Most of them have used vision-based approaches along with machine learning algorithms to detect and recognize static NSL.

There are two main approaches for HCI: vision-based and sensor-based. In the below table, we can see that the majority of papers have used computer vision and digital image processing techniques, which is a simple and cost-effective approach for interaction between humans and computers. Similarly, we can also observe that the type of gesture used in all the papers is static, which do not pose any movement. This implies that all the proposed systems are only suitable for study and research purposes. But when it comes to solving the real-world problem by translating the sign language practically, all of them will fail to meet the expectations as most of the sign language in the real world are performed dynamically and is dependent on various factors like lighting, environment, orientations, facial expression, body movements, and so on.

TABLE 10.1 Initial Overview of Papers.

Paper	Approach	Dataset
[1]	Vision Based	5 Static Classes [Custom Dataset]
[2]	Sensor Based	[Custom Dataset]
[3]	Vision Based	5/7/9 Static Classes [Custom Dataset]
[4]	Vision Based	35 Static Classes [Custom Dataset]

TABLE 10.2 Accuracy of Methods across Different Papers.

Paper	Methodology	Accuracy
[1]	Chain Coding+Shape Formation+KNN	56%
[2]	Random Forest	96.8%
[3]	CNN	95.4%
[4]	Chain Coding+Shape Formation+Data Structure	Not Mentioned

As shown in the above table, paper[2] has the highest accuracy among all the methods approached in the other papers. Similarly, the papers[1–3] used machine learning algorithms to produce accuracy as a metric for output while paper[4] used simple digital image processing techniques coupled with data structures for recognition and does not state any metrics as the final result. The accuracy obtained by paper[1] is 56% on the static sign, which concludes that it is not usable in a real-time application as there is large variation in lighting conditions and other aspects that may further decrease the accuracy. Similarly, for paper,[2] though it has a high accuracy, it is based on sensor gloves as a primary hardware constraint that makes it a hassle to use in real-time, and its availability to the differently abled people also might be something to think over. Though the paper[3] boasts an accuracy of 95.4%, it is also explained that the accuracy went down to 94.57% on increasing the number of classes to seven which further declined to 94.11% when the number of classes was increased to 9. It shows that the proposed methodology in paper[3] is unreliable as well as unscalable. Finally, paper[4] uses a premade database and data structure to store sign gestures which is not feasible if you consider the variations of signs and prevalence of some noises or other constraints while feeding input data to the system.

10.4 CONCLUSIONS

After deep study and analysis, we have collected the features, advantages, disadvantages as well as challenges for every paper. Now, our next goal is to minimize the limitations present in these papers. We tried to overcome those constraints by using different techniques, and at last, we came up with an idea of developing a NSL recognition system that is supposed to be effective, efficient, and implementable. In this system, we are thinking of generating video datasets on our own as no usable dynamic NSL datasets are available on the internet. We will be implementing a system that will work on dynamic gestures as all the above-mentioned methods only deal with finger-spelling and static gestures. Since dynamic sign language is applicable in real-world usage, we will implement a real-time system for NSL recognition by avoiding all the constraints of the reviewed papers.

KEYWORDS

- **Indian Sign Language**
- **British sign language**
- **convolutional neural network**
- **k-nearest neighbors**
- **random forest**
- **freeman chain code**

REFERENCES

1. Thapa, V.; Sunuwar, J.; Pradhan, R. Finger Spelling Recognition for Nepali Sign Language. In *Recent Developments in Machine Learning and Data Analytics: Advances in Intelligent Systems and Computing*; Kalita, J.; Balas, V.; Borah, S.; Pradhan, R.; Eds.; Vol, 740; Springer: Singapore, 2019. https://doi.org/10.1007/978-981-13-1280-9_22
2. Manandhar, S.; Bajracharya, S.; Karki, S.; Jha, A. Hand Gesture Vocalizer for Dumb and Deaf People. *SCITECH, Nepal* **2091,** *14* (1), 22–29. https://doi.org/10.3126/scitech.v14i1.25530
3. Mali, D.; Mali, R.; Sipai, S.; Panday, S. Two Dimensional (2D) Convolutional Neural Network for Nepali Sign Language Recognition, 2018, 1–5. DOI: 10.1109/SKIMA.2018.8631515.
4. Sunuwar, J.; Pradhan, R. Hand Gesture Recognition for Nepali Sign Language Using Shape Information. *Int. J. Comput. Sci. Eng.* **2015,** *3* (6), 129–135.

5. Huang, C-L.; Huang, W-Y. Sign Language Recognition Using Model-Based Tracking and a 3D Hopfield Neural Network. *Mach. Vis. App.* **1998,** *10*, 292–307.
6. Davis, J.; Shah, M. Visual Gesture Recognition. *IEE Proc. Vis. Image Signal Process. 1994*, *141* (2), 101106.
7. Chen, F-S.; Fu, C-M.; Huang, C-L. Hand Gesture Recognition Using a Real-Time Tracking Method and Hidden Markov Models. In *Institute of Electrical Engineering*, National Tsinghua University: Hsin Chu 300, Taiwan, March 2003.
8. Deora, D.; Bajaj, N. Indian Sign Language Recognition. In *1st International Conference on Emerging Technology Trends in Electronics, Communication and Networking (ET2ECN)*, 2012; pp 1–5.
9. Kirillov, A. Hand Gesture Recognition, 2008. http://www.codeproject.com/Articles/26280/Hands- Gesture-Recognition (accessed 9 Sept 2014).
10. Shamaie, A.; Sutherland, A. Accurate Recognition of Large Number of Hand Gesture. Machine Vision Group, Centre for Digital Video Processing School of Computer Applications, Dublin City University, Dublin 9, Ireland, 2003.
11. Hurlbut, H. M. A Lexicostatistic Survey of the Signed Languages in Nepal. SIL Electronic Survey Reports, June 2012; p 23.
12. Chai, X. et al. Sign Language Recognition And Translation with Kinect. Institute of Computing Technology, CAS, Microsoft Research Asia, Beijing, China, October 2013.
13. Quan, Y.; Jinye, P. Application of Improved Sign Languages Recognition and Synthesis Technology. *Industrial Electronics and Application*, 2008, ICIEA 2008, 3rd IEEE Conference, June 2008; pp. 1629–1634.
14. Starner, T.; and Pentland, A. Real-Time American Sign Language Recognition from Video Using Hidden Markov Models. In *SCV95*, 1995; p 265270.

CHAPTER 11

APRIORI-BASED ALGORITHMS WITH A DECENTRALIZED APPROACH FOR MINING FREQUENT ITEMSETS: A REVIEW

SATVIK VATS[1], SUNNY SINGH[2], and B. B. SAGAR[3]

[1]*Computer Science and Engineering, Graphic Era Hill University, Dehradun, India*

[2]*Data Scientist, NextGen TechEdge Solutions Pvt Ltd., India*

[3]*Computer Science and Engineering, Birla Institute of Technology, Mesra, Ranchi, India*

ABSTRACT

Big data in a distributed computational environment is the need of digital era, which requires analysis, monitoring, and control. In the present scenario of big data, space, and time complexity have raised out new challenges to frequent itemsets mining. The key to success of MapReduce framework is appropriate for distributed data processing on commodity nodes. Apriori algorithm is one of the most popular and widely used data mining algorithm. Which is applied to produce frequent itemsets. The present paper deals with the overview of several procedures, which is on parallel platform to improve the performance of the traditional Apriori algorithm. It also represents the advantages of MapReduce framework. And explore the new path of research in the future.

Advanced Computer Science Applications: Recent Trends in AI, Machine Learning, and Network Security. Karan Singh, PhD, Latha Banda, PhD & Manisha Manjul, PhD (Eds.)

11.1 INTRODUCTION

Data mining is a method that extract information's from the bulk amount of information obtained from several data analysis tools to make valid predictions.[1] These valid predictions are now used to set the strategy in industries. In today's digital era, the size of data in different fields is growing exponentially, which brings out new challenges toward data mining.

Big data is utilized to process the hefty size of information originated from different digital sources. Traditional data analysis tools are not appropriate to mine useful information from the Big data.[2] Big data is defined by 3Vs, that is, Volume, Variety, and Velocity. Now it is defined in form of 10Vs, by the addition of another 7vs that is Veracity, Validity, Variability, Volatility, Vulnerability, Visualization, and Value.[3–10] The MapReduce programming model provides the necessary platform to build the application that can be capable to run parallel in a distributed computational environment to deal with the Big data. Reliability of the data processing process is guaranteed by Hadoop framework. Frequent itemsets in minimal time are generated with help of improved Apriori algorithm[11] to run in a parallel fashion on the MapReduce programming model. A decentralized approach[12] is used to execute the application at different nodes in a distributed computation environment. Various methods have been anticipated to augment the competence of Apriori to deal with Big data problems.

The present paper deals with the all-major improvements of the Apriori method, which is proposed earlier. Simultaneously, it also provides a brief discussion on future research direction, based on our survey. The paper is prepared in four segments. Section 11.1 deals with the introduction. Section 11.2 include an summary of Association rule mining (ARM), Big data, Hadoop framework, and MapReduce-Apriori. Whereas Section 11.3 focuses on related work. Conclusion and future research direction are discussed in brief in Section 11.4.

11.2 BACKGROUND STUDY

11.2.1 ASSOCIATION RULE MINING (ARM)

ARM is utilized to identify frequent itemsets in transactional database. It is processed in two steps.[13]

- Finding of the frequent itemsets having support counts are more than or equal to minimum support.
- Generation of the association rules confidence is more than or equal to minimum support.

The whole ARM process is achieved by using the Apriori algorithm. Apriori algorithm consists of two phases based on two steps as given above. The first step performs to generate frequent itemsets in an iterative way, which is also responsible for the overall cost of the mining process. Decentralized approach to the Hadoop framework is used to reduce the cost of the mining process, which is the key to success to divide the whole problem into a subproblem.

11.2.2 BIG DATA

Today, different areas of digital world's are handled by the big data such as social network, healthcare,[34] finance, e-commerce,[35] education, web analytics,[14] character recognition,[15] etc. These areas generate three kinds of huge data:

- Structured data
- Unstructured data
- Semistructured data

Structured data is a term, which represent the data that can be accessed, processed, and stored in the form of fixed format. Which is represented as data stored in a relational database management system. Whereas, Unstructured data represent the data with an raw form data, for example, output produced by "Google Search Engine." Semistructured data consist of both the forms of data, for example, data in the XML file. Nowadays, it is become important to mine such a huge amount of data to extract useful information to help business decisions making scenarios to provide better quality services. The main challenge is to handle and analyze Big Data. A large number of the computational framework has been developed to analyze such a huge amount of data. MapReduce is one of the widely used frameworks that can be used to analyze Big data using an open-source Hadoop environment.

11.2.3 HADOOP FRAMEWORK

Parallel processing for large data sets on commodity nodes are easily configured and executed by the Apache Hadoop.[16,17] Hadoop Distributed File System (HDFS) and MapReduce is implemented in Google's File System.[18] Replicated files are stored across multiple nodes, which offer fault tolerance/ high availability.

11.2.4 MAPREDUCE-APRIORI

MapReduce helps in programming and fault tolerance.[19] It is required to make the whole system works on a decentralized approach using Big data in a distributed computational environment. Working of any algorithm on the MapReduce platform have to develop the whole algorithm in two main functions, named as map function and reduce function. Both functions are user-defined functions. Users define a map function that is used to produce a set of intermediate key-value pairs. Reduce function has the work of merging all intermediate values, which are associated with the same intermediate key. Figure 11.1 is given to show the basic concept of the MapReduce platform. Performance of MapReduce-Aprori depends on four major phases that are given below.

11.2.4.1 LOCAL ITEM FILTERING

Local item filtering is required to reduce the communication cost during uploading of data into HDFS. Items having temporal support less than minimum support are filtered out. After completing the local item filtering by scanning the data records are stored into the HDFS are in different computational nodes.

11.2.4.2 MERGING ITEMS INTO TRANSACTION

Data in HDFS are clustered in time slots to find their correlations. The overall process is conducted in a parallel fashion. Merging of the data records into transactions is done by the MapReduce pass, where each mapper has an input pair in key-value form. When mapper function finishes its job, MapReduce architecture grouped its outcomes and then submit to the reducer function as the key-value pair. Reduce function received the key-value pair and accordingly combines all the data items in similar time slot in a transaction with association of transaction ID.

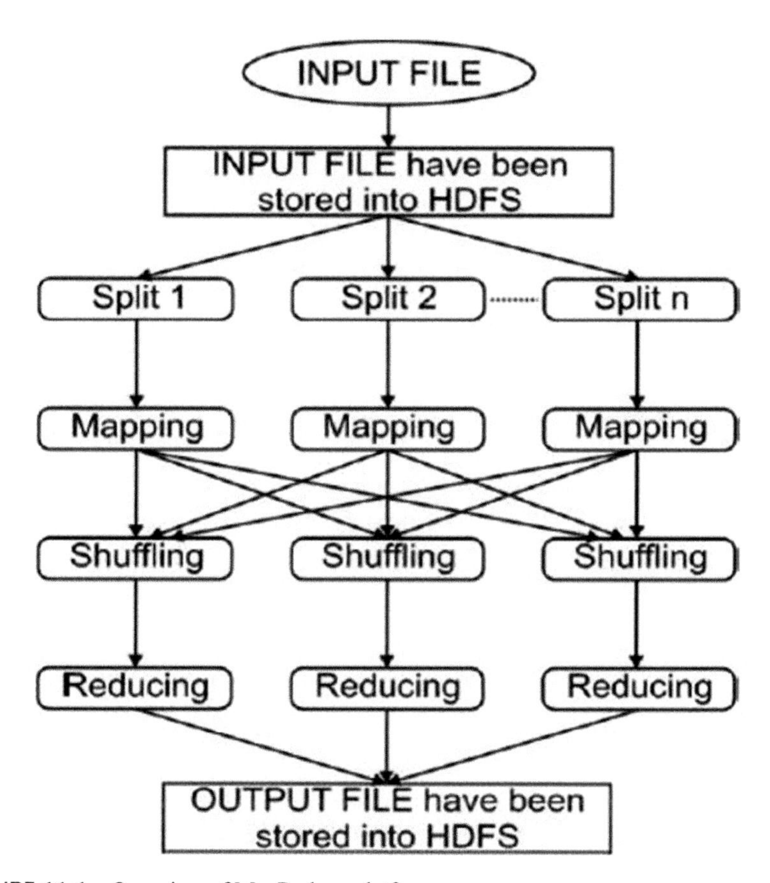

FIGURE 11.1 Overview of MapReduce platform.

Source: Adapted from Ref. 20.

11.2.4.3 Global Frequent Itemsets Mining

As the transactions are distributed at different nodes in a distributed computation environment, the MapReduce programming model is capable enough to parallel the global frequent itemsets mining process. In MapReduce-Apriori, first of all, each mapper loads all the candidate itemsets, which are generated in the first subset. The reducer function sums the values with the same key, which are based on a re-calculated lifetime of an itemsets. It also identify, if the total support count is not less than minimum support.

11.2.4.4 Global Monitoring of Association Rule Generation

The basic concept behind this is to discover subsets for each frequent itemset, to produce association rules. A pruning step is introduced here, which is based on the Apriori property to reduce the computation time. According to this all the subsets of frequent itemsets must be frequent.

Each mapper must read from a frequent item–itemsets along with its temporal support. Key-value pair of outputs were generated on frequent subitemsets that must be a subset of global frequent itemsets. Each reducer function has the task of calculating the confidence and support count based on the improved association rules.

11.2.5 ADVANTAGES OF DECENTRALIZED APPROACH FOR MINING ARE AS FOLLOWS

- To avoid centralization into one database.
- Overall mining time is considerably reduced.
- MapReduce need to enhance the computational capability of distributed environment to handle the Big data in today's scenario.
- Generation of frequent itemsets in parallel fashion to make the Apriori algorithm more efficient to deal with Big data.

11.3 RELATED WORK

QFP-growth method is proposed to search out the frequent itemsets based on FP-trees.[20,21] In this, a large database is divided into subdatabases having approximately similar in size and distributed uniformly over a distributed computational environment where FP-trees are constructed using the subdatabases. Finally, all FP-trees are merged to construct an FP-tree for a huge problem. The main drawback of this mining technique is that it is required to pass each child database to computing nodes to execute the mining operation. When the size of the database is large then it overhead the data transformation and communication cost in between the different nodes.

Ansari et al. [22] proposed a distributed tire-based algorithm named DTFIM. It is based on Message Passing Interface (MPI) to find out frequent itemsets in a distributed computational environment. It is an Apriori-like algorithm and has the iterative fashion to scan the database that further restricts its scalability and efficiency.

An Apriori-like algorithm is planned to explore the efficacy of the generation of frequent itemsets in a distributed computing environment using the Hadoop framework.[23] The main disadvantage of this method is that it produces a huge number of candidates itemsets after scanning the large database, many times that increase the time complexity.

Distributed Parallel Apriori algorithm is given by Yu et al.[24] It is an Apriori-like algorithm that works on a language-independent interface named as MPI over distributed computing nodes. This algorithm creates a table on each node where all the data records and transaction IDs are brought and then assigns a set of k-itemset to each node and corresponding candidate (k+1) itemsets are generated as a result of it.

MapReduce can process bulky datasets using several nodes in a distributed computational environment. A parallel application of the Apriori method configured on MapReduce is given by Li et al.[25] Findings of the present study clearly demonstrate that the planned algorithm can effectively executed huge datasets with scalability in a distributed computational environment. A decentralized approach for ARM is proposed for distributed system monitoring.[26]

Parallel Apriori algorithm is configured and executed on the distributed framework by Sudhakar et al.[27] It is categorized based on one-phase and k-phase, input-output Mapper, Combiner and Reducer used to implement it. It also discusses the various characteristics of the distributed Apriori system with its advantages and limitations on the MapReduce.

A parallel association rules method is introduced, which works on distributed environment for big data.[28] It implements both candidate generation steps and count over the MapReduce framework to speed up the whole process while the existing parallel Apriori technique only implements the count step. Experimental results express better performance for big data and records good speedup and scalability features.

Xian et al. in Ref.[29] proposed an iterative sampling-based frequent itemset mining algorithm to deal with the (i) space complexity and (ii) time complexity to handle both input data and intermediate results of big data. This algorithm samples the data into computationally subsets and finds out the frequent itemsets from these computationally subsets rather than process the whole data at once. The main advantage of this algorithm is that it can be parallelized with ease over the MapReduce framework. Scalability issues and massive outputted patterns responsible for out-of-memory problems can be handled by this algorithm.

For discovering frequent pattern, efficient mining algorithm is introduced by Lin et al.[30] which is capable to govern sufficient computational nodes to process the data and provides load balancing automatically. Moreover, efficiency of the overall system cannot be improved just by increasing the computational nodes because too several nodes create hefty network traffic. However, on the other hand, allocating a limited node can be reason the load to be imperfectly circulated. So that load balancing is required to improve the efficiency of the overall system.

Zhang et al.[31] proposed i^2 MapReduce, an incremental processing extension to MapReduce for mining big data. Supports both one-step and more sophisticated iterative computation. Help to reduce input-output overhead by saving computation states. This new computational environment combines an incremental engine, to execute an incremental program effectively over distributed system.

Ashburn et al.[32] proposed a methodology to evaluate the efficacy of a single node cluster concerning, to process huge data. In this experiment, a parallel Apriori system is used to search out the frequent itemset and the total time taken by the algorithm over MapReduce is to find out to evaluate the performance concerning the increasing number of nodes.

Solanki et al.[33] proposed a fuzzy ARM algorithm to produce the output in minimal time. Most of the method examine the database several times before to produce the final output which required high process cost, memory, and time to handle the large dataset. A tree-based algorithm can solve this problem but it is also suffering from the generation of a massive number of conditional FP-tree. Hence a fuzzy ARM method is the best choice to get the optimal frequent patterns in minimal time.

11.4 CONCLUSIONS AND FUTURE RESEARCH DIRECTION

MapReduce programming model is very lucrative for parallel processing using Big data in a distributed computational environment. In this chapter, we mainly concern about the parallelization of the Apriori algorithm using the MapReduce model on the Hadoop framework and also reviewed various proposed approaches to parallelize the Apriori algorithm or using an alternative solution to find the frequent itemsets in minimal time such as fuzzy algorithms. Although the MapReduce model is an efficient and scalable platform to solve Big data computational problems in an efficient way it may be difficult to port some problems on this framework such as problem consisting of iterative or looping function in itself.

In the future, the throughput of frequent itemsets mining approach can be improved by considering future research mainly in two dimensions. One dimension is to look at an alternative Apriori-like algorithm, which is capable enough to find out the frequent itemsets in minimal time, even in case of iterative-approach-based problem and by developing an algorithm that can support pipelining to speed up the computational process of the whole system. The pruning step of the Apriori algorithm needs to be improved further, considerably. The second dimension of future research focuses on the improvement of the MapReduce framework.

KEYWORDS

- **big data**
- **MapReduce**
- **Apriori**
- **decentralized approach**

REFERENCES

1. Ngai, E. W. T.; Xiu, L.; Chau, D. C. K. Application of Data Mining Techniques in Customer Relationship Management: A Literature Review and Classification. *Expert Syst. App.* **2009,** *36* (2), 2592–2602.
2. Chen, H.; Chiang, R. H. L.; Storey, V. C. Business Intelligence and Analytics: From Big Data to Big Impact. *MIS Quart.* **2012,** *36* (4), 1165–1188.
3. Agarwal, R.; Singh, S.; Vats, S. Implementation of an Improved Algorithm for Frequent Itemset Mining Using Hadoop. In *2016 International Conference on Computing, Communication and Automation (ICCCA)*; IEEE, 2016 April; pp 13–18.
4. Agarwal, R.; Singh, S.; Vats, S. Review of Parallel Apriori Algorithm on MapReduce Framework for Performance Enhancement. In *Big Data Analytics*; Springer: Singapore, 2018; pp 403–411.
5. Vats, S.; Sagar, B. B. Performance Evaluation of K-Means Clustering on Hadoop Infrastructure. *J. Discr. Math. Sci. Cryptogr.* **2019,** *22* (8), 1349–1363.
6. Vats, S.; Sagar, B. B. Data Lake: A Plausible Big Data Science for Business Intelligence. In *Communication and Computing Systems: Proceedings of the 2nd International Conference on Communication and Computing Systems (ICCCS 2018)*, December 1–2, 2018, Gurgaon, India; CRC Press, Oct 2019; p 442.
7. Bhati, J. P.; Tomar, D.; Vats, S. Examining Big Data Management Techniques for Cloud-Based IoT Systems. In *Examining Cloud Computing Technologies Through the Internet of Things*; IGI Global, 2018; pp 164–191.

8. Vats, S.; Sagar, B. B.; Singh, K.; Ahmadian, A.; Pansera, B. A. Performance Evaluation of an Independent Time Optimized Infrastructure for Big Data Analytics that Maintains Symmetry. Symmetry **2020,** *12* (8), 1274.
9. Vats, S.; Sagar, B. B. An Independent Time Optimized Hybrid Infrastructure for Big Data Analytics. *Modern Phys. Lett. B* **2020,** 2050311.
10. Agarwal, R.; Vats, S.; Singh, S. *Knowledge Discovery Using Big Data Analytics*; LAP LAMBERT Academic Publishing, Jun 2016; pp 1–84.
11. Wu, X. et al. Top 10 Algorithms in Data Mining. *Knowl. Inf. Syst.* **2008,** *14* (1), 1–37.
12. Wu, G. et al. A Decentralized Approach for Mining Event Correlations in Distributed System Monitoring. *J. Parallel Distrib. Comput.* **2013,** *3*, 330–340.
13. Asbern, A.; Asha, P. Performance Evaluation of Association Mining in Hadoop Single Node Cluster with Big Data. *Circ. Power Comput. Technol. (ICCPCT), 2015 Int. Conf.*; IEEE, 2015.
14. Kohli, S.; Kaur, S.; Singh, G.. A Website Content Analysis Approach Based on Keyword Similarity Analysis; 2012; pp 254–257. DOI: 10.1109/WI-IAT.2012.212.
15. Kaur, S.; Sagar, B. B. Brahmi Character Recognition Based on SVM (Support Vector Machine) Classifier Using Image Gradient Features. *J. Discr. Math. Sci. Cryptogr.* **2019,** *22* (8), 1365–1381. DOI: 10.1080/09720529.2019.1692445
16. Ma, B.; Liu, W.; Hsu, Y. Integrating Classification and Association Rule Mining. *Proceedings of the Fourth International Conference on Knowledge Discovery and Data Mining*, 1998.
17. Bhandarkar, M. MapReduce Programming with Apache Hadoop. *Parallel Distrib. Process. (IPDPS), 2010 IEEE International Symposium on*; IEEE, 2010.
18. Noll, M. G. Running Hadoop on Ubuntu Linux (Multi-Node Cluster). Apr 2013. http://www.michael-noll.com/tutorials/running-hadoop-on-ubuntu-linux-multi-nodecluster/. (accessed 15 Jun 2013) (2007).
19. Ghemawat, S.; Gobioff, H.; Leung, S-T. The Google File System. *ACM SIGOPS Operat. Syst. Rev.* **2003,** *37* (5).
20. Li, N. et al. Parallel Implementation of Apriori Algorithm Based on MapReduce. *Software Engineering, Artificial Intelligence, Networking and Parallel & Distributed Computing (SNPD), 2012 13th ACIS International Conference on*; IEEE, 2012.
21. Qiu, Y.; Yong-Jie, L.; Qing-Song, X. An Improved Algorithm of Mining from FP-Tree. *Machine Learning and Cybernetics, 2004. Proceedings of 2004 International Conference on*, Vol. 3; IEEE, 2004.
22. Ansari, E. et al. Distributed Frequent Itemset Mining Using Trie Data Structure. *IAENG Int. J. Comput. Sci.* **2008,** *35* (3), 377–381.
23. Yang, X. Y.; Zhen, L.; Yan, F. MapReduce as a Programming Model for Association Rules Algorithm on Hadoop. *Information Sciences and Interaction Sciences (ICIS), 2010 3rd International Conference on*; IEEE, 2010.
24. Yu, K-M. et al. A Load-Balanced Distributed Parallel Mining Algorithm. *Expert Syst. App.* **2010,** *37* (3), 2459–2464.
25. Li, L.; Zhang, M. The Strategy of Mining Association Rule Based on Cloud Computing. *Business Computing and Global Informatization (BCGIN), 2011 International Conference on*; IEEE, 2011.
26. Wu, G. et al. A Decentralized Approach for Mining Event Correlations in Distributed System Monitoring. *J. Parallel Distrib. Comput.***2013,** *3*, 330–340.

27. Singh, S.; Garg, R.; Mishra, P. K. Review of Apriori Based Algorithms on Map Reduce Framework. *Environment* 11: 18.
28. Zhou, X.; Huang, Y. An Improved Parallel Association Rules Algorithm Based on MapReduce Framework for Big Data. *Fuzzy Systems and Knowledge Discovery (FSKD), 2014 11th International Conference on*; IEEE, 2014.
29. Lin, M-Y.; Lee, P-Y.; Hsueh, S-C. Apriori-Based Frequent Itemset Mining Algorithms on MapReduce. *Proceedings of the 6th International Conference on Ubiquitous Information Management and Communication*; ACM, 2012.
30. Lin, K. W.; Chung, S-H. A Fast and Resource Efficient Mining Algorithm for Discovering Frequent Patterns in Distributed Computing Environments. *Future Gen. Comput. Syst.* **2015**.
31. Zhang, Y. et al. i2MapReduce: Incremental MapReduce for Mining Evolving Big Data; 2015.
32. Asbern, A.; Asha, P. Performance Evaluation of Association Mining in Hadoop Single Node Cluster with Big Data. *Circuit, Power and Computing Technologies (ICCPCT), 2015 International Conference on*; IEEE, 2015.
33. Solanki, S. K.; Patel, J. T. A Survey on Association Rule Mining. *Advanced Computing & Communication Technologies (ACCT), 2015 Fifth International Conference on*; IEEE, 2015.
34. Vats, S. et al. iDoc-X: An Artificial Intelligence Model for Tuberculosis Diagnosis and Localization. *J. Discr. Math. Sci. Cryptogr.* **2021,** *24* (5), 1257–1272.
35. Gupta, A., Lohani, M. C.; Manchanda, M. Financial Fraud Detection Using Naive Bayes Algorithm in Highly Imbalance Data Set. *J. Discr. Math. Sci. Cryptogr.* **2021,** *24* (5), 1559–1572.

CHAPTER 12

IMPACT OF ARTIFICIAL INTELLIGENCE AND THE INTERNET OF THINGS IN MODERN TIMES AND HEREAFTER: AN INVESTIGATIVE ANALYSIS

ANSHUL GUPTA, SUNIL KR. SINGH, and MUSKAAN CHOPRA

CSE, CCET, Panjab University, Chandigarh, India

ABSTRACT

Innovations in Man-made Intelligence have a long history that is effectively and continually evolving and growing. This chapter discusses the Artificial Intelligence (AI) framework and architecture that permits individuals to improve their performance over the long haul. Further, a profound learning and prescient investigation on the working and applications of the Internet of Things (IoT) is brought out and the major talk of, how both of the said technologies are interrelated, has been incorporated. In view of this, the proposed research also focuses on the positive impact of AI and the IoT in the times of the COVID-19 Pandemic. Besides, this chapter revolves around the count and percent share of the present sectors in which AI and the IoT, have been utilized and forecasts the results using effective Data Visualizations and Analysis.

Advanced Computer Science Applications: Recent Trends in AI, Machine Learning, and Network Security. Karan Singh, PhD, Latha Banda, PhD & Manisha Manjul, PhD (Eds.)

12.1 INTRODUCTION

It is asserted that Artificial Intelligence (AI) is assuming an expanding part in the exploration of the executives science and operational examination zones. The term insight alludes to the capacity to secure and apply various abilities and information to take care of a given issue. AI is the investigation and improvements of astute machines and programming that can reason, learn, accumulate information, impart, control, and see the items. Computerized reasoning was first proposed by John McCarthy in 1956 in his first scholastic meeting on the subject. Alan Turing had the option to placed his theories and inquiries into activities by testing whether "machines can think"? After arrangement of testing (later was called as Turing Test) incidentally, it is conceivable to empower machines to think and learn much the same as people. Major AI zones are Expert Systems, Natural Language Processing, Speech Understanding, Robotics and Sensory Systems, Computer Vision and Scene Recognition, Intelligent Computer-Aided Instruction, Neural Computing. Man-made brainpower has the favorable circumstances over the characteristic insight as it is more perpetual, reliable, and less expensive.[29] AI is a super-set of Machine learning, which further is a super-set of Deep Learning[17] that have also been portrayed in Figure 12.1 along with the broader expansions of the said terms. IoT is a tremendous idea incorporating an excessive number of sensors, actuators, and data processing capacities interconnected by the Web.[8] In this manner any IoT-empowered gadget can detect its surrounding elements, convey, stash the data and act likewise. The Internet of Things (IoT) and AI play a fundamental contribution due to the increase in the developing requirements. AI has become an ideal answer to deal with numerous associated IoT components, it's limitless processing and learning capacities.[21] Few instances of existing IoT administrations with the working of AI behind them are Voice assistants (Alexa, Siri, Google Assistant),[8] Smart Decisions, Smart Cities, and Smart Devices.[21]

The overall description of the chapter is as follows: First, Section 1 covers the introduction, terms, and terminologies. Second, Section 2 confers about the literary work in AI and how it is involved with the IoT. Next, Section 3 discusses the architecture and the cycle followed while building an AI-based product and the working model of IoT. Section 4 ponders upon these emerging technologies and their impact during COVID times. Section 5 describes analyses and provides their future aspects through Visual Analysis. Section 6 concludes the chapter.

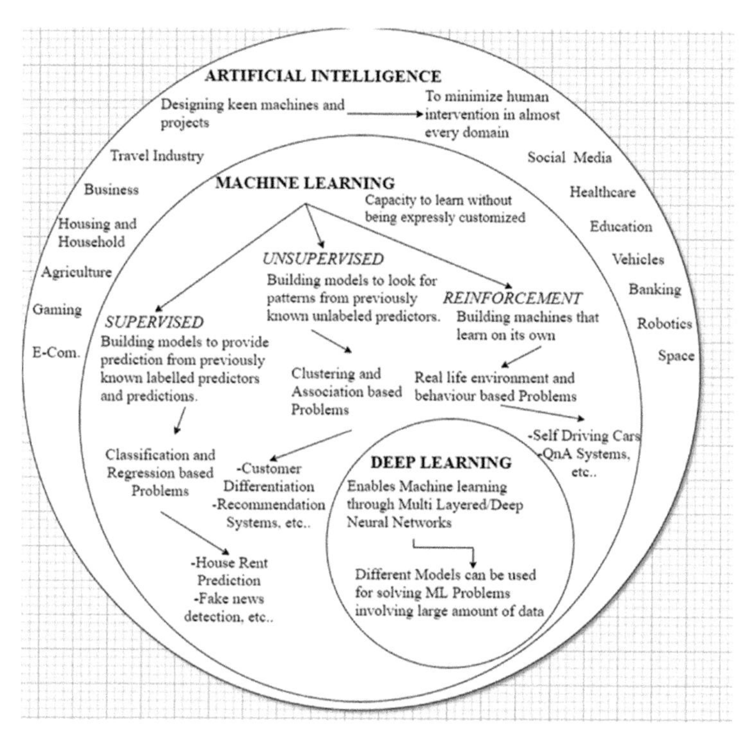

FIGURE 12.1 Broader domains of artificial intelligence, machine learning, and deep learning.

12.2 LITERARY WORK

Artificial or manmade intelligence alludes to the capability of PC-controlled machines/robots toward performing assignments that are practically like individuals or learns to work and perform like humans.[1,6,28] For this situation, Artificial knowledge is utilized to create different robots that have human scholarly attributes, conduct, gaining from past experience, have capacities to detect, and capacities to making forecasts.[24] As a rule, AI is fundamentally known as the capacity or capability of advanced mechanics to choose, take care of issues and reason in a quick and robust manner.[27] In present times, the applications and use cases of this newly evolved technology are unending, may it be Gaming,[31] Robotics,[5,18,] Biological findings,[11] Neuroscience,[14] detection of faulty software,[10] enhancement in business activities ,[7,19,] Social Media,[23] Sentiment,[3] Depression, and Disease Analysis,[13] and other such fields as shown in Figure 12.1.

IoT can be referred to as a system of innovations or modules, which consist of certain developmental innovations.[9] This assists the system in conveying and cooperating along individual's information. The cycle includes getting, moving, and broadcasting information down the organization besides having no human associations or individual-PC contribution. This information gained by gadgets, sensors, or modules can be put away in the cloud and it will in general be made available for continuous investigation. IoT gathers these immense measures of information from various sources. With the assistance of information science and applying examination, AI changes over this aggregate information into applications. So the entire cycle includes gathering and handling information. IoT can be called as the information "provider" while AI or Machine Learning[12] can be considered as information "miner."[21] The cycle happens as follows:

(1) IoT sensor supplies a large number of discrete information.
(2) "Miner" or AI recognizes the connections in them
(3) Concentrate significant understanding from these factors.

IoT would create immense measure of information because of the quick developing of gadgets and sensors. These informations would be a ton accommodating for different things, for example, foreseeing regular disasters, mishaps, helps specialists getting continuous data from clinical hardware, improved efficiency across ventures, prescient support on gear and apparatus, make savvy homes with associated machines and give basic correspondence between self-driving cars. These huge informations are significant just when it is changed into important worthy data inside a given time period.[25] This is the place where AI becomes an integral factor. Computer-based intelligence gathers the information and concentrates the significance from it by applying analytics.

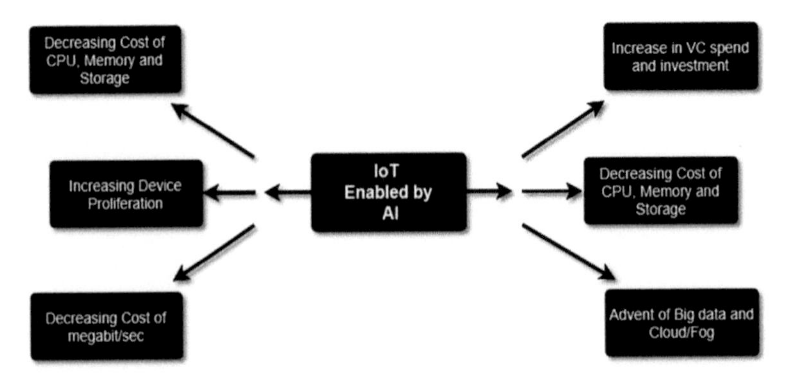

FIGURE 12.2 Artificial intelligence enabled IoT.[21]

12.3 ARCHITECTURE OF AI AND IOT

12.3.1 A BASIC MACHINE LEARNING CYCLE

Inside aspects of AI, depict that almost every data science project deals with similar cycle as shown in Figure 12.3 whereby the steps include:

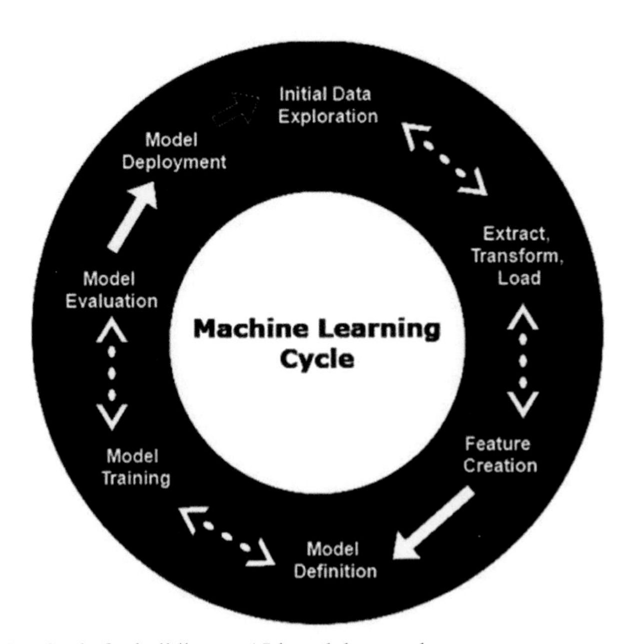

FIGURE 12.3 Cycle for building an AI-based data product.

12.3.1.1 Data Exploration

This is the initial step of the overall process in which the effective collection of data is performed and an adequate investigation can assist cut with bringing down your enormous informational collection to a sensible size where one can zero in the endeavors on examining the most pertinent information. It is both a workmanship and a science. There is the study of delving into and preparing the information for effective results consisting of the following,

(1) Gather the data in raw form using web scraping for real-time entries or data sets provided on the internet based on the requirements.
(2) Get to know the quality of data through inputs, and visualizations.

(3) Look for other resources or ask the proprietors if the data is having numerous inconsistencies.

12.3.1.2 ETL Process

Extract, Transform, and Load commonly known as the ETL process is a significant advancement in changing the information from the source framework into that part of information which is appropriate for examination. In conventional data warehousing, this cycle incorporates,

(1) Getting to the online exchange,
(2) Preparing the framework data sets for the exchange,
(3) Changing the information from an exceptionally standardized data model into a useful schema, and
(4) Putting away the information to an information stockroom.

In the projects involving AI, this progression is normally a lot less difficult as the information shows up in a traded design (for instance, JSON or CSV). However, some of the time de-standardization needs to be performed on the obtained data. The outcome generally winds up in a mass stockpiling like Cloud Object Store.

12.3.1.3 Feature Creation

This errand changes input segments of different relations into extra segments to improve model execution. A subset of those highlights can be made in an underlying undertaking (for instance, one-hot encoding of unmitigated factors or standardization of mathematical factors). Some others require business understanding or various cycles to be thought of. This assignment is one of those profiting the most from the exceptionally iterative nature of this strategy. It includes the making of new characteristics that can catch the significant data in an informational collection considerably more effectively than the first ascribes. Three general techniques and their importance are as follows:

(1) Feature Extraction

The formation of another arrangement of highlights from the first crude information is known as highlight extraction. Think about a bunch of photos, where each photo is to be characterized by whether it contains a human face. The crude information is a bunch of pixels, and in that capacity, is not appropriate for some kinds of arrangement calculations. Nonetheless, if the information is prepared to give more elevated level highlights, for example, the presence or nonattendance of particular sorts of edges and regions that are exceptionally connected with the presence of human faces, at that point a lot more extensive arrangement of characterization procedures can be applied to this issue. This strategy is profoundly area explicit.

(2) Feature Development

Sometimes the highlights in the first informational collections have the vital data, yet it is not in a structure appropriate for the information mining calculation. In this circumstance, at least one new highlight built out of the first highlights can be more valuable than the first highlights. Example: isolating mass by volume to get the thickness.

(3) Planning data to new space

An entirely unexpected perspective on the information can uncover significant and fascinating highlights. Consider, for instance, time-arrangement information, which frequently contains intermittent examples. In the event that there are just a solitary intermittent example and very little commotion, at that point the example is effortlessly recognized. On the off chance that, then again, there are various occasional examples and a lot of clamor is available, at that point these examples are difficult to identify. Such examples can, in any case, frequently be identified by applying a Fourier change to the time arrangement to change to a portrayal in which recurrence data is express.

12.3.1.4 Model Definition

This undertaking characterizes the AI or profound learning model. Since this is an exceptionally iterative technique, different emphases inside this undertaking or including upgrading and downgrading pipelines are conceivable. It is also suggested to begin with basic models first for pattern creation after those models are assessed. As a whole, models in AI are numerical calculations that are "prepared" utilizing information and human master contribution to reproduce a choice a specialist would make when given that equivalent data. It endeavors to repeat a particular choice cycle that a group of specialists would make in the event that they could survey all accessible information.

12.3.1.5 Model Definition

This errand prepares the model. A model reproduces a choice cycle to empower mechanization and comprehension. Simulated models include numerical calculations that are "prepared" utilizing information and human master contribution after previous processes to recreate a choice as that of a professional. Therefore, this undertaking is separate from model definition and assessment for different reasons. In the first place, preparing is a computationally extraordinary errand that may be scaled on PC groups or GPUs. Consequently, a compositional cut is some of the time unavoidable. For instance, model definition occurs in Keras, however preparing occurs on a Keras model fare utilizing Apache SystemML on top of Apache Spark running on a GPU group.[15] Also, in hyper-parameter tuning and hyper-parameter space investigation, the prospect of "Model Evaluation" can be essential for this resource.

12.3.1.6 Model Evaluation

Here the assessment of the model's presentation is performed. Having said that, various measurements should be applied, for instance, cross entropy for a multiclass grouping issue, confusion matrix in case of a classification problem. It's imperative to isolate the data gathered into a preparation, test, and validation (if cross-approval is not utilized) and monitor the outputs of various component designing, model definition, and other parameters. Overall, it plans to gauge the speculation exactness of a model on future (concealed/out-of-test) information. Techniques for assessing a model's exhibition are partitioned into two classes: in particular, Holdout and Cross-approval. These two techniques utilize a test set (i.e., information not seen or utilized by the model in any case) to assess model execution.

12.3.1.7 Deploying the Model

The final step of the cycle is to send the model for deployment after enormously training the model. Conveying an AI model, known as model deployment, just intends to incorporate an AI display and coordinate it into a current creation domain where it can take in an info and return a yield. It becomes important to note that the deployment relies intensely upon the utilization case, particularly, on the partner's assumption on devouring the information item. Along these lines, substantial methods of arrangement include:

(1) A REST endpoint permitting scoring (and preparing) of the model (for instance, supported by a docker holder running on Kubernetes)
(2) A web or versatile application

12.3.2 *WORKING STRUCTURE OF IOT*

The architecture of the IoT has five layers that are responsible for the ideal working of an IoT system. The first and bottom-most layer is the perception layer which consists physical objects such as WSN and Sensors responsible for conveying data to Network layer, which is responsible for transmission by the means of wired/remote mediums, next is the middleware layer which handles data and processes it, then the application layer utilizes the processed data for applications and the last and top most layer—the business layer that deals with system management and builds business models, visualizes data of application layer by the use of graphs and flowcharts. The architecture is also shown in Figure 12.4.

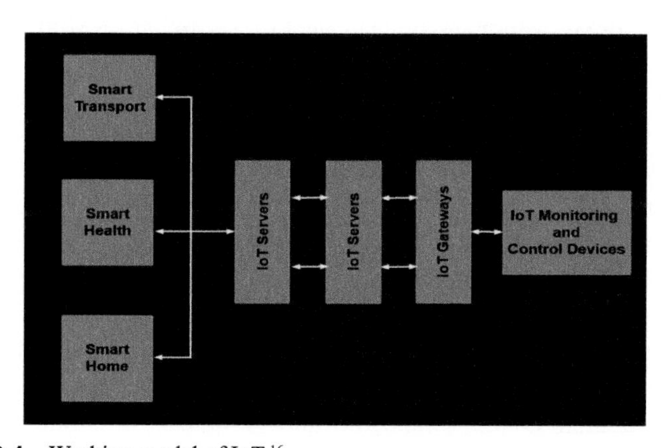

FIGURE 12.4 Working model of IoT.[16]

12.4 ROLE OF AI AND IOT IN ALLEVIATING THE AFFECT OF THE COVID-19 OUTBREAK

As Corona proceeds with its disruption around the globe, the world is under the massive pressure of smashing finances. Lamentably, huge number of individuals are as yet under a consistent danger of disease, with the circumstance not liable to improve in the near future.[30] Notwithstanding, a huge number of innovative methodologies are arising to manage the effects of this outbreak. Computerized advances, including IoT, AI, Blockchain, and cutting edge drones have been at the front line and are playing a major role.[4] As per the WHO and the CDC, computerized advancements can assume a fundamental part in improving general wellbeing reaction to the COVID-19 pandemic. In the accompanying segments, we investigate the adequacy of the previously mentioned advancements in mitigating the tragic effects of the COVID-19.

12.4.1 ROLE OF AI

AI has ended up being a milestone innovative headway. Whenever utilized appropriately, proves to be a profoundly powerful instrument against the COVID-19. BlueDot's AI model uses a few AI and Natural Language Processing instruments to search for proof of arising sicknesses. This model helped BlueDot to follow the COVID-19 spread.[3] The Fast finding of the COVID-19 can permit government to take compelling reaction measures to restrict the illness' further spread. The lack of testing packs around the world, notwithstanding, made it difficult for the specialists to do mass symptomatic testing. AI is transforming the process of COVID-19 screening and determination. To restrict the openness of the front-line staff to COVID-19 patients, a few emergency clinics, air terminals, and clinical focuses have embraced the utilization of cameras with AI-based multisensory innovation. Voice recognition is probably the least complex innovation utilized to distinguish COVID-19 patients. Voice recognition stages can act as a screening test to choose who should be tried. The way to building up an effective treatment against COVID-19 is to comprehend the actual infection. Since infections cannot duplicate without anyone else, they depend on host cells to produce duplicates of their DNA. DeepMind and Google's AI organizations have received the utilization of its AlphaFold framework to foresee the design of the proteins related with the Coronavirus.[3] These expectations can help researchers in better understanding the general construction of the infection, and therefore, in building up another medication to treat the COVID19. The questionable occasions following the flare-up of the COVID19 have reared a few fantasies and fear-inspired notions. Much deception has been getting out and about via web-based media stages. To control the proliferation of this counterfeit news and give checked data, innovation organizations like Google, Youtube, and Facebook have utilized the utilization of AI procedures.

12.4.2 ROLE OF IOT

The Internet of Medical Things are the medical services of IoT, that is, a combination of clinical equipment and programmed applications offering major benefits on medical services associated with the medical care IT frameworks. Inferable from their capacity to gather, dissect, and send the wellbeing information effectively, the medical services area has utilized the extraordinary capability of technology advances. The clinical associations and government bodies are hoping to use these advances to lessen the work-load on the frontline workers.

(1) IoT Buttons To keep up high precautions and cutoff the number of medical clinic procured contaminations, a few medical clinics have introduced IoT buttons operating on battery.[3] These buttons are used for quick sending in any facility, independent of their size, to give alerts to the administration, notice them of any sterilization or support issue that may be a danger to public security. A momentous element of these buttons is their freedom on outside framework, that is, their capacity to adhere to any surface.

(2) Telemedicine, The act of utilizing IoT advancements to encourage distant patient observing is known as telemedicine or telehealth, this training permits clinicians to assess, analyze, and treat patients without requiring any actual collaboration with them. The benefits of embracing telehealth strategies have been twofold:

    (1) it has diminished the weight on the exhausted medical clinic staff,
    (2) it has decreased the danger of radiation of the infection from the contaminated people to the medical care workforce.

Telemedicine can go far in conceivably disburdening clinics of clinical waste and prepare for them to receive all the more earth maintainable practices. In India, keeping as a top priority its latent capacity and developing prominence, the Ministry of Health and Family Welfare, along with NITI Aayog and Board of Governors Medical Council of India, has given tele-medicine rules.[22]

12.4.3 UTILIZATION OF TRACKING APPLICATION

The utilization of versatile applications has arisen as a noticeable methodology in the battle against the COVID-19. A few government and private associations around the planet have just built up certain applications utilizing a wide assortment of advances, including Bluetooth, Global Positioning System, and Geographic Information System, while a few others are currently doing as such (refer Fig. 12.5).based media stages. To control the proliferation of this counterfeit news and give checked data, innovation organizations like Google, Youtube, and Facebook have utilized the utilization of AI procedures.

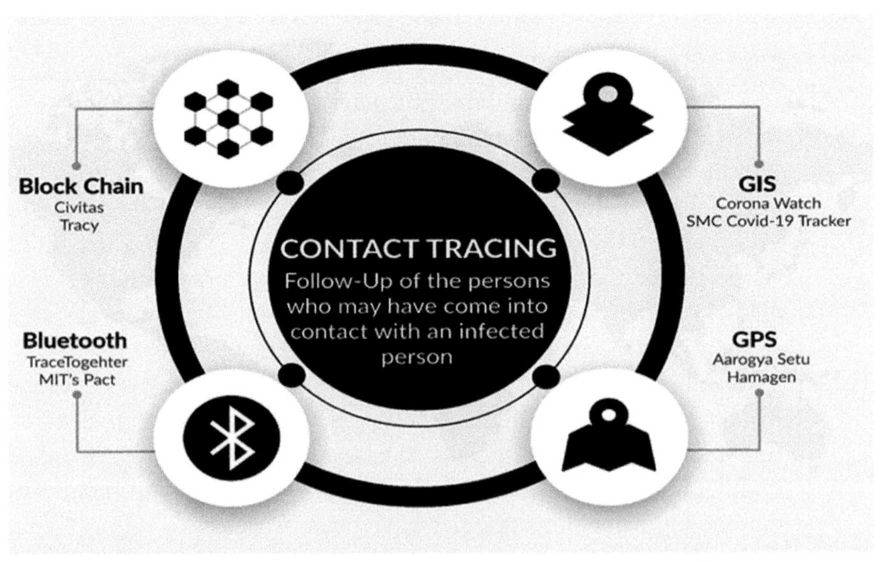

FIGURE 12.5 Tracking and contact detection softwares for COVID-19.[3]

12.5 RESULTS AND DISCUSSION

12.5.1 COMPARATIVE ANALYSIS

Building an AI and IoT product is pointed toward making human existence simpler. With the presentation and effective usage of AI arrangements, numerous ventures having unending applications (refer Fig. 12.6) in various countries[26] (also refer Fig. 12.7) are and will profit by expanded productivity which in any case will lead to great financial development rates. Moreover, man-made brainpower openings will focus on inventive, human-focused methodologies and estimating rise in the materialness of automated innovations like IoT (refer Fig. 12.9) to gain different benefits and ease-of-use. Development of AI frameworks for the products delivering Internet of Things facility will help the whole world to surmise the accessible representative designs.[20] The authors believe that the said technologies in different areas on the planet might be because of different examination advancements and are likely to benefit every second person on the earth. Overall, the work comprised on a database of 3465 companies[2] whose visual analysis is shown in Figures 12.6 and 12.7. Next, figuring out the applicability of IoT in different domains, those having the most percentage share at present, have been shown in Fig. 12.8.

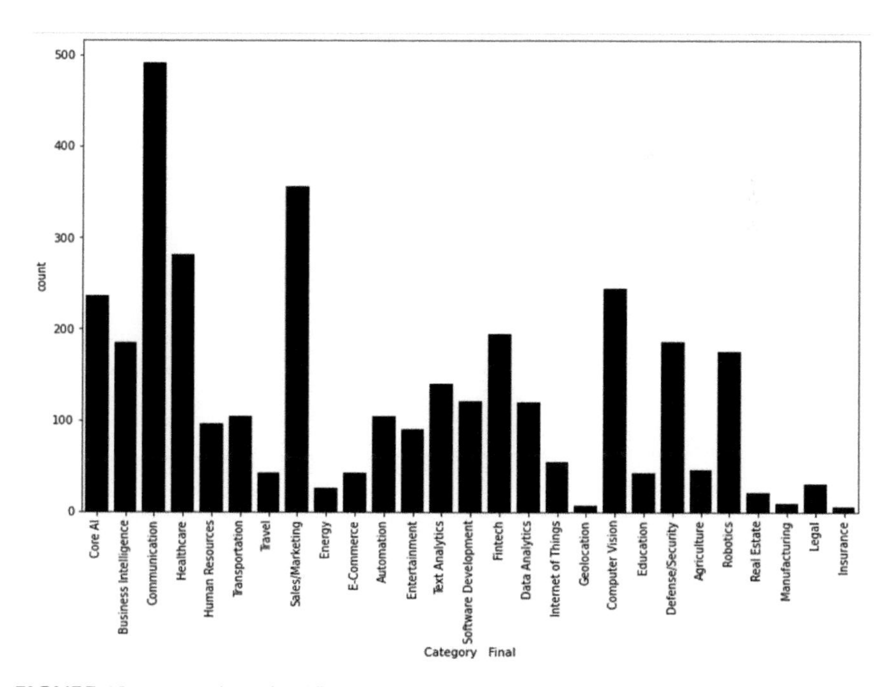

FIGURE 12.6 Number of artificial intelligence companies in various industries.

12.5.2 FUTURE PERSPECTIVE ON IOT

The authors analyzed that the estimated rise of connected IoT devices will rise up to 65 billion by 2032, which was calculated based on these devices from the past and the results have been portrayed in Figure 12.9.

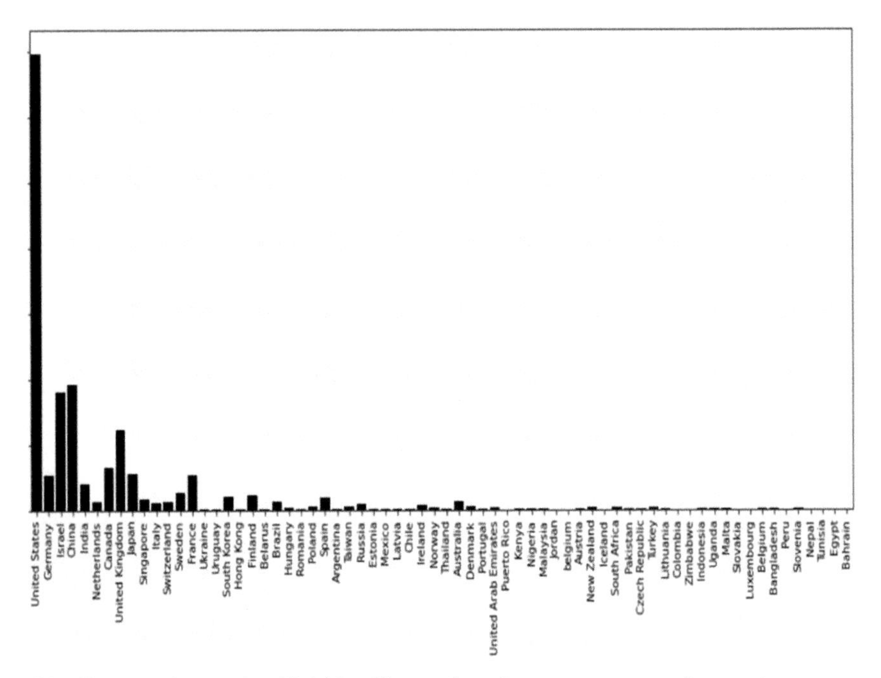

FIGURE 12.7 Count of artificial intelligence based startups across various nations

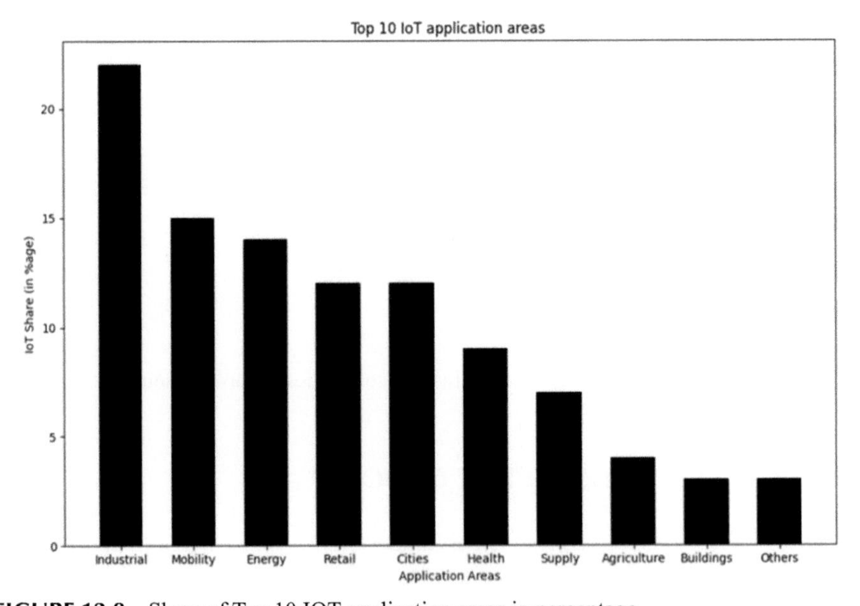

FIGURE 12.8 Share of Top 10 IOT application areas in percentage.

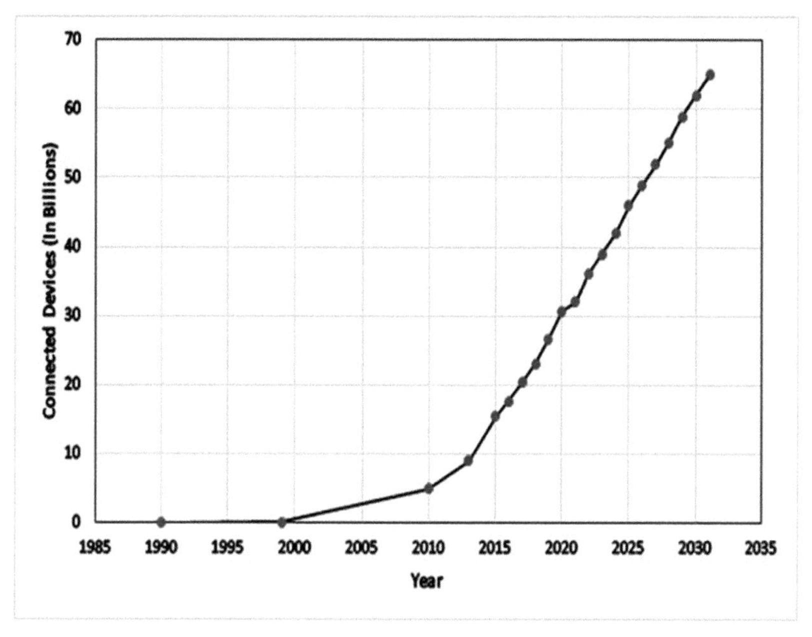

FIGURE 12.9 Exponential curve showing the estimated rise of connected IOT devices by 2032.

12.6 CONCLUSIONS

AI is programming that can reason, learn, accumulate information, impart, control, and see the items. With the assistance of IoT, AI changes over the aggregate information into applications. AI alludes to the capability of PC-controlled machines/robots toward performing assignments that are practically like individuals. IoT is a network of innovations or sensors that contains some development innovation implanted into it. As per the WHO and the CDC, computerized advancements like AI and IoT act as a fundamental part in improving general well-being reaction to the COVID-19 pandemic. Moreover, Artificial Intelligence openings will focus on inventive and human-focused methodologies. The estimated rise of connected IoT devices provides an insight of how this impact of technology can double the number of connected useful devices than present by 2032.

KEYWORDS

- **artificial intelligence**
- **machine learning**
- **internet of things**
- **deep learning**
- **COVID-19 pandemic**
- **coronavirus**
- **visualizations**
- **connected IoT devices**

REFERENCES

1. Alsedrah, M. *Artificial Intelligence*, 2017.
2. Asgard. The global artificial intelligence landscape, 2018.
3. Campos, V.; Salvador, A.; Giro-i Nieto, X.; Jou, B. Diving Deep into Sentiment: Understanding Fine-Tuned CNNS for Visual Sentiment Prediction. In *Proceedings of the 1st International Workshop on Affect & Sentiment in Multimedia*; Association for Computing Machinery, ASM '15: New York, 2015; pp 57–62.
4. Chamola, V.; Hassija, V.; Gupta, V.; Guizani, M. A Comprehensive Review of the Covid-19 Pandemic and the Role of IoT, Drones, AI, Blockchain, and 5G in Managing Its Impact. *IEEE Access* **2020,** *8*, 90225–90265.
5. ET Government. Indian Robotics Solution Launches Corona Combat Drone to Fight Covid-19, 2020.
6. Feuerstein, R. *The Dynamic Assessment of Cognitive Modifiability: The Learning Propensity Assessment Device : Theory, Instruments and Techniques*; ICELP Press, 2002.
7. Fragkiadaki, K.; Levine, S.; Felsen, P.; Malik, J. Recurrent Network Models for Human Dynamics, 2015.
8. Ghosh, A.; Chakraborty, D.; Law, A. Artificial Intelligence in Internet of Things. *CAAI Trans. Intell. Technol* **2018,** 3.
9. Gupta, M.; Singh, S. K. The Internet of Things: An Overview of the Awareness, Architecture & Application. *Int. J. Latest Trends Eng. Technol* **2019,** *12*, 19–24.
10. Gupta, P.; Sahai, P. A Review on Artificial Intelligence Approach on Prediction of Software Defects. *IJRDASE* **2016,** *9*, 4.
11. Holland, J. H. *Adaptation in Natural and Artificial Systems: An Introductory Analysis with Applications to Biology, Control and Artificial Intelligence*; MIT Press: Cambridge, 1992.
12. H¨oller, J.; Tsiatsis, V.; Mulligan, C.; Avesand, S.; Boyle, D. From Machine-to-Machine to the Internet of Things—Introduction to a New Age of Intelligence, 2014.
13. Islam, M. R.; Kabir, A.; Ahmed, A.; Kamal, A.; Wang, H.; Ulhaq, A. Depression Detection from Social Network Data Using Machine Learning Techniques. *Health Information Science and Systems* **2018,** *6*, 8.

14. Kang, Q.; Huang, B.; Zhou, M. Dynamic Behavior of Artificial Hodgkin-Huxley Neuron Model Subject to additive noise, *IEEE Transactions on Cybernetics* **2016,** *46,* 2083–2093.
15. Kienzler, R. The Lightweight IBM Cloud Garage Method for Data Science, 2019.
16. Kumar, S.; Tiwari, P.; Zymbler, M. Internet of Things Is a Revolutionary Approach for Future Technology Enhancement: A Review, *J. Big Data* **2019,** *6* (1), 111.
17. LeCun, Y.; Bengio, Y.; Hinton, G. Deep Learning. *Nature* **2015,** *521* (7553), 436–444.
18. Lenz, I.; Lee, H.; Saxena, A. Deep Learning for Detecting Robotic Grasps, 2014.
19. Milford, M.; Lowry, S.; Sunderhauf, N.; Shirazi, S.; Pepperell, E.; Upcroft, B.; Shen, C.; Lin, G.; Liu, F.; Cadena, C.; Reid, I. Sequence Searching with Deep-Learnt Depth for Condition- and Viewpoint- Invariant Route-Based Place Recognition, In *2015 IEEE Conference on Computer Vision and Pattern Recognition Workshops (CVPRW)*, 2015; pp 18–25.
20. Mu¨ller, V. C.; Bostrom, N. Future Progress in Artificial Intelligence: A Survey of Expert Opinion. In *Fundamental Issues of Artificial Intelligence*; Mu¨ller, V., Ed.; Springer, 2016l pp 553–571.
21. nasscom. AI in IoT for a Better Future, 2020.
22. ND, S. How Telemedicine Is Silently Making India's Healthcare Ecosystem Future-Ready, 2021.
23. Nigl, A.; Grey, D. AI and Machine Learning Applications for Social Media Platforms final, 2018.
24. Nilsson, N. J. *Principles of Artificial Intelligence*; Morgan Kaufmann Publishers Inc: San Francisco, CA, USA, 1980.
25. Ranger, S. 5G: What It Means for IoT, 2020.
26. Rao, D. A. S.; Verweij, G. Sizing the Prize, PWC's Global Artificial Intelligence Study: Exploiting the AI Revolution, What's the Real Value of AI for Your Business and How Can You Capitalise?
27. Russell, S.; Dewey, D.; Tegmark, M. Research Priorities for Robust and Beneficial Artificial Intelligence. *AI Magaz* **2015,** *36* (4), 105–114.
28. Shabbir, J.; Anwer, T. Artificial Intelligence and Its Role in Near Future. CoRR, abs/1804.0**1396,** 2018.
29. Turan, M.; Pilavcı, Y.; Jamiruddin, R.; Araujo, H.; Konukoglu, E.; Sitti, M. A Fully Dense and Globally Consistent 3d Map Reconstruction Approach for GI Tract to Enhance Therapeutic Relevance of the Endoscopic Capsule Robot, 2017.
30. Vaishya, R.; Javaid, M.; Khan, I. H.; Haleem, A. Artificial Intelligence (AI) Applications for Covid-19 Pandemic. *Diab. Metabol. Synd.* **2020,** *14* (4), 337–339.
31. Weddle, C. Artificial Intelligence and Computer Games, 2006.

CHAPTER 13

INTELLIGENT POST-LOCKDOWN MANAGEMENT SYSTEM FOR PUBLIC TRANSPORTATION

HARI MOHAN RAI[2], BARNINI GOSWAMI[1], SHREYA MAJUMDAR[1], and KAJAL GUPTA[1]

[1]*Department of Computer Science and Engineering, Krishna Engineering College, Ghaziabad, India*

[2]*Faculty of Electronics and Communication, Krishna Engineering College, Ghaziabad, India*

ABSTRACT

This chapter describes a unique approach to solve the issue of maintaining social distance while traveling due to the COVID 19 outbreak and also to make public transportation function like it used to function early. The COVID 19 pandemic is a worldwide pandemic because of which it became risky to allow the public transportation without proper mechanism to maintain social distancing. So, to resolve this problem we came up with an idea of making an intelligent application to schedule the timings of transportation, avoiding over occupancy of public transport, providing them the shortest route to reach their desired destination, providing them proper guidelines, and also providing them with the information of the nearest hospitals for any emergency. It will find the shortest possible route from the source to the

Advanced Computer Science Applications: Recent Trends in AI, Machine Learning, and Network Security. Karan Singh, PhD, Latha Banda, PhD & Manisha Manjul, PhD (Eds.)

destination and seats will be booked following the social distancing and our web app will notify people with an alert as an alarm.

13.1 INTRODUCTION

The outbreak of Coronavirus pandemic, 2019 (COVID-19) has created a global health crisis that has had a deep impact on the way we perceive our world and our everyday lives. COVID-19 is highly contagious and mortality is also very high; this makes the disease all the more dangerous. One of the major preventions is to maintain social distancing, which would smash the chain of the expansion of the disease. Lockdown has been implemented to implement this idea, but we cannot keep lockdown in a country for a very long time; otherwise, the country's economy will drastically decline. Post lockdown many things will change around us. And after the daily routine resumes post lockdown, the public transportation system will play a crucial role, as it is most commonly used by people and to prevent the further spread of COVID-19 cases, social distancing needs to be maintained at all the public places. So there has to be a proper management for public transportation which can allow its use, without further producing more COVID-19 cases.

To solve this problem we have been implementing a solution, by building an intelligent app to schedule the timings of various transports, avoiding the over occupancy of buses/railway stations etc.[1]

This app also finds the possible shortest route from a particular source to the destination for a customer and monitors whether the people are maintaining social distance or not by using the Bluetooth frequency manipulation technique.[2]

13.1.1 SCOPE AND OBJECTIVES

The goals of the proposed work are as follows:

a) It will help in preventing people from getting vulnerable to COVID-19.
b) It will provide the shortest route for the traveler's desired destination thus saving time.
c) This app will be used to make public transportation possible while maintaining social distancing.
d) It will help in avoiding over occupancy of public transport.
e) It is user friendly so many people can easily use it.

13.1.2 RELATED WORKS

This section includes methods that formulated the problem, addressed a central or related problem, used a similar methodology as our work to a similar problem, and also how our work is inspired by their work.

By clearly describing previous work, we can better describe the current limitations and the need for a new methodology. It also gives an opportunity to demonstrate our area of work and will help others to relate our current work to other scientific areas. We collected, analyzed, and coded the author-assigned keywords of other research papers to start a discussion on the topic "Intelligent post lockdown transportation management system." Presently, how to maintain social distancing while traveling is a boiling area of research. Managing traffic, safer routes, and also telling the public about the shortest route to reach out their destination have been used in designing many applications in the past.[3] Some of those existing approaches from the past are discussed in this section.

Existing work in the field motivated our design idea to tackle the needs of a safer traveling environment for the public. A handful of solutions have been suggested pertaining to the issue in discussion.

Transit app was designed for aggregating and mapping real-time public transit data, crowdsourcing user data to determine the actual location of buses and trains. This app was first released for iPhone users; then, it was launched for the android users as well. It offers upcoming departure times for all nearby transit lines and alerts for various types of transportation where available, including both bus and train. If it is found that public transit is not cooperating, people can easily request an Uber or grab the closest bicycle.

Whether you ride the train, subway/underground /tube, bus, light rail, ferry or metro, use bikes, ride-sharing like Uber, getting the best urban mobility information is critical.[4]

Moovit[5] guides you from point A to B in the easiest and most efficient way. Get train and bus times, maps, and live arrival times with ease so you can plan your trip with confidence. Find critical alerts and service disruptions for your favorite lines. Get step-by-step directions of optimal route bus, train, metro, bike, or a combination of them.[6]

GoTo is the First ever Real-time Bus Booking Platform for Intercity Travels.[7] It is focused on saving your precious time which is generally wasted waiting for buses at the Boarding point, with No Clue of bus arrival. Though on booking operators claim that you can track the bus, in most of the cases you are unable to get its actual expected time of arrival.

All the above-mentioned applications help people to book tickets, reach their destination in time, provide them with a scheduled time table, and notify them of numerous modes of transportation. There are several such applications like Chalo, UTS, and NextBus. So, after exploring all the previous works we have come up with an application that in addition to existing features provides new and innovative ways to implement social distancing in public transport.[8–10]

After exploring several design options, we eventually decided to develop a website that will have features from several of our brainstormed and existing ideas named COVID TRAVEL COVER. By integrating the features of distinct applications together in one platform, COVID TRAVEL COVER offers a more holistic and effective solution than the existing applications by providing them with a variety of tailor-made features as per user's requirements. Another highlight is that the currently available applications fail to provide a solution to the COVID-19.

13.1.3 NOVELTY

a) We all are aware of Arogya Setu application that keeps the record of people's health conditions and makes us aware about the nearby COVID-19 patients, but it does not monitor whether the social distancing is being maintained or not which is equally important to stop the spreading of the virus. So, to cover those gaps our application would monitor whether the social distancing is being maintained by the public or not in public transportation and additionally this application is also going to notify public as soon as they violate social distancing.
b) Our application will check the user's authenticity and accountability by asking for their details to enter while they login to our app.
c) This application will provide services based on the preference of people's urgency or purpose of their work.

13.2 EXPERIMENTAL METHODS AND MATERIALS

13.2.1 ARCHITECTURAL DESIGN

The block diagram of the proposed methodology of our work is presented in Figure 13.1. The main idea is to design an App which will smartly manage the social distancing and decide the route of the transport based on crowding and shortest route. First, the App will ask to sign up, which means that registration will take place for all passengers and after signing up, an authentication page will be launched, which will ask for the ID proof of the customer, so that our app could be used in the desired way. Then the customer has to enter his address details in Address Page and then destination details in Destination Page. Ticket booking page would be launched to book the tickets and it would show the available timings and seats observing the social distancing method. The app would contain a page of Guidelines, another feature.

13.2.2 REGISTRATION AND LOGIN

This would be the first step for a new user. This step would enable the user to create an id and for maintaining the authenticity every user would be required to upload the credentials of any address proof and workplace or institutional details (employee id /student id card, etc.)

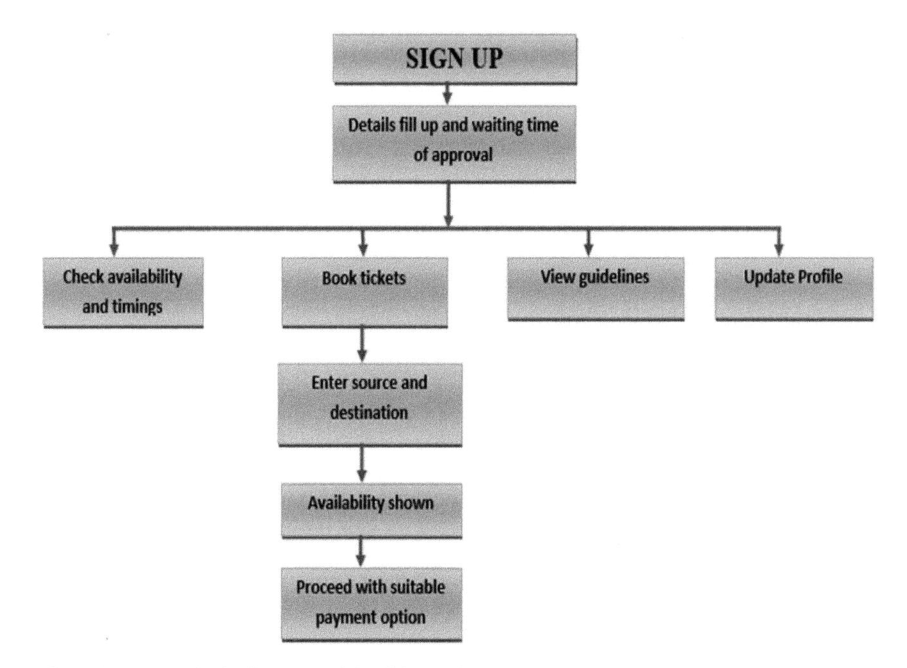

FIGURE 13.1 Block diagram of Covid travel cover.

13.2.2.1 TICKET BOOKING

Whenever a user needs to travel he or she will have to book a ticket online through our app at least 5 h before they intend to start their journey. So, there will be a booking page where the users can mention the purpose, places, and timings of their travel and will receive the tickets on the grounds of availability.

13.2.2.2 GUIDELINES PAGE

There will be a page to which any user can refer to, any time for viewing the safety guidelines to be taken to ensure they are having least risk of infection.

13.2.2.3 FACILITIES NEARBY

This page will be showing all the hospitals, medical shops, and other health-care and emergency facilities nearby even when the users are in transit.

13.2.2.4 ALERT

There will be an alert on the user's screen if they are not following the social distancing criteria (using Bluetooth proximity), so that they can be aware and can safeguard themselves.

13.2.3 METHODOLOGY

The flow chart of the proposed method used to find the shortest path is shown in Figure 13.2.

13.2.3.1 SSSSCHEDULING

- ***Shortest Path Algorithm***

If there will be more than one person requesting for the same path, they will be allocated ticket through the shortest route on first come first served basis.[11–14]

- ***Traffic Management***

The app will try to improve the travel of people in a way that most people are able to go through the best possible route and also avoiding traffic congestion in all routes as long as possible.[15,16]

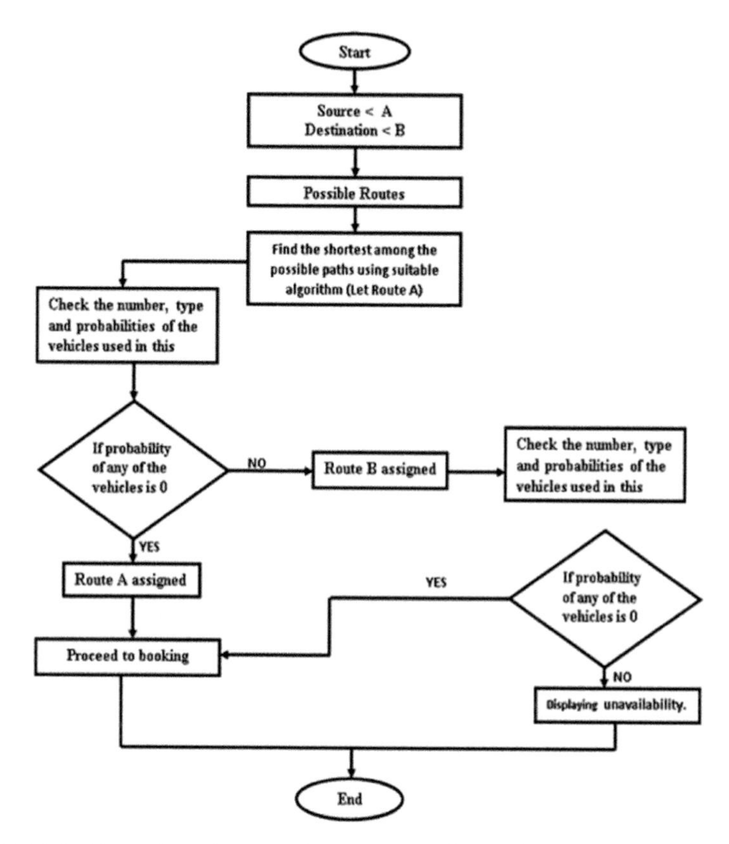

FIGURE 13.2 Flow chart of proposed methodology for finding shortest path.

- ***Timetable Generation***

While booking tickets the users would be able to see the preset timings of the transport availability which would help them make decisions about the finest hour to start their journey.[17] As the architecture is designed, now the implementation of this comes: We will design our frontend using HTML, CSS, and JAVASCRIPT in BOOTSTRAP. It will contain: Login Page, Categories Page, Ticket Booking Page, Guidelines Page, and Payment Page.

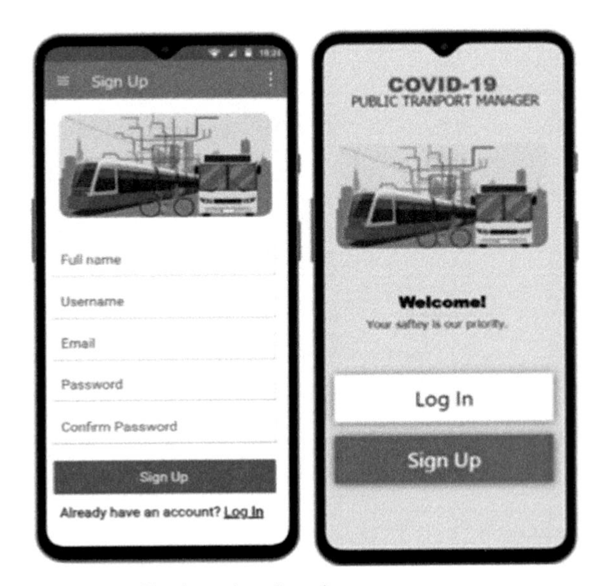

FIGURE 13.3 Prototype of login and registration page.

Backend we will code on working on pages mentioned above and all pages would be made dynamic. We will use Java for backend coding. (Atom/Visual studio code). We will code a scheduling algorithm to schedule the timings and routes of the buses and another algorithm to manage the traffic for the ease of the working of the buses. A feature would be coded in the backend which would notify the person who would violate the social distancing rules using the Bluetooth detection technique/algorithm. By this we will be able to identify data items and concepts and find relations between them.

13.2.3.2 PROTOTYPE OF UI

- The front page or Welcome page—It would contain login and signup button.
- Sign up page—for a new user.
- Login page—for existing users.
- Authentication Page—I will ask for ID proof of the customer, so that the authenticity of the customer can be verified and he/she can use the category feature in our app without any issues.
- Address Page will ask the address from where the customer will board or the home address.

- Destination page will look for the address of where the customer wants to go most often maybe on a daily basis (if any); therefore, his usual destination like office address, college address, or hospital address, etc. Figure 13.4 shows the pictorial view of prototype UI for authentication and user address details.
- Ticket booking page, the user will have to come to this page whenever he/she wants to book tickets for their journey. The user will be required to enter the source and destination which can be the same as entered while registration or different according to the requirements; the underlying algorithm will then find the shortest possible route and availability of seats considering social distancing and the book tickets.[18]

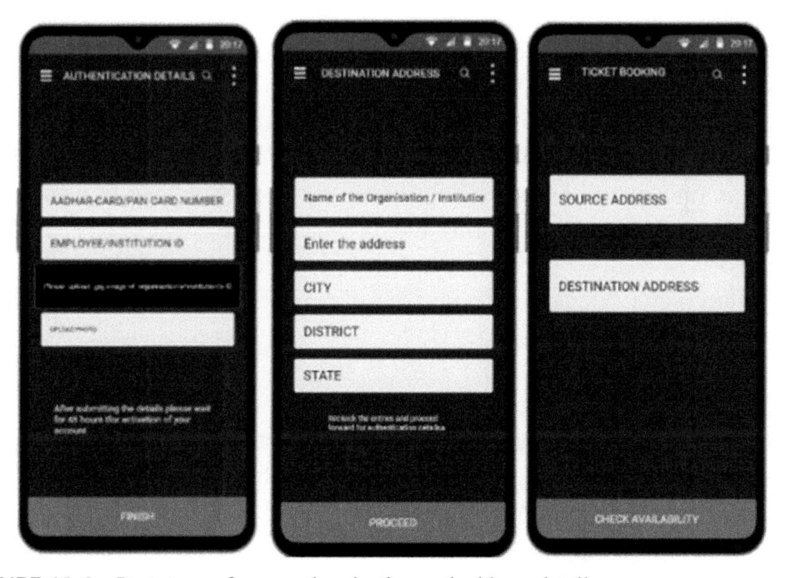

FIGURE 13.4 Prototype of user authentication and address details.

13.3 RESULTS AND DISCUSSION

13.3.1 SCHEDULING MANAGEMENT

A basic scheduling according to the guidelines from the government will have to be maintained on a daily basis; on the basis of the additional demand by the citizen additional services can be employed.[1]

13.3.1.1 TRANSPORT DISTRIBUTION

Essential facilities will be facilitated by most frequent bus timings. The commuters of school, colleges, companies, and factories usually travel during the morning and evening so frequency of buses has been kept highest in those periods.

Assuming that we have 100 buses available in a particular area. The bus services will be provided for citizens from 6:00 am to 12:00 pm. We have divided the buses into three categories based on the type of service of its commuters. Category A: Essential services (doctors, police, and army officials, etc.), Category B: Less essential services (college students, office workers etc.), Category C: Least essential services (people who want to go to salons, marketing, etc.). The distribution of buses has been done as follows:

Category A–30, Category B–50, Category C–20. The primary fixed schedule will be facilitated by the data obtained from the users (after the initial launch of the app). As people would be filling their daily requirement of public transport in a particular time interval that will help us determine the number of buses efficiently.

Let us understand with the help of an example. Suppose in a particular time slot (say 8:00 am to 9:00 am), five people (P1, P2, P3, P4, P5) belonging to category B want to go from the same pincode (say Code1) to the same destination (say Code2). As the number of people is considerable so we will analyze the possible routes from Code1 to Code2. Let us assume in this case there are two possible routes (R1 and R2). In every possible route we will now observe the number of stops for a bus and users willing to board from those stops, respectively. Now we will observe the total number of commuters in both the routes. Let us say, in our example, number of commuters along R1 = 20 and number of commuters along R2 = 40. If the maximum occupancy of a normal bus is kept 25 people per bus keeping social distancing in mind, then R1 will be allotted one bus and R2 will have two buses(with a gap of 15 min between them) running between 8:00 am to 9:00 am. This will avoid the rush of people in bus stops also.

13.3.2 RESCHEDULING CRITERIA

Now if suddenly due to some reason there is a demand for buses in timings other than the ones already being implemented or in areas where there are no frequent buses scheduled then an additional bus can be rescheduled accordingly, provided that there should be a minimum number of people willing to avail that service. All the available buses will not be run daily as the buses once used for commuting will have to be sanitized and cleaned as preventive measures for the COVID-19 outbreak. Similarly, employees should also follow precautionary measures from time to time. Similarly, the scheduling of auto-rickshaws, trains can also be done.

13.3.3 TICKET ALLOCATION

We will ask the user to enter the purpose of travel while booking tickets and we will also have the work details of the user beforehand which the user was asked to fill during the registration itself. A chart of scheduled timings of various transports would be available all the time for the ease of the users.

When a user requests for booking ticket the shortest possible path (analyzing the factor manipulation for availability of seats and traffic congestion already present in that route) will be allocated to him or her keeping social distancing in mind using algorithms like Dijkstra Algorithm and A* algorithm.[19,20]

To better understand the working of the application after this let us assume a scenario that a doctor and a college student might have requested to book a ticket at the same time during the same time through the same route. Let the purposes entered by the doctor and college student be as follows:

Doctor: Heart surgery of a scheduled patient

Student: Attending regular classes

Now the ticket allotment algorithm will first compare the two purposes on the basis of urgency and associate priority accordingly to both of them (lowest index being the highest priority). On the basis of priority it will check for availability in possible routes and vehicle options along those routes to complete the journey. First, the seat would be allotted to the doctor then only the algorithm will check availability for the student as doctor would have been associated with the highest priority at that time. After the allocation of tickets the allotted seats would be displayed on the user's screen, simulating the actual arrangement of seats on that particular vehicle, so that there is no confusion regarding the seating pattern sanctioned for maintaining social distancing by the concerned authorities. Also, the guidelines would be mentioned so that the users do not fall into the trap of negligence of not paying attention. The payment would be done by an electronic gateway that could be linked to the users' bank, e-wallets or the user can choose to use various cards for the payment.

13.3.4 MAINTAINING SOCIAL DISTANCING

The users would be required to keep their Bluetooth on especially during travel. We can keep a standard distance closer to which the social distancing criterion would be assumed to be violated.

Let us suppose that 1 meter is the accepted closest distance with a view to keeping social distancing in mind. Assume that the frequency exchanged between two devices (device A and device B) 1 m apart is ‘x’ MHz. We are aware of the fact that distance is inversely proportional to frequency detected, that is, as the distance between the devices increases the frequency between them would decrease. Using the above-mentioned principle we can devise an algorithm in a way that if the frequency input in any of the devices is more than the expected minimum frequency (i.e., observed frequency $> x$), then an alert would be popped up on the user's screen as a warning that social distancing has been violated and immediate action is required.

13.3.5 ADVANTAGES

- The app will have two user interfaces with distinguished utility—(a) General Public, (b) Healthcare Officials.
- The authentication of the user would be guaranteed by ensuring that the user should upload id proofs like Aadhar card, PAN card, etc.
- The app would be implementing the idea of frequency observation using Bluetooth to maintain social distancing by popping an alert whenever a person would be violating the minimum distance from other people.
- User-friendly
- A chart or table showing available buses, auto-rickshaw, metro, and local trains with timings in the nearby areas will always be displayed for the user's convenience (listing on the basis of pin codes).
- Users will be able to book tickets or monthly passes for both short- and long-route distances through this app.
- A visualization of seating arrangement will be provided beforehand so that the users can easily get acquainted with the newly implemented patterns.
- The app will also be able to suggest nearby medical shops, hospitals, etc. whenever necessary.

- To minimize the rush on roads categorization of buses has been done on the basis of essentiality of services.
- Guidelines for maintaining social distancing inside the vehicle like modified sitting arrangement will be available with the officials as well as the commuters.
- There will be a complaint registration form where anyone can put up their issues with the app or some incidents happening around them so that assistance can be provided accordingly.

13.4 CONCLUSIONS

Our work has provided a dependable solution to tackle the growth of infections of COVID-19 and would facilitate proper public transportation management post lockdown. Once the lockdown ends, there needs to be elaborate planning to manage all the people using public transport as offices, factories etc. will start functioning. So, to ensure everyone's wellbeing and safety we are introducing an intelligent post lockdown management system for public transport which will help in avoiding over occupancy of public transport and will help people to maintain social distancing which will lower their risk of getting infected by COVID-19. In our application, we will have ticket allocation on the basis of urgency of the purpose of travel; this will help the doctors and other health workers get to their destination in time as in tough times of the COVID-19 pandemic these people hold the highest priority. Through our ticket booking system we will try to reduce the transit time for users as much as possible, thus saving time and comparatively reducing the risk of the individual getting infected. We will also strive to avoid over-occupancy of the public vehicles and ensure social distancing by allowing the commuters to board only the allowed seats as available seats would be arranged in a pattern to facilitate social distancing. All the money transactions would be done through e-currency, thus reducing the chances of infection being spread through physical currency. All the users will be able to view the details of the nearest available healthcare facility in case they would need it. An algorithm to have dynamic bus scheduling will be present to ensure maximum resource utilization.

This proposed work is very much useful in maintaining social distancing, intelligently scheduling and rescheduling of transport services, and tracking the status of passenger.

KEYWORDS

- **COVID-19**
- **transportation**
- **Bluetooth**
- **social distancing**
- **safety**
- **ticket allocation**
- **schedule**
- **authentication**

REFERENCES

1. Nguyen, Q. T.; Ba, N.; Phan, T. Scheduling Problem for Bus Rapid Transit Routes. *Adv. Intell. Syst. Comput.* **2015,** *358*, 1–415.
2. Effects of COVID-19 on Transportation Demand, 2020. https://www.teriin.org/article/effects-covid-19-transportation-demand (accessed 23 May 2020).
3. Yu, B.; Lam, W. H. K.; Tam, M. L. Bus Arrival Time Prediction at Bus Stop with Multiple Routes. *Transp. Res. Part C Emerg. Technol.* Dec. **2011,** *19* (6), 1157–1170.
4. M. & N. Transit. Transit: Bus & Subway Times, 2019. https://play.google.com/store/apps/details?id=com.thetransitapp.droid&hl=en (accessed 23 May 2020).
5. Moovit: MaaS Solutions & the #1 Urban Mobility App. https://moovit.com/ (accessed 24 May 2020).
6. Matheus, R.; Janssen, M.; Maheshwari, D. Data Science Empowering the Public: Data-Driven Dashboards for Transparent and Accountable Decision-Making in Smart Cities. *Gov. Inf. Q.* Feb **2018,** 0–1.
7. Sharma, H. V. et al. Home—GoTo Bus. https://gotomobility.in/, 2020. https://www.gotobus.in/ (accessed 24 May 2020).
8. Cats, O.; Loutos, G. Real-Time Bus Arrival Information System: An Empirical Evaluation. *J. Intell. Transp. Syst. Technol. Planning, Oper.* **2016,** *20* (2), 138–151.
9. Shalaby, A.; Farhan, A. Prediction Model of Bus Arrival and Departure Times Using AVL and APC Data. *Public Transp.* **2004,** *7* (1), 41–61.
10. Cats, O.; Koutsopoulos, H. N.; Burghout, W.; Toledo, T. Effect of Real-Time Transit Information on Dynamic Path Choice of Passengers. *Transp. Res. Rec. J. Transp. Res. Board* Jan **2011,** *2217* (1), 46–54.
11. Tan, Y. A New Shortest Path Algorithm Generalized on Dynamic Graph for Commercial Intelligent Navigation for Transportation Management, 2019.
12. Ferone, D.; Festa, P.; Napoletano, A.; Pastore, T. Shortest Paths on Dynamic Graphs: A Survey. *Pesqui. Operacional* **2017,** *37* (3), 487–508.
13. Tan, Y. snatchagiant/Shortest-paths-algorithm-over-dynamic-graph-for-transportation-management.

14. Lu, J.; Dong, C. Research of Shortest Path Algorithm Based on the Data Structure. In *2012 IEEE International Conference on Computer Science and Automation Engineering*; Beijing, 2012; pp 108–110.
15. Zhu, W.; Li, R. Research on Dynamic Timetables of Bus Scheduling Based on Dynamic Programming. In *Proceedings of the 33rd Chinese Control Conference CCC 2014*; 2014; pp 8930–8934.
16. Aceto, G.; Ciuonzo, D.; Montieri, A.; Persico, V.; Pescapé, A. MIRAGE: Mobile-app Traffic Capture and Ground-truth Creation. In *2019 4th International Conference on Computing, Communications and Security (ICCCS)*; Rome, Italy, 2019; pp 1–8.
17. G. A. N. S. and S. A. P. S. Al Perumal, Tabassum, M.; Norwawi, N. M. Development of an Efficient Timetable System using AngularJS and Bootstrap 3. In *2018 8th IEEE International Conference on Control System, Computing and Engineering (ICCSCE)*; Penang, Malaysia, 2018; pp 70–75.
18. Xu, T.; Shi, F.; Liu, W. Research on Open-Pit Mine Vehicle Scheduling Problem with Approximate Dynamic Programming. In *Proc.—2019 IEEE Int. Conf. Ind. Cyber Phys. Syst. ICPS* ***2019***, no. 71790614; 2019; pp 571–577.
19. Jasika, N.; Alispahic, N.; Elma, A.; Ilvana, K.; Elma, L. et al. Dijkstra's Shortest Path Algorithm Serial and Parallel Execution Performance Analysis. In *2012 Proceedings of the 35th International Convention MIPRO, Opatija*, 2012.
20. Bogdan, P. Dijkstra Algorithm in Parallel- Case Study. In *Proceedings of the 2015 16th International Carpathian Control Conference (ICCC)*; Szilvasvarad, 2015; pp 50–53.

CHAPTER 14

SMART WALKING STICK (FOR THE BLIND)

PRABHPREET SINGH SANDHU[1], DEEPTI SAHU[1], and LATHA BANDA[2]

[1]*Sharda University, Greater Noida, Uttar Pradesh, India*

[2]*ABES Engineering College, Ghaziabad, Uttar Pradesh, India*

ABSTRACT

Visually challenged people face a lot of problems while walking on the roads finding hindrances in their path, which is not safe for them. This walking stick provides solution for this problem. We come up with a solution in this chapter, embedded in a stick comprising many sensors like ultrasonic to detect obstacles of the way and pot holes. Water sensor is placed at the bottom to avoid water on the roads and smoke sensor to guard from fire or heat. A buzzer and alarm system is there to warn the person when obstacle is detected. Arduino Nano is used as a microcontroller in the system for all the working. It is economic, responsive, power efficient, and foldable.

14.1 INTRODUCTION

People who are visually challenged are the people who are not able to recognize any minor detail with their eyes. The horizontal extent of the visual field with both eyes open less than or equal to 20 degrees and acuity of vision is 6/60, people with these symptoms are considered blind.[1] These people need digital devices which help them in coping the disability. As explained in Reference [2], around 10% of people have unusable eyesight so that they

Advanced Computer Science Applications: Recent Trends in AI, Machine Learning, and Network Security. Karan Singh, PhD, Latha Banda, PhD & Manisha Manjul, PhD (Eds.)

can move around safely and on their own. These embedded devices are manufactured to tackle such issues.

Active or passive sensors can be used to keep the information about hindrances present on the road. Sensors that only receive signal are passive ones. Radiations emitted from natural energy sources are detected by these sensors. Active sensors can detect near and far objects and give actual measurement between an obstacle and a blind person.

We tried overcoming some disadvantages in our product:

- The smart walking stick is developed to detect stairs, obstacles, and other hindrances and gives warning (Figure 14.1).
- Our product is aimed to be practiced at lower costs and is comfortable to use in position.
- The buzzer and alarm are used to give warning to the blind person and are also user friendly in public places.
- The embedded GPS also provides the geographical location of the blind person to the registered number of their guardian.
- It gives warning of obstacle hitting in an average distance of ≤50 cm before.
- "Find my stick" feature is also added to search for the stick with a remote by pressing a button on it and the stick will make a beep sound to follow it.

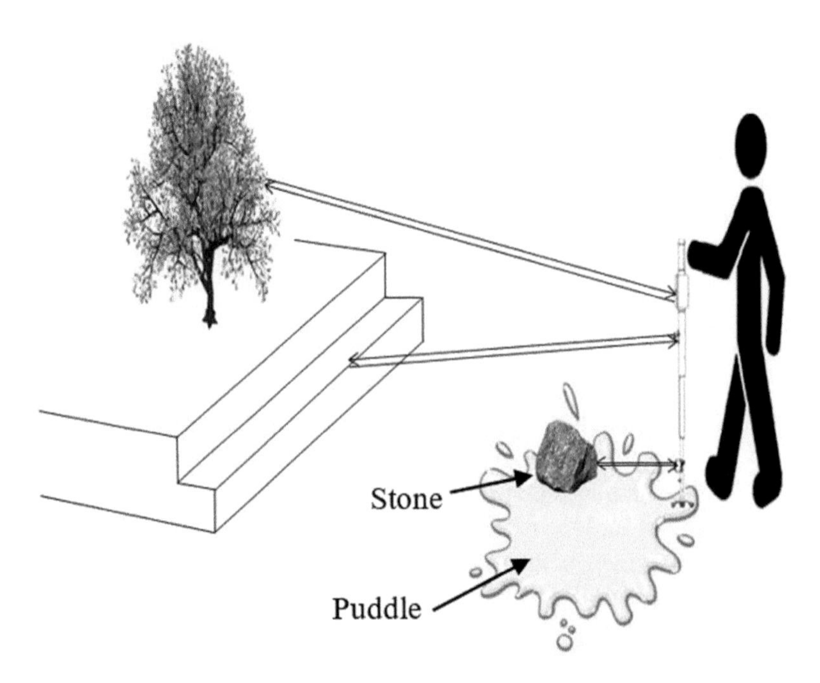

FIGURE 14.1 Working of smart stick for blind.

14.2 PROPOSED WORK

There are a plenty of reasons to design a product like smart stick for blinds, that is: (I) the person just has to handle a stick and not any bulky system. (II) The system detects outdoor as well as indoor obstacles like pot holes, stairs, footpaths, sidewalks, etc.

In Figure 14.2, an integrated embedded system is illustrated that makes ultrasonic sensor (HC-SR04) work for sensing obstacle for the blind person from a height till ground level in the range of 50 cm ahead;ultrasonic sensor provides real-time values and shares it with Arduino Nano (Microcontroller). Then after sensing any obstacle it gives a beep alarm warning to the blind person. Smoke sensor detects heat or fire, water sensor is for the detection of water, GPS detects live location along with finding my stick feature, and a rechargeable lithium-ion battery provides power to whole circuit.

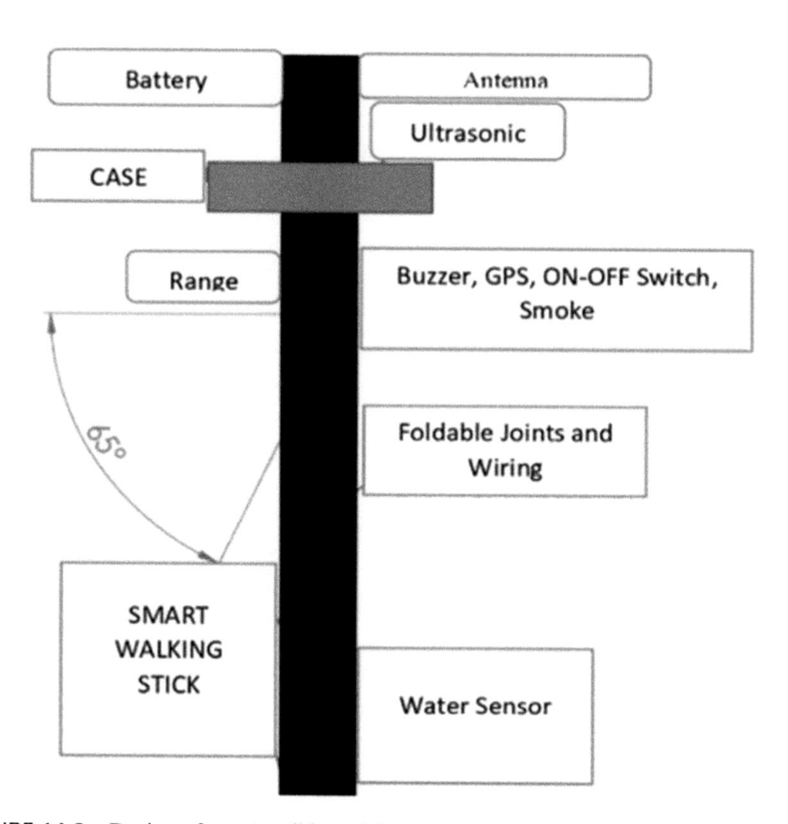

FIGURE 14.2 Design of smart walking stick.

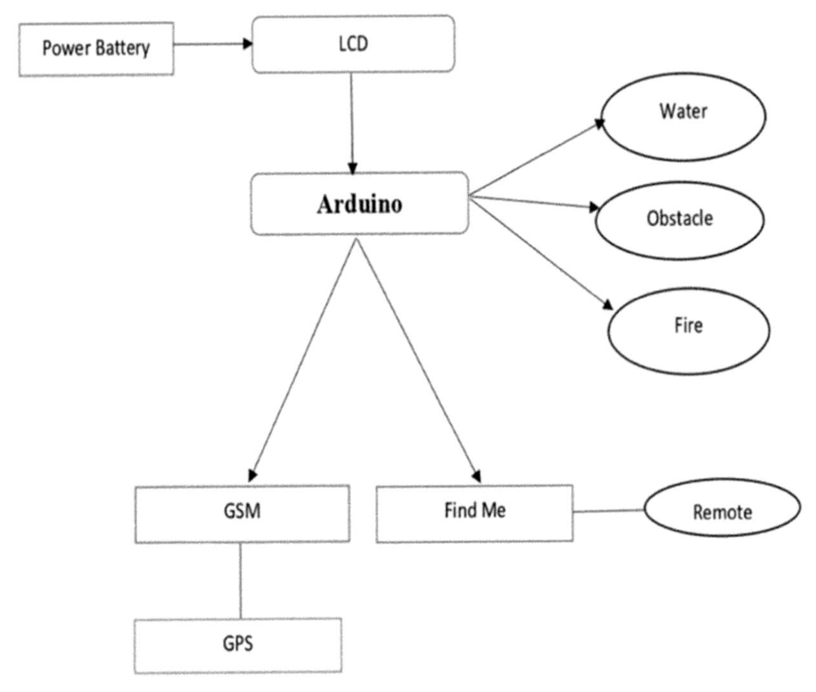

FIGURE 14.3 Block diagram of smart walking stick.

14.2.1 APPLIED SENSORS

The various factors on which sensors are selected are detection of range, condition of atmosphere, cost efficiency, types of obstacles, precision range of measurements, and frequency of transmission as shown in Table 14.1.

TABLE 14.1

	GSM&GPS	Smoke	Water	Ultrasonic
Principle	Transmissionand receptioncellular signal and location	Detection of heat and fire	Detection of water and wet surface	Obstacle Sensing
Range	Geographical location	From 30 to 100cm	20 cm	From 2 to 400 cm
Affect from Atmosphere	YES	YES	YES	NO
Cost	High	Low	Low	Low

14.2.1.1 ULTRASONIC SENSOR

Unlike laser one ultrasonic sensors are more accurate for close obstacles, when an object is very much closer, laser sensor (less than 10 cm) is not able to give exact reading. At 40 kHz an ultrasonic pulse is sent out and it passes through air and object or obstacle and reflects or bounces back to the sensor. The calculation of time and the speed of sound for the calculation of distance are given. We use ultrasonic sensor at the middle position of the stick so that it can catch the objects that are at height as well as on the level of ground.

Working of an ultrasonic sensor relies upon the following factors:

- (TOF)Time of flight: The arrival of an echo that depends on the distance of an obstacle and delay amount between emissions of sound is directly proportional to distance.
- Size of beam: Size of an obstacle depends upon reflected wave amount. Sound waves of the obstacles having dimensions greater than beam size will be reflected to the receiver. Then portion of ultrasonic sound wave will be reflected to the receiver and the remaining will be lost.

14.2.2 ARDUINO NANO MICROCONTROLLER ATMEGA328

Researchers in (17, 18) used the microcontroller. It has three disadvantages: (1) It has memory 32 kB (2 kB used by bootloader). (2) It has no internal oscillator so we will need an external crystal as a clock source. (3) Clock speed at 16 MHz. The Arduino Nano microcontroller ATmega328 is a mini, complete breadboard. It has similar functionality as Arduino Uno, but in a different package. It is only short of a DC power jack but works with a mini USB efficiently. It can be easily mounted on any surface due to its small size and light weight and can be used with little power due to its less power consumption. It also has a blue color placed on the top to show the power indication.

14.2.3 BUZZER ALARM AND WARNING

Many researchers[5,6,14] use different buzzer frequencies and audio patterns to warn the impaired person about different hindrances and obstacles. The device uses preloaded and different alarming patterns for all the installed sensors uniquely to avoid getting confused with sensors. In the proposed system, we have given different alert warning tones depending on the distances to aware the user.

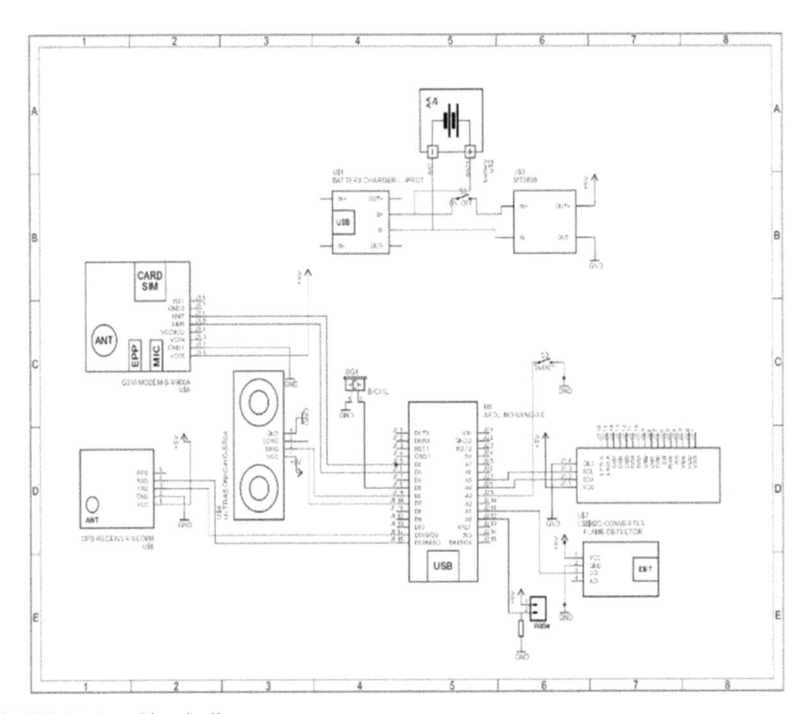

FIGURE 14.4 Circuit diagram.

14.2.4 WATER SENSOR

To determine wet surfaces and to detect water in the way the water sensors are being used in the device; it is also used to sense water spread, water pits, and puddles. The moment water gets into the contact of the sensor, it sends the information to the microcontroller and after that it generates attention warning that will come out from the buzzer as a medium.

14.2.5 GSM AND GPS

The GPS and GSM modules work together to give the live location of the person when the hazard button situated on the stick is pressed. A sim card is being installed in the GSM module and on pressing the button it will generate the notification and a message is being sent to the number of the guardian of the blind person that is being stored in the code and a live location link is being sent through the GPS module.

14.2.6 FIND MY STICK

Frequency modulation (FM) module along with wireless remote communication is provided to a blind person if the stick is far away from him. Frequency waves are being generated in the circuits of RF transmitter and "carrier signal," information is added by modulating the carrier signal. The generated signal is then given to antenna. From atmosphere the RF receiver receives the same frequency signal, by altering the Magnetic and Electric fields from antenna. The circuits of the receiver then give the information of signal to the carrier, and give an audio after amplifying it.

14.3 RESULTS AND COMPARISON

Smoke sensor, ultrasonic sensor, water sensor, GSM, and GPS are tested in an individual manner as well as in an integrated one. Checking ultrasonic sensors on various obstacles is must. We have calculated and checked the sensor at different values and distance for the results as shown in Table 14.2.

TABLE 14.2 Comparison and Result of Ultrasonic.

Distance (cm)	Analog value calculated (mV)	Analog value measured (mV)	Error
5	25	23	1 mv
10	50	48	1.9 mv
20	100	98.1	2.3 mv
30	150	145.6	3.8 mv
40	200	194.8	4.6 mv
50	250	246.3	5.8 mv
75	375	369	9.1 mv
100	500	491	11 mv

TABLE 14.3 Comparison of Other Devices.

Devices	Detection range	Time response	Power consumption	Stairs and waterdetection	Portable
Proposed Stick	High	Fast	Medium	Yes	Yes
Warning System	High	Medium	Low	Yes	Yes
GSM & GPS	High	Fast	Medium	NA	Yes
Ultrasonic	Low	Fast	Low	Yes	Yes

For getting the actual results the testing of the model has been done at the ground level or real world. Many experiments have been done by generating different circumstances and varying the types and pattern of obstacles. Along with the system the test for the battery has also been conducted in which the power lasts for about 2 h and 30 min which is a good backup considering the functionality provided.

System that is proposed can be compared with existing technologies as well. Some parameters can be evaluated:

- Starting with the range that is the first parameter, we have kept our system in between a low range of 2 m to a maximum of 75 m.
- Coming to second should be response time that our system is providing a very high response time of 0–100 ms.
- The third parameter is durability, light in weight, portable, and foldable to carry it anywhere easily.
- Fourth parameter that is important is the consumption of power of the system on which we have worked really well to get a decent power output or backup.

14.4 CONCLUSIONS

The Smart Walking Stick is a future aiding device for new generation which will grant a decent help to the visually challenged people around the world, giving near to accurate results for the user in the water pits with a good range of obstacle detection.

It is really economical, portable, reliable, and robust with a power-efficient capabilities system which also gives a real-time navigation experience to the user giving security aspects as well as quick response for every sensor enabled and also finding the stick when lost with a click on a remote. Blind people's concern was on top priority while developing the following product.

14.5 FUTURE SCOPE

We can collaborate with nursing homes, hospitals, surgical item manufacturers for the production and get it available to the needy ones easily and at a very reasonable price.

KEYWORDS

- **ultrasonic sensor**
- **electronic travel aids**
- **blind navigation**
- **visually challenged**
- **smoke sensor**
- **water sensor**

REFERENCES

1. World Health Organization. Visual Impairment and Blindness. Fact sheet N 282. Oct 2014.
2. National Disability Policy: A Progress Report—October 2014, National Council on Disability, Oct 2014.
3. Terlau, T.; Penrod, W. M. K'Sonar Curriculum Handbook, June 2008. http://www.aph.org/manuals/ksonar.pdf
4. Whitney, L. Smart Cane to Help Blind Navigate, 2009. http://news.cnet.com/8301-17938_105-10302499-1.html.
5. Hans du Buf, J. M.; Barroso, J.; Rodrigues, M. F.; Paredes, H.; Farrajota, M.; Fernandes, H.; Jos, J.; Teixeira, V.; Saleiro, M. The Smart Vision Navigation Prototype for Blind Users. *Int. J. Digital Content Technol. Its App* May **2011,** *5* (5), 351–361.
6. Ulrich, I.; Borenstein, J. The Guide Cane-Applying Mobile Robot Technologies to Assist the Visually Impaired. *IEEE Trans. Syst. Man Cybern. A* **2001,** *31* (2), 131–136.
7. Meijer, P. An Experimental System for Auditory Image Representations. *IEEE Trans. Biomed. Eng.* Feb **1992,** *39* (2), 112–121.
8. Nie, M.; Ren, J.; Li, Z. et al. Sound View: An Auditory Guidance System Based on Environment Understanding for the Visually Impaired People. In *Proceedings of the 31st Annual International Conference of the IEEE Engineering in Medicine and Biology Society: Engineering the Future of Biomedicine (EMBC '09)*; IEEE, September 2009; pp 7240–7243.
9. Balakrishnan, G.; Sainarayanan, G.; Nagarajan, R.; Yaacob, S. Wearable Real-Time Stereo Vision for the Visually Impaired. *Eng. Lett.* **2007,** *14* (2).
10. Fajarnes, G. P.; Dunai, L.; Praderas, V. S.; Dunai, I. CASBLiP—A New Cognitive Object Detection and Orientation System for Impaired People. In *Proceedings of the 4th International Conference on Cognitive Systems*; ETH Zurich, Switzerland, 2010.
11. Hoyle, B.; Withington, D.; Waters. D. UltraCane. June 2006. http://www.soundforesight.co.uk/index.html.
12. Kee, E. iSONIC Cane for the Virtually Impaired, 2011. http://www.ubergizmo.com/2011/01/isonic-cane-for-the-virtually- impaired/
13. Farcy, R.; Leroux, R.; Damaschini, R.; Legras, R.; Bellik, Y.; Jacquet, C.; Greene, J.; Pardo, P. Laser Telemetry to Improve the Mobility of Blind People: Report of the 6

Month Training Course. In *ICOST 2003 1st International Conference on Smart homes and health Telematics Independent Living for Persons with Disabilities and Elderly People*; Paris, 2003; pp 24–26.

14. Kumar, A.; Manjunatha, M.; Mukhopadhyay, J. An Electronic Travel Aid for Navigation of Visually Impaired Person. In *Proceeding of the 3rd International Conference on Communication Systems and Networks*, 2011; pp 1–5.
15. Bouhamed, S. A.; Kallel, I. K.; Masmoudi, D. S. New Electronic White Cane for Stair Case Detection and Recognition Using Ultrasonic Sensor. *Int. J. Adv. Comput. Sci. App.* **2013,** *4* (6).
16. Design of Non-Weighing Type Desert Plant Lysimeter Observation System Based on PIC18. In *Information Management, Innovation Management and Industrial Engineering (ICIII), 6th International Conference on IEEE* **2013,** *3*, 42–44.
17. Mahmud, N.; Saha, R. K.; Zafar, R. B.; Bhuian, M. B. H.; Sarwar, S. S. Vibration and Voice Operated Navigation System for Visually Impaired Person. In *Informatics, Electronics & Vision (ICIEV), International Conference on IEEE*, 2014; pp 1–5.
18. Al-Fahoum, A. S.; Al-Hmoud, H. B.; Al-Fraihat, A. A.. A Smart Infrared Microcontroller-Based Blind Guidance System. Active and Passive Electronic Components, 2013.
19. Su, B.; Wang, L. Application of Proteus Virtual System Modelling (VSM) in Teaching of Microcontroller. In *E-Health Networking, Digital Ecosystems and Technologies (EDT), 2010 International Conference on*, Vol. 2; 2010; pp 375–378.
20. Direct Mode ISD 1932, Jan 5, 2009. https://www.futurashop.it/...PDF_ENG/7300-ISD1932.pdf
21. Bhatlawande, S. S.; Mukhopadhyay, J.; Mahadevappa, M. Ultrasonic Spectacles and Waist-Belt for Visually Impaired and Blind Person. *Communications (NCC), National Conference on*; IEEE, 2012.
22. Dada, E.; Shani, A.; Adekunle, A. Smart Walking Stick for Visually Impaired People Using Ultrasonic Sensors and Arduino. *Int. J. Eng. Technol.* **2017,** *9*, 3435–3447. DOI: 10.21817/ijet/2017/v9i5/170905302
23. Muhammad, K.; Sajjad, M.; Baik, S. W. Dual-Level Security Based Cyclic18 Steganographic Method and Its Application for Secure Transmission of Key Frames During Wireless Capsule Endoscopy. *J. Med. Syst.* **2016,** *40*, 1–16.
24. Wu, J.; Cheng, B.; Wang, M.; Chen, J. Energy-Aware Concurrent Multipath Transfer for Real-Time Video Streaming over Heterogeneous Wireless Networks. *IEEE Transactions on Circuits and Systems for Video Technology*, 2017.
25. Panda, R.; Chowdhury, A. R. Multi-View Surveillance Video Summarization via Joint Embedding and Sparse Optimization. *IEEE Transactions on Multimedia*, 2017.
26. Viola, P.; Jones, M. Rapid Object Detection Using a Boosted Cascade of Simple Features. In *Computer Vision and Pattern Recognition, 2001. CVPR 2001. Proceedings of the 2001 IEEE Computer Society Conference on*, 2001; pp I–I.
27. Hua, Z.; Zhou, Y. Image Encryption Using 2d Logistic-Adjusted-Sine Map. *Inf. Sci.* **2016,** *339*, 237–253.
28. Machkour, M.; Saaidi, A.; Benmaati, M. A Novel Image Encryption Algorithm Based on the Two- Dimensional Logistic Map and the Latin Square Image Cipher. *3D Res.* **2015,** *6*, 1–18.

CHAPTER 15

DYNAMIC NEPALI SIGN LANGUAGE RECOGNITION

KANCHAN SAPKOTA, SANTOSH KHANDAL, SAILESH RANA, YUGARAJ TAMANG, and PANKAJ SHARMA

Department of CSE, Sharda University, Greater Noida, Uttar Pradesh, India

ABSTRACT

It has always been difficult to communicate and interact with those people who are unable to speak or listen. Human translators somewhat try to bridge the communication gap between the deaf-mute community and those who do not know how to read and use the sign language. However, they are limited in number and are not available everywhere, all the time. So, to solve this problem, we can use various computer science technologies to detect and classify the sign language gestures. This chapter proposes a system to detect and recognize dynamic Nepali Sign Language (NSL) in real time using a deep learning technique with the help of computer vision. The proposed approach takes video input from the user, extracts its frames, and classifies the sequence of images using a combined model of Convolutional Neural Network (CNN) and Long Short-Term Memory (LSTM). We have used InceptionV3, a transfer learning approach to extract spatial features and LSTM, a type of Recurrent Neural Network (RNN) to recognize the temporal features. The dataset is collected manually by capturing videos using a smartphone for five different classes.

Advanced Computer Science Applications: Recent Trends in AI, Machine Learning, and Network Security. Karan Singh, PhD, Latha Banda, PhD & Manisha Manjul, PhD (Eds.)

15.1 INTRODUCTION

Nepali Sign language (NSL) is a visual language used by physically impaired individuals in Nepal who cannot speak or hear for communication in their day-to-day conversational activities.

They communicate by performing signs using their hand shapes, orientation, and movement of hands. It requires the usage of fingers, hands, face, and sometimes head. To perform a sign language gesture accurately, there are five main factors to be considered: hand shapes, orientation, location, movement, and facial expression. However, despite having this mode of communication, there is still a huge communication gap between the deaf-mute people and the hearing-speaking public. It is not actively used by most of the nonhearing and nonspeaking people in Nepal. It is because the majority of deaf people are born in the family and society where there is not a single NSL user. They only get a chance to learn this language at the primary level of their education. Hence, deaf-mute people do not have the opportunity to learn and acquire Nepali Sign Language under normal situations. To some extent, there are instructors and translators who teach and interpret sign language respectively, but they are scarce in number and cannot be available all the time and everywhere. In this scenario, there is an inevitable need for science and technology. With its advancement, we can think of developing a system that can translate gesture signs into human understandable text and vice versa. This can help bridge the conversational gap between the two groups.

15.2 EXISTING SYSTEMS

Hand gesture recognition for sign language classification is an active and demanding research area. It is not an easy field to deal with, as working with hand gesture recognition in real time poses various factors that can affect the efficiency of the system such as lighting conditions, gesture performance, gesture momentum, variations in language structure, and so on. Many different researches have been done regarding sign language recognition in different countries such as the USA, the UK, and India; however, this field is still in its rudimentary stage in Nepal. Some researches that have been done in this field of NSL are stated below:

1. Nepali Sign Language Translation Using Convolutional Neural Network used a CNN model to build a Nepali Sign Language

translation system to identify and predict seven different signs statically. The data acquisition was done through a one-megapixel camera and a red glove was used for hand segmentation, which was then fed into a CNN model consisting of a total of two convolutional layers, a max-pooling layer and finally one fully connected layer. First, they developed a model for five classes which gave an accuracy of 95.4%. Later, another model was created for seven classes and due to this addition of two new signs, it was noticed that the model's accuracy was reduced by 0.83%

2. Hand Gesture Vocalizer for Dumb and Deaf People used glove controller to recognize sign alphabets and some words, and then translate sign language into text and voice. In this system, classification is done using the Random Forest Algorithm. Random Forest Algorithm is a supervised classification algorithm, which is operated by constructing a large number of decision trees at operating or training time. This algorithm creates some sets of decision trees from different subsets of the training data and overall votes are collected from all sets of decision trees to select the output class. When the testing was done in Nepali sign language, it was able to give an accuracy of 96.8%.
3. Hand Gesture Recognition for NSL Using Shape Information was obtained using digital image processing. Image acquisition is done by using a four-megapixel camera and manually preprocessed according to the need and hand segmentation is done to detect the concerned part. Erosion and dilation and fill operation are applied to improve the quality of the segmented image. Freeman chain code is applied to segmented images to prevent lossless data and feed in the image database. Freeman chain code draws a boundary from the starting point of the image until a blob is formed or the starting point is met. They have fed 13 vowels and 49 alphabet sets in the database. Classification was done by simply finding the maximum count of the matching sequence between the shape numbers of the database with the hand gesture to be classified.
4. This chapter has proposed a system to recognize NSL where the hand gesture is recognized using shape information. Here, Vertex Chain Code (VCC) and Freeman Chain Code are used as the feature extractors to identify distinguishable information from the hand gesture which is later used by the K-Nearest Neighbor (KNN) for classifying the gestures.

The accuracy is determined by dividing the total number of correctly predicted gestures by the total number of tested gestures. The accuracy of recognition using the radial approach for sampling along with the Freeman chain code was 56%. Similarly, when the grid line method was applied for sampling along with Freeman chain code, the accuracy was 48% and for the radial approach with the VCC method, the accuracy was 44%. Hence, the methodology where radial approach was used for sampling and Freeman chain code for extracting features was found to be comparatively better than the other two approaches.

15.3 DATA ACQUISITION

Since the research work on Nepali Sign Language recognition is still in its initial phase, there are no video datasets of NSL on the Internet. Hence, we generated a dataset of five different signs (father, food, promise, tea, wife) from the NSL by ourselves.

The videos were recorded using smartphones in landscape mode with a minimum frame rate of 30 fps. Each sign is performed 20 times by a single signer in different lighting conditions, background and speed of signing so that we can achieve variations in the dataset. We generated 80 videos per class (66 for training and 14 for testing).

15.4 PREPROCESSING

To introduce variability and eventually achieve better generalization, we have augmented the training video dataset using various operations like flipping the frames horizontally, rotating, increasing (multiply), and decreasing contrast (add). Applying such augmentation allowed the dataset size to increase from 66 videos per class to 330 videos per class allowing the overall dataset size to increase from mere 330 total videos to 1650 total videos.

FIGURE 15.1 Sample frames for a video with multiply augmentation.

Frames are extracted from all the video files belonging to both train and test datasets using a command-line tool called "FFmpeg" and then saved in the same directory. Simultaneously, metadata in CSV format is created to contain video file details like whether it belongs to a train or test dataset, its class name, filename, and a total number of frames in that video. This file is later used while extracting features from video frames using transfer learning. Now, while extracting features, the number of frames is down sampled to 40 for each video file to avoid overfitting and reduce computational cost. Now, distinguishable information is extracted from each frame using the InceptionV3 model and saved into a NumPy array. Eventually, this array is used for training the LSTM model.

15.5 METHODOLOGY

In this system, we are predicting sign language dynamically, which means that we have to classify the gesture from a set of sequential frames. For this task, a model is needed to learn two aspects:

- Features of the images
- Features of the sequence (temporal or time-related features)

1. For extracting features from images:

Convolution Neural Networks (CNNs) is the best choice for this task. They are excellent at extracting distinguishable information from the input frames through the application of relevant convolution filters. For our system, we have opted for a transfer learning approach. This supervised learning technique helps to get the final layer of a pre-built model which leads to a remarkable drop in training time as well as the amount of dataset required. A popular model used in transfer learning is Inception V3. It has 48 layers in a network and is trained on the ImageNet dataset having 1000 classes.

We have fed 40 frames for each video to the InceptionV3 architecture whose top classifying layer is removed, and got a feature vector of shape (40, 2048) for each video that will be used as an input for LSTM architecture.

2. For extracting time-related features from sequences:

We have used LSTM, which is an improved version of RNN, to extract the temporal features of image sequences.

The features we obtained from Inception-V3 are passed to this LSTM architecture. After that, we defined some of our own layers. We tried with dense layers with "relu" activation with some dropouts between multiple layers. After trying with different layers, we settled down to a 256-wide

LSTM layer along with batch-normalization and 0.4 dropout followed by another LSTM layer with 128 units. After that, a 16 dense layer along with ReLU as an activation function is used and a dropout of value 0.5 is applied.

The last layer consists of a dense layer with nodes equal to the number of labels along with a SoftMax activation function which basically provides a class as output in terms of probability.

We have used "categorical_crossentropy" as a loss function and "Adam" as an optimizer with a learning rate of 1e-5 and decay of 1e-5. Similarly, "accuracy" and "top_k_categorical_accuracy" are used as metric functions for training and testing. Finally, we compiled and trained our model using these parameters.

15.6 EXPERIMENTAL RESULT

In this project, we have trained our model on Google Colaboratory with the help of GPU provided by Google. The number of epochs used while training was 100 with the patience value of 10. A batch size of 32 was set for maintaining the weight of the LSTM model. The model was trained with an Adam optimizer with a learning rate of 1e-5 and categorical cross entropy as a loss function. It took around 10 min to train the model on Google Colaboratory.

First, we trained the model on a small dataset of 66 videos each class where we got an accuracy of 98% on training data and an accuracy of 20% on validation data which indicated that our model was overfitting. Then we increased the dataset size by applying different augmentation operations to 330 videos per class. Then we got 89% training accuracy, and 59% validation accuracy which is a huge improvement compared to the previous result.

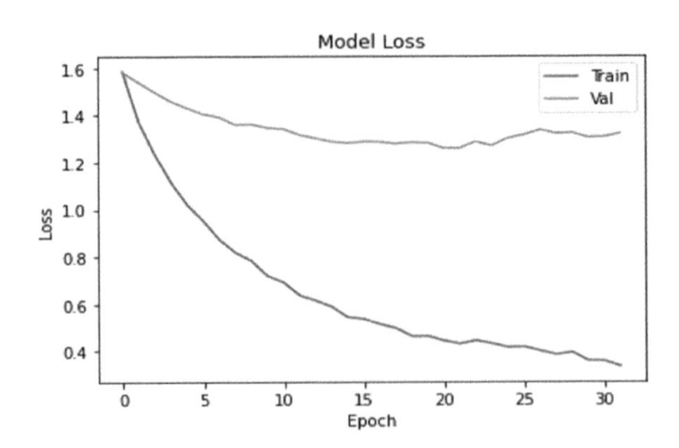

FIGURE 15.2 Validation curve.

15.7 CONCLUSIONS

The proposed system has given considerable outcome for the recognition of Nepali sign language gestures. We have taken both spatial and temporal features of the videos into account while training our model. That is why, it is successful to classify the sequence of video frames into its actual class.

Even though the model implemented using CNN-LSTM architecture gave a validation accuracy of 60% with a loss of 1.32% which is very less compared to that of other state-of-the-art techniques, it is not bad considering our limitations like unavailable datasets, lack of high-end devices, and insufficient knowledge of other better techniques.

KEYWORDS

- **Nepali Sign Language (NSL)**
- **convolutional neural network (CNN)**
- **long short-term memory (LSTM)**
- **recurrent neural network (RNN)**
- **InceptionV3**
- **computer vision**
- **spatial features**
- **temporal features**

REFERENCES

1. Mali, D.; Mali, R.; Sipai, S.; Panday, S. Two Dimensional (2D) Convolutional Neural Network for Nepali Sign Language Recognition, 2018; 1–5. DOI: 10.1109/SKIMA.2018.8631515.
2. Manandhar, S.; Bajracharya, S.; Karki, S.; Jha, A. Hand Gesture Vocalizer for Dumb and Deaf People. *SCITECH, Nepal* **2019,** *14* (1), 22–29. https://doi.org/10.3126/scitech.v14i1.25530
3. Sunuwar, J.; Pradhan, R. Hand Gesture Recognition for Nepali Sign Language Using Shape Information. *Int. J. Comput. Sci. Eng.* **2015,** *3* (6), 129–135.
4. Thapa, V.; Sunuwar, J.; Pradhan, R. Finger Spelling Recognition for Nepali Sign Language. In *Recent Developments in Machine Learning and Data Analytics. Advances in Intelligent Systems and Computing*; Kalita, J., Balas, V., Borah, S., Pradhan, R., Eds., Vol. 740; Springer: Singapore. https://doi.org/10.1007/978-981-13-1280-9_22

5. Huang, C-L.; Huang, W-Y. Sign Language Recognition Using Model-Based Tracking and a 3D Hopfield Neural Network. *Mach. Vis. App.* **1998,** *10*, 292–307.
6. Davis, J.; Shah, M. Visual Gesture Recognition. *IEE Proc. Vis. Image Sign. Proces.* **1994,** *141* (2), 101106.
7. Starner, T.; Pentland, A. Real-Time American Sign Language Recognition from Video Using Hidden Markov Models. In *SCV95*, 1995; p 265270.
8. Chen, F-S.; Fu, C-M.; Huang, C-L. Hand Gesture Recognition Using a Real-Time Tracking Method and Hidden Markov Models. *Institute of Electrical Engineering, National Tsinghua University*, Hsin Chu 300, Taiwan, March 2003.
9. Deora, D.; Bajaj, N. Indian Sign Language Recognition. In *1st International Conference on Emerging Technology Trends in Electronics, Communication and Networking (ET2ECN)*, 2012; pp 1–5, 2012.
10. Kirillov, A. Hand Gesture Recognition, 2008. http://www.codeproject.com/Articles/26280/Hands-Gesture-Recognition (accessed 9 Sept 2014).
11. Shamaie, A.; Sutherland, A. Accurate Recognition of Large Number of Hand Gesture. Machine Vision Group, Centre for Digital Video Processing School of Computer Applications, Dublin City University, Dublin 9, Ireland, 2003.
12. Hurlbut, H. M. A Lexicostatistic Survey of the Signed Languages in Nepal. SIL Electronic Survey Reports, June 2012; p 23.
13. Chai, X. et al. Sign Language Recognition and Translation with Kinect. Institute of Computing Technology, CAS, Microsoft Research Asia, Beijing, China, October 2013.
14. Quan, Y.; Jinye, P Application of Improved Sign Languages Recognition and Synthesis Technology. Industrial Electronics and Application, 2008, ICIEA 2008, 3rd IEEE Conference, June 2008; pp 1629–1634.

CHAPTER 16

PERFORMANCE EVALUATION OF A MULTIBAND EMBROIDERED FRACTAL ANTENNA ON THE HUMAN BODY

SHRUTI GITE and MANSI SUBHEDAR

Electronics and Telecommunication Engineering, Mumbai University, Pillai HOC College of Engineering and Technology, Rasayani, India

ABSTRACT

Several recent technologies are emerging that integrate the wearable system with antenna technology. The wearable antennas embedded into textiles are the most promising and fully integrative ones in the field of wireless body area networks (WBAN). In this chapter, a fully textile wearable antenna operating at 2.4 GHz with Minkowski fractal design is presented. The antenna is fabricated using a pure silver conductive thread on the polyester substrate using the embroidery technique. A 3D full-wave electromagnetic simulation tool is used to simulate the design. The antenna is placed onto the human body to obtain a specific absorption rate (SAR) result. Simulation results are demonstrated using different antenna parameters like S11 return loss, VSWR, Gain, and Directivity. Results showcase the multiband antenna performance for various applications with four resonant frequencies 2.68, 4.06, 4.32, and 4.46 GHz. The SAR value for the simulated embroidered textile fractal antenna is 1.32 W/kg when placed along with the realistic human male torso model in HFSS.

Advanced Computer Science Applications: Recent Trends in AI, Machine Learning, and Network Security. Karan Singh, PhD, Latha Banda, PhD & Manisha Manjul, PhD (Eds.)

16.1 INTRODUCTION

The usage of wearable devices has been rising with regard to growing demand for wearable electronics. They have become a part of our daily lifestyle. We very much rely on these wearable devices as they will not only keep us in touch with our social circle but also provide additional features of monitoring daily activities. Thus, these devices are enhancing the performance of our day-to-day tasks by providing us with some accountability for it. There is a wide range of applications for wearable antennas that receive a tremendous response from the market. Some of the body area network application areas include health monitoring, military, navigation, RFID tracking, security, emergency, intelligent child protection, smart garments, transportation system, astronaut space monitoring, social networking, etc. Effective antenna design is crucial for wireless body area network communication. This involves improved performance characteristics of antenna like multiband, wideband, compact, low cost, ease of fabrication, and so on. The designed wearable devices must be comfortable to wear, compatible in size, as well as flexible for best performance. It should not be bulky in size so as to avoid tangling with garments as well as appearance of antenna should be aesthetic and ornamental when placed on the body.

To meet the above requirements for a reliable wearable antenna, fractal geometry is adopted for the proposed antenna. Fractal geometry fulfils all these requirements due to its properties like self-similarity, space filling, and ability of miniaturization. It is desirable that the wearable antenna will be functional at more than one band and hence the multiband wearable antenna is being proposed along with fractal geometry. This work focuses on the body area network applications mainly the e-health monitoring sector. The proposed antenna can be integrated onto the aprons of doctors as well their assistants to monitor the health condition of patients anytime from anywhere even outside the hospitals. This can ensure 24 h health monitoring of patients under experts as well as it can prove to be beneficial in life-endangering emergency situations.

Impact of antennas on the human body is an essential aspect of antenna design as wearable antennas function in close proximity to the human body. The man's body tissues captivate the electromagnetic energy emitted by the antennas to produce heat. Such exposure to electromagnetic waves can lead to several health hazards and also cause permanent damage to functioning of body parts. The measurement known as Specific Absorption Rate is used to determine how much electromagnetic radiation is absorbed by the human body (SAR). This work presents SAR analysis on the human body.

16.2 LITERATURE REVIEW

As the wearable antenna market is rapidly expanding, various wearable antennas have already been developed. These antennas are fabricated on different types of dielectric substrates. A plethora of flexible type of substrates are developed for wearable antennas like paper,[1,2] textiles, plastic,[3] and so on.[4,5] Several studies are available demonstrating wearable antennas showcasing its design and fabrication applicable to various fields.[6–9] Implementation of certain textile materials for body area applications has also been studied.[10,11] Amaro et al. published wearable antenna design with simple planar microstrip geometry on felt fabric which shows significant loss of bandwidth.[12] Khaleel et al. showcased a simple monopole antenna fabricated on artificial magnetic conductor ground plane using the simple printing technique.[13] The work presented in[14,15] shows a flexible wearable antenna fabricated using the inkjet printing method. This method also involves ink compatibility issues when combined with different types of substrates.

Elias et al. presented a single-band dipole antenna and its performance was studied using CST Microwave Studio. The outcomes showed that the SAR values were affected due to antenna position.[16] A rectangular patch and planar inverted-F antenna is implemented on a jean textile substrate (thickness $h=1$ mm, dielectric constant $\varepsilon r=1.7$, loss tangent $\tan \delta=0.025$) using an adhesive copper sheet. This antenna is being analyzed for SAR results that depict the on-body SAR value of 2.85 W/kg, not permitted for public exposure.[17] An SAR analysis of a planar inverted-F antenna utilizing a mobile phone on a human head and hand model is provided, revealing a nearly 40% reduction in the antenna's radiation efficiency.[18] A wearable planar dipole antenna with a single band was built, resulting in high values of specific absorption rate. To compensate this, an additional EBG layer is to be added that causes increase in the size of antenna.[19] Two different wearable antennas are being investigated for SAR results by Ramli et al. These antennas show the use of additional reflectors in antenna structure to reduce back radiation and attain satisfactory SAR value.[20]

16.3 PROPOSED ANTENNA DESIGN AND METHODOLOGY

The emergence of the wearable technologies is enhancing the lifestyle of humankind. Several recent technologies are emerging which integrate wearable system with antenna technology. Furthermore, the most promising and fully integrative wearable antennas are the ones that are embedded into textiles. Wearable systems are rapidly gaining importance and appeal as a result of a variety of benefits and applications. They can be used to sense biological information from the human body and send it wireless, especially for body-cantered applications in healthcare. The biological data obtained by the sensors is communicated to the control device by the sensors. The data gathered by the control device can then be remotely sent to a bodyworn wearable antenna for further patient diagnosis. Flexible substrates may be advantageous for wearable antenna applications, given the high demand for them. As a result, the use of various flexible substrates for such applications is increasing.

16.3.1 PROPOSED DESIGN

There has been rapid and revolutionary advancement in the biomedical field in recent years. Due to this, today the doctors can monitor the patient health condition wirelessly in real time from anywhere. To support this an efficient body area network communication is needed. This can be done by designing a reliable wearable antenna that will enhance the communication between the doctors and patients. To match the growing speed of technology, a wearable antenna is proposed which is being attached to the pocket of doctor's apron. This will ensure the effective communication between doctors and patients as they will be continuously updated with the health parameters. This will ensure that the patient is safe as they can be rescued and diagnosed if a medical emergency occurs. Figure 16.1 depicts an apron where the proposed antenna is to be integrated.

FIGURE 16.1 Medical apron for integrating proposed antenna.

Figure 16.2 depicts the proposed antenna approach, which is further explained in the following subsections.

16.3.2 MATERIAL SELECTION

Dielectric Substrate material: The parameters to be considered for selecting a fabric for antenna design are flexibility, availability, cost effectiveness, comfort, softness, long-lasting properties, and ease of production. Textile materials is a good choice for antenna substrate as they have very low dielectric constant for boosting antenna impedance along with reduction of surface losses. Because the performance of the antenna is based on these two parameters, a thorough evaluation of the substrate material, particularly the permittivity and loss tangent, is critical. The substrate of antenna is made of fully textile polyester material.

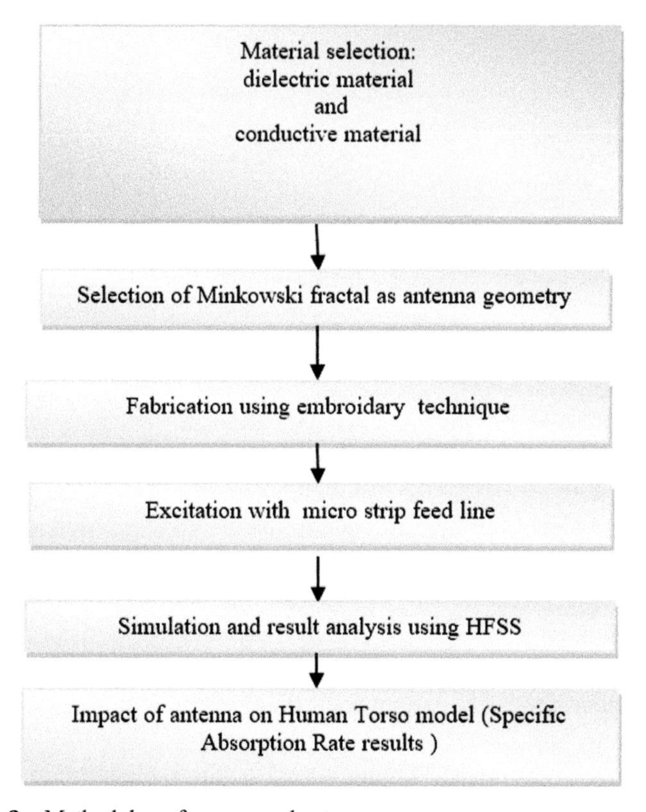

FIGURE 16.2 Methodology for proposed antenna.

Conducting Materials. The suggested antenna's conducting portion is made of pure silver conductive thread. It allows the electric current to pass through the antenna; hence, it can also be called an "E-textile" or "electro textile." For E-textiles, the most important parameter is conductivity. Conductivity (σ) is inversely proportional to the surface resistivity (R_s) of the material. For the selected thread the surface resistivity is less than 300 Ω/m, so a good amount of conductivity is obtained. E-textiles can be considered to be the best option for textile wearable antenna applications when compared with the design issues of copper adhesive patches or sheets that are not suitable for heat exposure as well as washing. They are not aesthetic in look and also damage the antenna performance on bending.

Ekatrina[12] introduced a detailed study on queues pending with customers and their relationship with impatience caused. Our study takes these scores into consideration.

16.3.3 MAJOR CONTRIBUTION

Realizing following points with the smart apron, further leverage to healthcare industry can be obtained:

- Early diagnosis: Doctors can detect early symptoms of a life-threatening disease or circumstance based on the patient's clinical parameters by using smart apron.
- Remote access: Doctors can monitor patients remotely and in real even though they are out of reach of hospital and patient. Thus smart apron will allow the doctors to treat the patients by accessing them remotely.
- Patient's Data: The patient's data medical statistics and history is recorded real time and stored in database during each visit to hospital. This will enable us to track the overall health and treatment of a patient. This will result in a more perfect and detailed diagnosis analysis by the doctor.
- Customized record: The doctor can maintain the database of an individual patient in real time and can have access to his or her complete medical history. This can be helpful for doctors to customize the treatment for every patient accordingly depending on the needs of the patient.
- Flawless Diagnosis: Having access to the patient's current data and previous reports digitally will provide clear condition of the patient to

the doctor. This will further help the doctor to compare, analyze, and give the optimum treatment.

- Reminder: Patients sometimes fail to take regular medicines, treatment and ignore the instructions given by doctors. So in such cases smart apron can act as a reminder as it will update doctors if patients fail to take treatment on time.
- Affordable healthcare: Usage of smart apron can save a lot of time and transport facility for a patient to visit the hospital again and again for the treatment.
- Patient engagement: Due to continuous and effective monitoring of health condition by doctors, the patients are satisfied and relieved and are engaged into the treatment process. This will enable a faster recovery of patients.

16.3.4 FABRICATION

The proposed antenna is a wearable antenna embedded in smart textile. To achieve better efficiency, flexibility of an antenna is an important aspect. Wearable antennas fabricated on epoxy-based substrates lack this aspect of design. In screen printing fabrication, conductive ink is used onto textile substrates that tend to permeate through the fabric during the process. There is also a possibility of oxidation reaction on the surface of substrate that results in degradation of antenna performance.

To overcome these issues during fabrication, a better wearable antenna fabrication method is chosen for the proposed work. Embroidery is an alternate fabrication approach for fractal wearable antennas that has never been explored before. The proposed antenna uses a pure silver conductive thread onto polyester substrate, fabricated using the embroidery technique. The fabrication is done on Brothers Embroidery machine F440E. The fabrication technique of the proposed antenna is depicted in Figure 16.3.

There is a need of fully integrative wearable antennas. As a result, the proposed antenna is constructed with a polyester fabric substrate and a silver conductive thread patch. The silver conductive thread and the textile substrate claims the antenna to be fully textile wearable antenna. The antenna design is Minkowski fractal designed at 2.4 GHz frequency. The antenna is being designed to operate at various frequency bands like GPS, LTE, Bluetooth, and WiMAX. The proposed antenna measures $44.17 \times 52.72 \times 1.6\ mm^3$. The antenna is being fed using a microstrip feed line. The fractal geometry enables the miniaturization of the antenna along with multiband characteristics.

FIGURE 16.3 Embroidery machine for the fabrication of proposed antenna.

16.4 EVALUATION OF ANTENNA AND IMPACT ON HUMAN TORSO MODEL

16.4.1 ANTENNA UNDER TEST (AUT)

The Antenna Under Test (AUT) is Minkowski fractal that needs to be evaluated for the study of overall antenna performance. The suggested antenna is created using a finite-element finite-difference 3D electromagnetic simulation tool. The commercial Ansys HFSS simulation software was used to simulate the antenna's performance. The results are obtained in terms of antenna parameters. The AUT is illustrated in Figures 16.4 and 16.5.

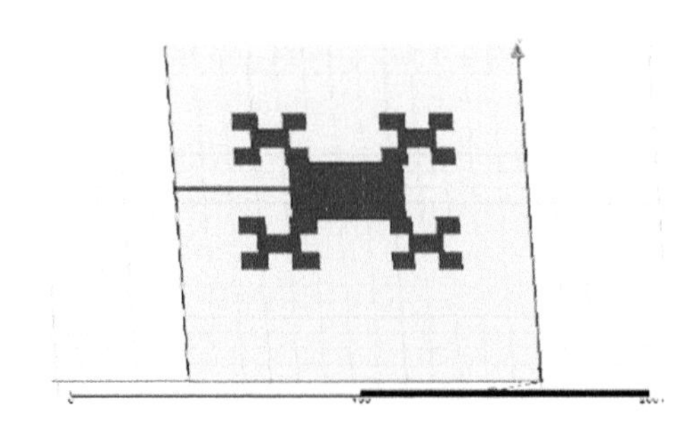

FIGURE 16.4 HFSS top view of AUT.

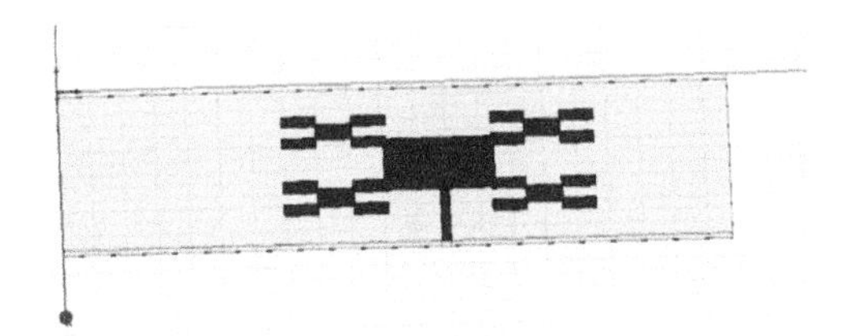

FIGURE 16.5 HFSS side view of AUT.

16.4.2 IMPACT OF HUMAN TORSO MODEL

Specific absorption rate analysis. When AUT is placed near the human body, some of the power is reflected back and absorbed by the human body; this value can be quantified and termed Specific Absorption Rate (SAR). It is given by the equation

$$SAR = (\sigma E^2)/(2\rho) \tag{16.1}$$

where σ and ρ are conductivity and density of human body respectively and E is the rms value of electric field strength. To obtain accurate results in terms of Specific Absorption rate (SAR), AUT is evaluated using the ANSYS Human Body Model (HBM). The geometrical and material properties of HBM are the same as the human body. This software has both adult human models for male as well as female with fine accuracy (up to millimeters) in dimensions. It is frequency dependent and functions from around 10 Hz to 10 GHz. The male HBM that is utilized for the proposed work has around 300 body parts including organs, muscles, and bones. The HBM can be used to analyze the antenna design with three levels of accuracy by changing the resolution. Figures 16.6 and 16.7 show AUT placed in chest position of HFSS male phantom model.

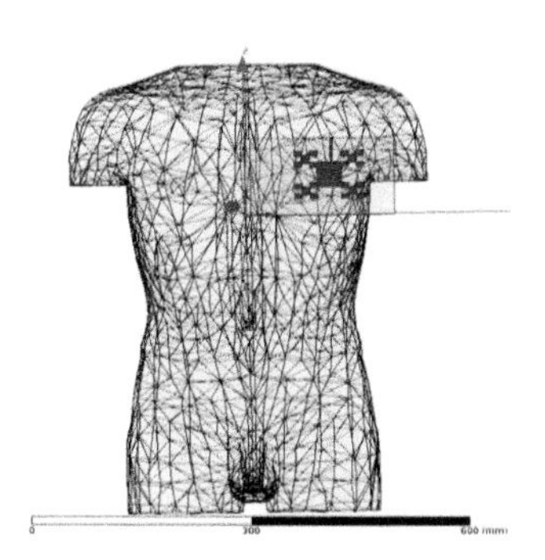

FIGURE 16.6 Front view of human torso model.

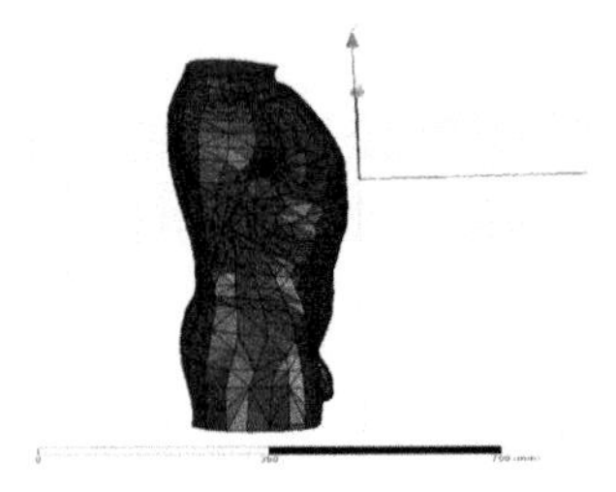

FIGURE 16.7 Side view of human torso model.

16.5 SIMULATION RESULTS

The simulated results can be seen in Figures 16.8–16.15. The various antenna parameters that are observed to study the performance of the proposed antenna are S11 return loss, VSWR, Gain, and Directivity. The proposed antenna resonates into four different resonant frequencies that are 2.68, 4.06, 4.32, and 4.46 GHz. This can be observed from Figures 16.8–16.10 respectively.

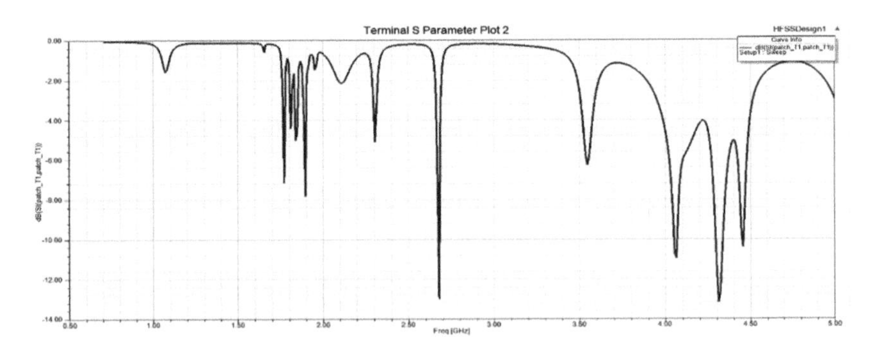

FIGURE 16.8 S11 return loss graph of the proposed antenna.

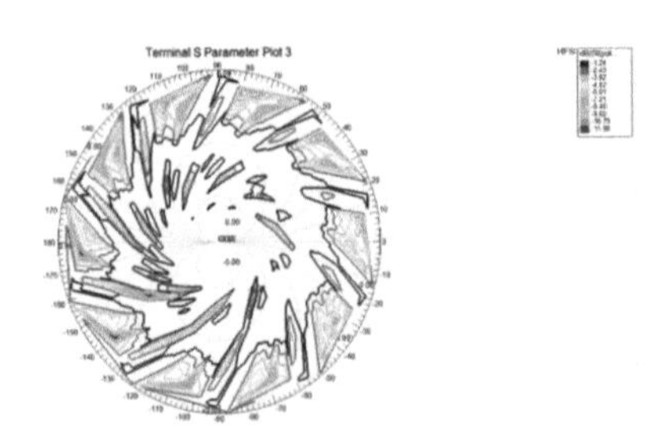

FIGURE 16.9 S11 Smith contour plot.

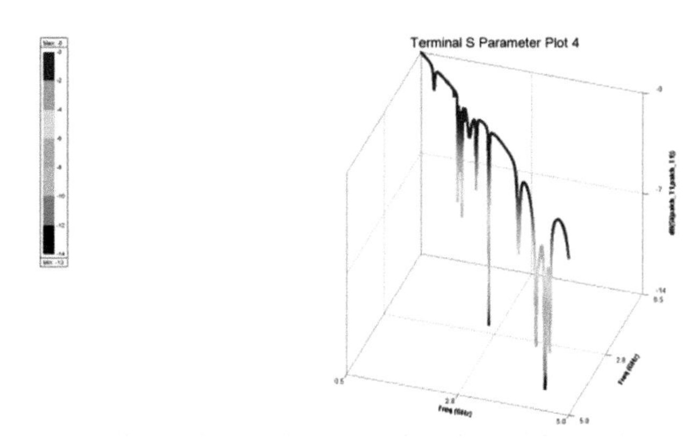

FIGURE 16.10 S11 3D rectangular plot.

The VSWR results for the proposed antenna can be seen in Figures 10.11–16.13, respectively. The 3D plots of gain and directivity of the proposed antenna can be seen in Figures 16.14 and 16.15, respectively.

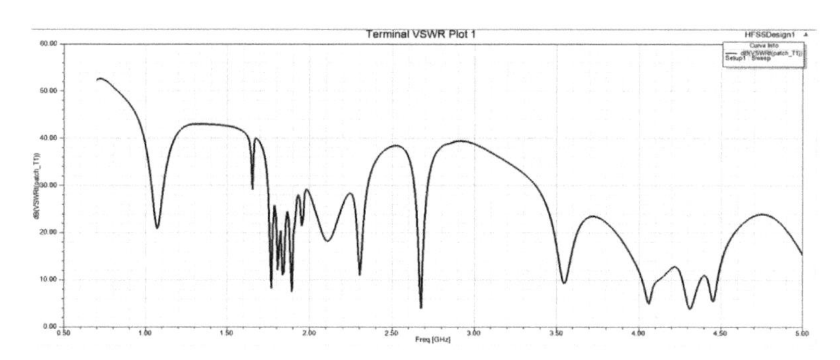

FIGURE 16.11 VSWR graph of the proposed antenna.

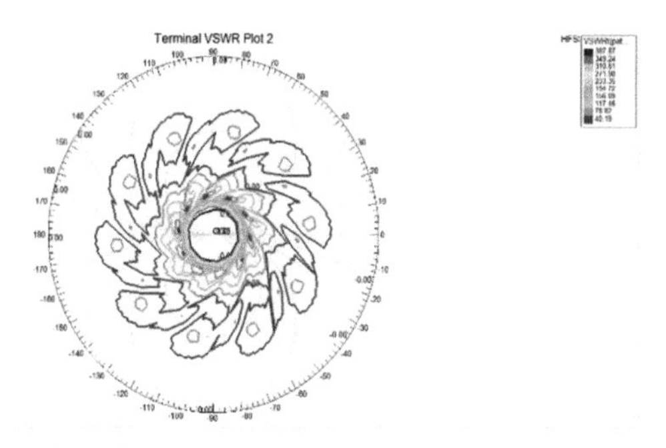

FIGURE 16.12 VSWR Smith Contour plot.

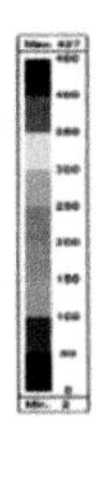

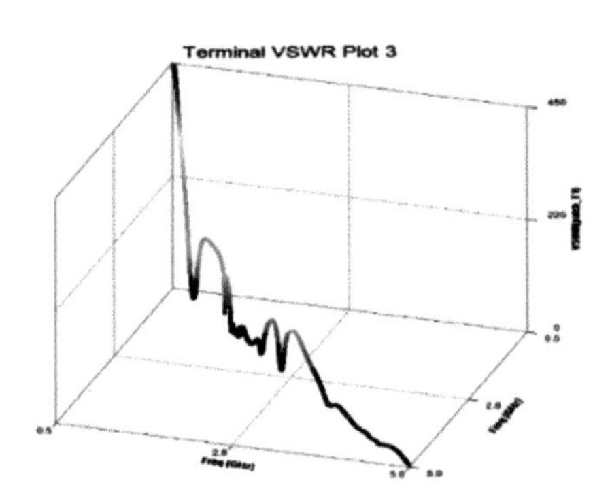

FIGURE 16.13 VSWR 3D rectangular plot.

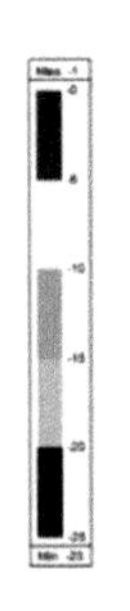

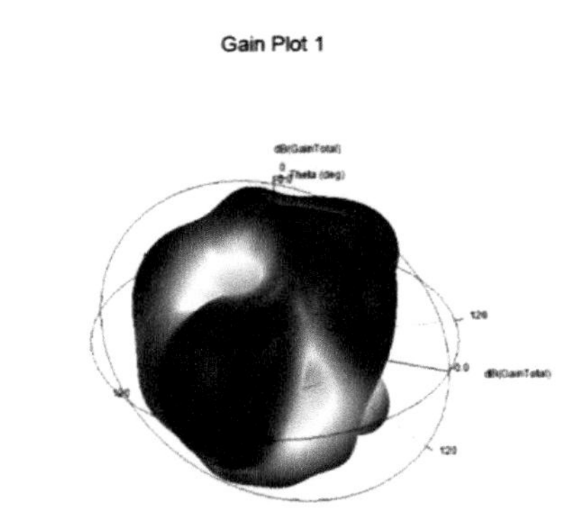

FIGURE 16.14 3D polar gain plot.

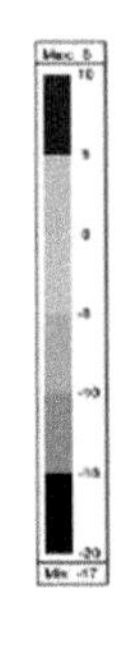

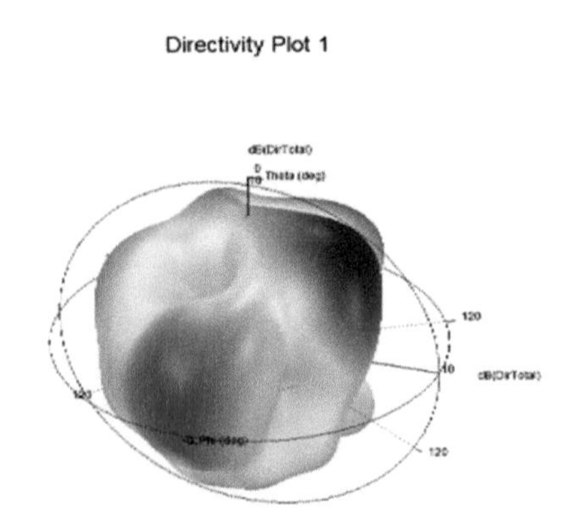

FIGURE 16.15 3D polar directivity plot.

It can be observed that at 1.7 cm distance, the wearable textile antenna safely gives the SAR value as 1.32 W/kg. Evaluation of SAR of AUT shows a satisfactory safety margin. It does not exceed the permitted limit of 2 W/kg averaged over10 g of tissues.

16.6 CONCLUSIONS

This chapter demonstrates a wearable textile antenna suitable for different applications demanding a compact, flexible, low cost, and efficient antenna. The proposed antenna operational at 2.4 GHz uses a fractal geometry that enables the miniaturization of an antenna. The antenna is fabricated onto a fully textile polyester substrate. The radiating element of the antenna is the highly conductive silver thread embroidered onto the fabric. This E-textile antenna is tested for its performance and also for the impact on the human body when placed in the vicinity of it. Simulated results indicate multiband characteristics and better gain. The proposed antenna resonates into four different resonant frequencies that are 2.68 GHz, 4.06 GHz, 4.32 GHz, and 4.46 GHz. The SAR results are obtained using a phantom model of male human torso that indicates a 1.32 W/kg value. The proposed antenna can be potentially used as safe, flexible, and efficient wearable antenna for several Wireless Body Area Networks (WBAN) applications.

KEYWORDS

- **body area network**
- **fractal design**
- **Minkowski**
- **high-frequency structure simulator (HFSS)**

REFERENCES

1. Leng, T.; Huang, X.; Chang, K.; Chen, J.; Abdalla, M. A.; Hu, Z. Graphene Nanoflakes Printed Flexible Meandered-Line Dipole Antenna on Paper Substrate for Low-Cost RFID and Sensing Applications. *IEEE Antennas Wireless Propagation Lett.* **2016,** *15,* 1565–1568.

2. Abutarboush, H. F.; Farooqui, M. F.; Shamim, A. Inkjet-Printed Wideband Antenna on Resin-Coated Paper Substrate for Curved Wireless Devices. *IEEE Antennas Wireless Propagation Lett.* **2016,** *15*, 20–23.
3. Durgun, A. C.; Reese, M. S.; Balanis, C. A.; Birtcher, C. R.; Allee, D. R. Venugopal, S. Design, Simulation, Fabrication and Testing of Flexible Bow-Tie Antennas. *IEEE Trans. Antennas Propagation* Dec **2011,** *59* (12), 4425–4435.
4. Hamouda, Z.; Wojkiewicz, J. -L.; Pud, A. A.; Kone, L.; Belaabed, B.; Bergheul, S.; Lasri, T. Dual-Band Elliptical Planar Conductive Polymer Antenna Printed on a Flexible Substrate. *IEEE Trans. Antennas Propagation* **2015,** *63* (12), 5864–5867.
5. Janeczek, K.; Jakubowska, M.; Koziol, G.; Mlozniak, A.; Arazna, A. Investigation of Ultra-High-Frequency Antennas Printed with Polymer Pastes on Flexible Substrates. *IET Microwaves, Antennas Propagation* **2012,** *6* (5), 594–554.
6. Hong, S.; Kang, S. H.; Kim, Y.; Jung, C. W. Transparent and Flexible Antenna for Wearable Glasses Applications. *IEEE Trans. Antennas Propagation* **2016,** *64* (7), 2797–2804.
7. Khaleel, H. R.; Al-Rizzo, H. M.; Rucker, D. G.; Mohan, S. A Compact Polyimide-Based UWB Antenna for Flexible Electronics. *IEEE Antennas Wireless Propagation Lett.* **2012,** *11*, 564–567.
8. Zahran, S. R.; Abdalla, M. Novel Flexible Antenna for UWB Applications. In *IEEE AP-S International Antenna and Propagation Symposium Digest*; IEEE: Vancouver, Canada, July 2015; pp 147–148.
9. Abbasi, Q. H.; Ur Rehman, M.; Yang, X.; Alomainy, A.; Qaraqe, K.; Serpedin, E. Ultrawideband Band-Notched Flexible Antenna for Wearable Applications. *IEEE Antennas Wireless Propagation Lett.* **2012,** *12*, 1606–1609.
10. Kiourti, A.; Lee, C.; Volakis, J. L. Fabrication of Textile Antennas and Circuits with 0.1 mm Precision. *IEEE Antennas Wireless Propagation Lett.* **2016,** *15*, 151–153.
11. Sun, Y.; Cheung, S. W.; Yuk, T. I. Design of a Textile Ultra-Wideband Antenna with Stable Performance for Body-Centric Wireless Communications. *IET Microwaves, Antennas Propagation* **2014,** *8* (15), 1363–1375.
12. Amaro, N.; Mendes, C.; Pinho, P. Bending Effects on a Textile Microstrip Antenna. In *IEEE International Symposium on Antennas and Propagation (APSURSI)*; IEEE, 3–8 July 2011; 978-1-4244-9561-0.
13. Khaleel, H. R. Design and Fabrication of Compact Inkjet Printed Antennas for Integrate Within Flexible and Wearable Electronics. *IEEE Trans. Components Packag. Manuf. Technol.* **2014,** *4* (10), 1722–1728.
14. Whittow, W. G.; Chauraya, A.; Vardaxoglou, J. C.; Li, Y.; Torah, R.; Yang, K.; Beeby, S.; Tudor, J. Inkjet-Printed Microstrip Patch Antennas Realized on Textile for Wearable Applications. *IEEE Antennas Wireless Propagation Lett.* **2014,** *13*, 71–74.
15. Ahmed, S.; Tahir, F. A.; Shamim, A.; Cheema, H. M. A Compact Kapton-Based Inkjet-Printed Multiband Antenna for Flexible Wireless Devices. *IEEE Antennas Wireless Propagation Lett.* **2015,** *14*, 1802–1805.
16. Elias, N. A.; Samsuri, N. A.; Rahim, M. K. A.; Othman, N.; Jalil, M. E. Effects of Human Body and Antenna Orientation on Dipole Textile Antenna Performance and SAR. *IEEE Asia-Pacific Conference on Applied Electromagnetics (APACE)*; IEEE: Malaysia, 2012.
17. Gil, I.; Fernández-García, R. SAR Impact Evaluation on Jeans Wearable Antennas. In *11th European Conference on Antennas and Propagation (EUCAP)*, 2017.

18. Hossain, Md. I.; Iqbal, M. R.; Faruque, M.; Islam, T. Investigation of Hand Impact on PIFA Performances and SAR in Human Head. *J. Appl. Res. Technol.* **2015,** *13*, 447–453.
19. Khan, A.; Bashir, S.; Ullah, F. Electromagnetic Bandgap Wearable Dipole Antenna with Low Specific Absorption Rate In *International Conference on Computing, Mathematics and Engineering Technologies*; iCoMET, 2018.
20. Ramli, M. N.; Soh, P. J.; Rahim, H. A.; Jamlos, M. F.; Giman, Ezzaty, F. N.; MohdHussin, F. N.; Lago, H.; Lil, E. V. SAR for Wearable Antennas with AMC Made Using PDMS and Textiles. In *32nd URSI GASS*, 19–26 August ; Montreal; 2017.

CHAPTER 17

LEARNING DEEP REPRESENTATION USING A FULLY CONVOLUTIONAL AUTOENCODER FOR AUTOMATED FLOOR PLAN IMAGE RETRIEVAL

SAYALI DESHPANDE[1], RASIKA GURURAJ KHADE[2], KRUPA N. JARIWALA[2], and CHIRANJOY CHATTOPADHYA[3]

[1]*Department of Computer Engineering, Pune Institute of Computer Technology, Pune, Maharashtra, India*

[2]*Department of Computer Engineering, Sardar Vallabhbhai National Institute of Technology, Surat, Gujarat, India*

[3]*Department of Computer Science and Engineering, Indian Institute of Technology Jodhpur, Jodhpur, Rajasthan, India*

ABSTRACT

The real estate industry is going digital by transforming its entire project workflow, such as planning, modeling, and designing. One of the pivotal requirements of the early design stage of any infrastructure design is coming up with a floor plan. For an experienced architect also, referring to the existing model to get inspirations and ideas is very common. Searching through a database of floor plan images manually and identifying the plan of interest is a cumbersome process. This chapter presents a system for the retrieval of similar floor plan images from a large-scale database under the query-by-example paradigm that is proposed. The proposed method uses

Advanced Computer Science Applications: Recent Trends in AI, Machine Learning, and Network Security. Karan Singh, PhD, Latha Banda, PhD & Manisha Manjul, PhD (Eds.)

deep representation learning to extract essential features from the floor plan images. These features are then used for similarity measurement for image retrieval. We contribute to the research in two ways. First, by proposing a new query-by-example-based floor plan retrieval framework for a large-scale dataset. Second, a fully convolutional auto-encoder architecture is proposed for discriminative feature representation and fast retrieval. An experimental study, followed by qualitative and quantitative comparison with state-of-the-art methods, authenticates our proposed floor plan image retrieval system.

17.1 INTRODUCTION

Real-estate industries have started adapting to online platforms for searching, buying, and selling properties. It is common for potential buyers to browse through properties on the Internet and shortlist the ones that satisfy their requirements. Digitization has not only affected the real-estate industry in its commerce but also has transformed the project workflow. A large portion of the planning, modeling, and design is carried out using different software tools. Thus, a lot of digital data is generated and used by architects and designers. This data has an immense potential to be effectively used to aid buyers, sellers, designers, architects, and other stakeholders of the real estate industry.

Keeping that in mind, researchers have also started contributing in this domain. A variety of work has been conducted on one key aspect of any property, that is, building floor plans. Examples include the construction of 3D floor plans,[5] development of navigation systems in historical buildings and monuments,[6,7] automatic rendering systems to convert the textual description into real-world images,[9] and automatic interpretation and retrieval of floor plan images.[1,14,15,17,18] However, for most of these applications, experiments were conducted on small and relatively simple datasets.

Room layout segmentation, adjacent room detection, and graph spectral embedding feature for unique floor plan layout representation were proposed.[16] In Ref. [11], a system to detect rooms in the architectural floor plan by using room and door hypothesis is discussed. An automatic system for the detection and labeling of rooms from architectural floor plans has been introduced in Ahmed et al.[2] In Aoki et al.,[3] a system for the interpretation of hand-sketched floor plans into a computer-aided design format is presented. Later in graph-based methods were introduced[10,12]. A deep learning-based floor plan retrieval framework was proposed in,[16] where the experiments were conducted on a relatively smaller dataset, ROBIN.

This chapter focuses on one of the aspects of automated floor plan analysis, which is automated floor plan retrieval. We propose an unsupervised deep representation learning technique for transforming the floor plan images into a more compact form, suitable for querying and retrieval. A fully convolutional autoencoder is used for this purpose. We conduct our experiments on the BRIDGE[6,7] and ROBIN[16] datasets. The proposed technique significantly outperforms the traditional unsupervised techniques in the task of retrieval of floor plan images on this real-world, complex dataset. The critical contributions are (1) A new query-by-example-based floorplan retrieval framework; (2) A fully convolutional auto-encoder architecture for feature extraction and dimensionality reduction of floor plan images.

17.2 MATERIALS AND METHODS

17.2.1 PROPOSED FRAMEWORK

We have proposed a methodology using a convolutional autoencoder for the effective retrieval of the floor plan images. The first phase is the training of the autoencoder using the BRIDGE dataset. In this phase, the raw dataset images are first preprocessed to make them uniform and suitable for training. These preprocessed images are used for training the autoencoder. After the model has been trained, it is used for feature extraction in the second phase. The feature vectors thus obtained are stored for future use. The third phase is the retrieval of the most similar floor plans from the dataset, given a query image. Whenever a query image is provided, it is first preprocessed using the preprocessing method used in phase one. Then, features are extracted from this image using the trained autoencoder. Thus, we obtain the query feature vector. The distance between the query feature vector and each of the feature vectors stored in the database is calculated. The images corresponding to the feature vectors giving minimum distances are returned as the output of retrieval. This process is outlined in the block diagram shown in Figure 17.1.

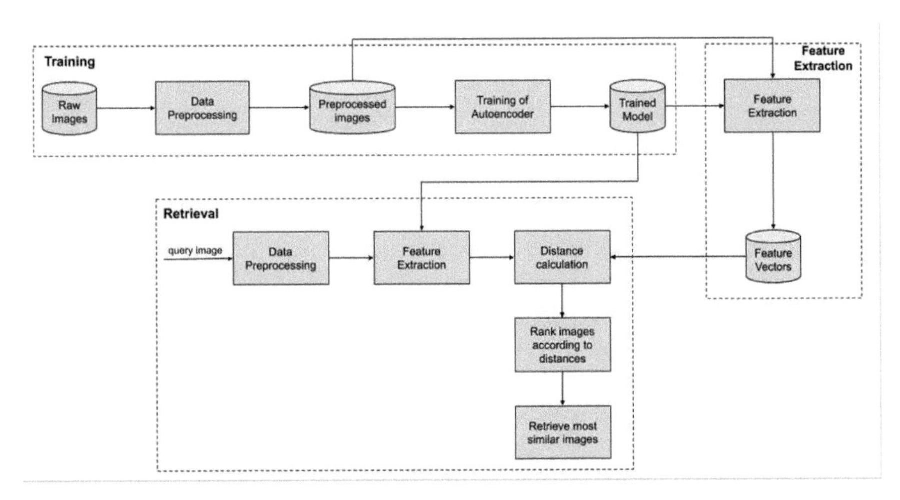

FIGURE 17.1 Block diagram of the proposed framework.

17.2.2 DATA PREPROCESSING

Our proposed framework is evaluated on the BRIDGE dataset. This dataset contains many images of different sizes and shapes. Thus, to maintain uniformity, all images are resized to size 256 × 256 as a part of data preprocessing. To have all images centered around a common mean, zero-mean normalization is used. Let N be the total number of pixels in an image I. Let xi represent an individual pixel of an image. The value of each pixel (xi) in an image (I) is updated as follows:

$$x_i = \frac{\left(x_i - \mu(I)\right)}{\sigma(I)} \tag{17.1}$$

Here, the mean pixel intensity of image I is subtracted from each individual pixel value xi. Then feature scaling is performed, to bring all the pixel values on the same scale (between −1 and 1). This is achieved by dividing each pixel value in every image by the standard deviation of pixel intensity. The mean pixel value of the resultant image is 0 and the standard deviation is 1.

17.2.3 AUTOENCODER ARCHITECTURE

As shown in Figure 17.2, the proposed autoencoder consists of 20 layers. The input layer is placed as the first layer, which is the 256×256×1 input feature map. This feature map is then passed to a convolutional layer followed by a max-pooling layer. This gives a feature map of dimension 28×128×8. This is further passed through two convolutional layers followed, and then through a max-pooling layer. Afterward, the resultant feature maps are then passed through another pair of convolution-max pool layers. The part of the architecture up to this point is called the encoder. The output of the encoder is a feature map of dimensions 16×16×16. The architecture of the decoder is just the mirror of the architecture up to the encoder. Instead of the max-pooling layers, upsampling layers are used in the decoder. All convolutional kernels are of size 3×3. The stride of the max-pooling and upsampling layers is 2.

17.2.4 TRAINING OF THE NETWORK

For training, we pass the input images to the network with a batch size of 64. The input is forward propagated through the network. The rectified linear unit (ReLU) activation function is used for each of the convolutional layers. The activation for the output layer is sigmoid. The mean squared error between the output of the last layer and the input image is calculated. This error is back-propagated and the weights are updated using Adam optimization. The training is carried out for 200 epochs.

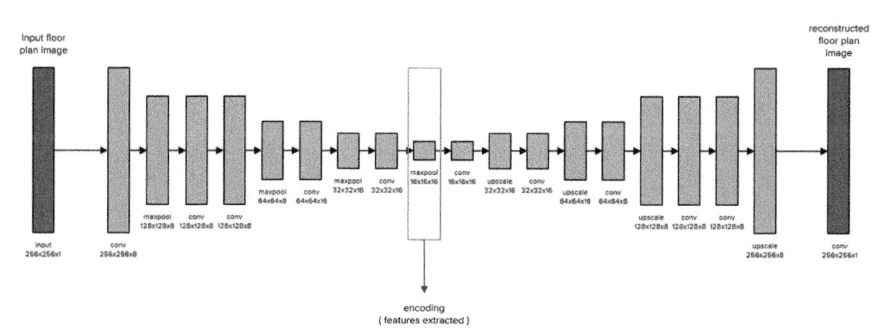

FIGURE 17.2 Network architecture of the autoencoder.

17.2.5 IMAGE RETRIEVAL

The feature maps of size 16×16×16 obtained from the trained encoder are used for image retrieval. The feature maps corresponding to each image in the database are unrolled to form feature vectors, each of length 4096. The input query image is also converted into a 4096-dimensional feature vector after preprocessing it and passing it through the trained encoder. Then, the distance between the features of the query image and those of the images in the database is calculated. These distance values are sorted in an ascending order and the corresponding images are given ranks accordingly. The first "*n*" images are retrieved as a result.

We have used cosine distance as a measure of dissimilarity between the vectors as it gives more relevant results. Let F(q): F1(q), F2(q), ...FL(q) be the set of encoded features of the query image q and x be a sample image of the dataset. Then the cosine similarity between features of q and x is given as

$$CosineSimilarity = \frac{\overrightarrow{f(q)}.\overrightarrow{f(x)}}{\overrightarrow{f(q)}.\overrightarrow{f(x)}} \quad (17.2)$$

$$Cosine\ Distance = 1 - Cosine\ Similarity \quad (17.3)$$

17.3 RESULTS

We performed our experiments on the BRIDGE dataset to demonstrate how our proposed framework performs with respect to other state-of-the-art. To train the autoencoder, we used the NVIDIA Quadro P4000 graphics card with 8GB GDDR5 memory.

17.3.1 DATASET DETAILS

The publicly available datasets for floor plan analysis are CVC-FP,[8] FPLAN-POLY,[13] SESYD,[4] and ROBIN.[16] CVC-FP has 122 scanned floor plans, while FPLAN-POLY contains 42 vectorial floorplan images. The numbers of floor plans in SESYD and ROBIN are 1000 and 510, respectively. ROBIN is a structured dataset with 510 images and has three categories of layout namely layout with three, four, and five rooms. These datasets are quite small and simple, having a little variation. Architectural floor plans are much more complex and extremely varied. Very recently,[6,7] a novel dataset named BRIDGE was introduced. It contains 13k unlabeled, unstructured, and complex floor plan images. All the images are web-scraped and compiled together. We conduct our experiments on the BRIDGE[6,7] andROBIN[16] datasets. A few samples of these two datasets are shown in Figure 17.3.

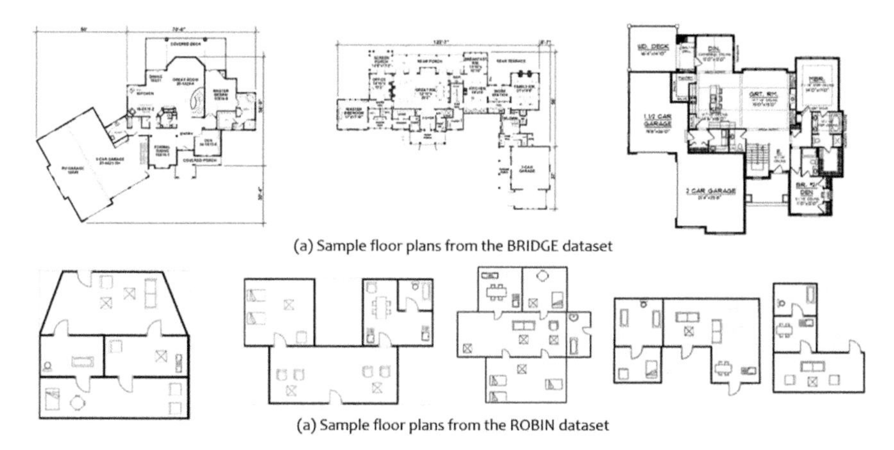

(a) Sample floor plans from the BRIDGE dataset

(a) Sample floor plans from the ROBIN dataset

FIGURE 17.3 Sample floor plan images from the two datasets used for experiments.

17.3.2 EXPERIMENTAL SETUP

During the experiments, we split a dataset into training and testing sets. All the samples in the dataset are resized to 256 × 256. This is done for training as well as retrieval (testing). We have taken into consideration the deep model depicted schematically in Figure 17.2. For a quantitative evaluation of our model and comparison with other techniques, ground truth labels would be required. However, manually labeling the floor plan images for the BRIDGE data set is difficult due to the size of the data set and the complexity of the floor plans. Thus, k-means clustering is used to structure the data set. The cluster labels thus obtained are considered ground-truth labels for recording performance measures. We compare the performance of the proposed technique with the following techniques: (1) feature extraction by pretrained models like VGGNet, ResNet, and Inception; (2) using handcrafted features like HOG, SIFT, SURF, BRISK, and ORB. Table 17.1 lists out the comparative results.

TABLE 17.1 Comparison Feature Dimensions Obtained from Various Techniques.

Feature extraction technique	Size of feature vector obtained for an image
VGGNet	25,088
ResNet	100,352
Inception	51,200
HOG	34,020

TABLE 17.1 *(Continued)*

Feature extraction technique	Size of feature vector obtained for an image
SIFT	8192
SURF	4096
BRISK	1024
ORB	512
Ours	4096

17.3.3 VISUALIZATION OF RESULTS

The proposed framework efficiently retrieves similar images from the BRIDGE dataset. The top 10 rank-ordered retrieval results for four different queries on BRIDGE are shown in Figure 17.4. As we can see, the retrieved images have a similar outer shape as that of the query images.

For all the results shown in Figure 17.4a and b, we have used the features extracted from the trained model of the autoencoder. The top-ranked retrieved result appears very similar to the query image. Only the internal details slightly differ. Because of less distance (Cosine distance) value between the retrieved layout and the query layout, the retrieved images are efficiently tank ordered. The subsequent results differ more in the distance values and are thus, ranked lower. In the case of Figure 17.4a, the top nine results belong to the query class, because the global layouts of all the floor plans are identical. The individual plans differ in terms of the number of objects present in the room and their position. The 10th sample belongs to a different category. In contrast, Figure 17.4b shows the failure cases, where the 6th and 10th rank result is incorrectly retrieved (highlighted by "red") by our proposed framework.

Figure 17.5 denotes the qualitative results for the ROBIN dataset. From the two example queries it can be observed that our proposed technique is able to retrieve the floor plans correctly. In Figure 17.5b, the 10th retrieved sample is erroneously retrieved. In the next subsection, we present the quantitative evaluation.

17.4 PERFORMANCE EVALUATION

Quantitatively, we evaluated the performance of the proposed image retrieval framework by considering the Precision-Recall (PR) metric as the evaluation method. We plotted the average precision and average recall scores by varying the number of images retrieved. Given a query floor plan image, the retrieved images should have a similar layout as that of the query image. Thus, the contour of the layout is a measure of similarity among the samples of the dataset and hence influences the class belongingness of a given floor plan. Figure 17.6a shows that our proposed technique (the "green" curve) significantly outperforms the traditional techniques on BRIDGE.

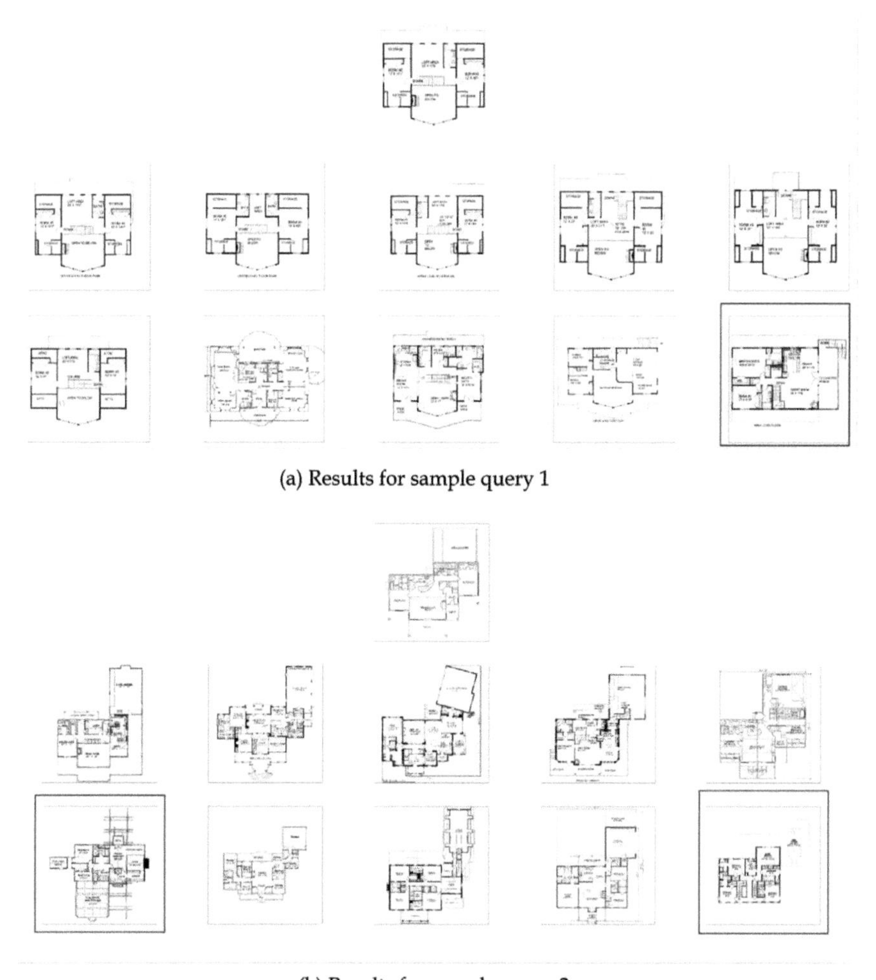

(a) Results for sample query 1

(b) Results for sample query 2

FIGURE 17.4 Top 10 rank-ordered images retrieved using our proposed framework on two different query images of the BRIDGE dataset. The red bounding box denotes an erroneously retrieved sample.

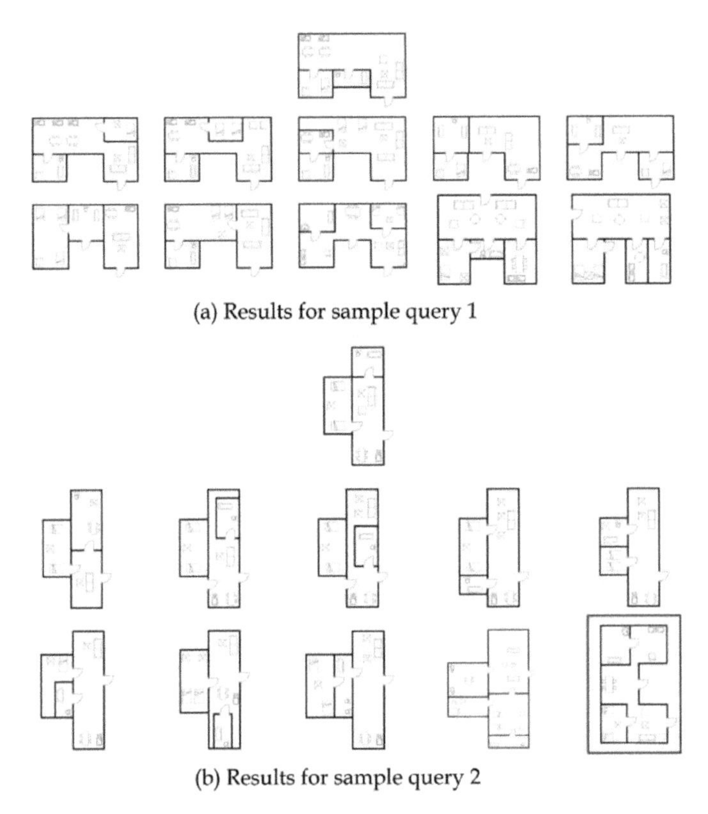

(a) Results for sample query 1

(b) Results for sample query 2

FIGURE 17.5 Top 10 rank-ordered images for two different queries on ROBIN. The red bounding box denotes an erroneously retrieved sample.

We have also compared our results with other existing methods in the literature. Table 17.2 shows the comparison of the Mean Average Precision (MAP) values between ours and the other existing techniques. The MAP value of our proposed technique (0.603) is the best on theBRIDGE dataset. However, it can be observed from Figure 17.6b and Table 17.2 that our proposed method is not the best in the case of the ROBIN dataset.

One possible reason behind this is that the ROBIN dataset is quite small, with only 510 images. Since our method involves training the autoencoder before feature extraction, this number of images is not adequate for it. Thus, our proposed method is more suitable for large datasets. Whereas for the methods that use pretrained networks or hand-crafted features, there is no training phase, making them more suitable for smaller datasets. The performance of our proposed method can be improved by data augmentation and by tweaking the autoencoder parameters for training it on a small dataset like ROBIN.

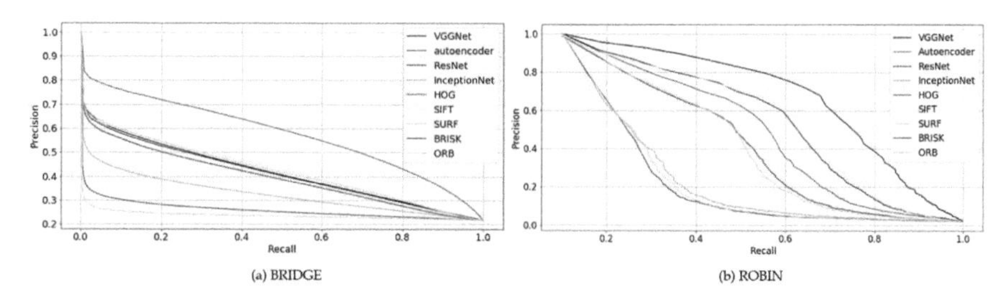

(a) BRIDGE (b) ROBIN

FIGURE 17.6 Precision-Recall (PR) plot, comparing the result of our proposed technique (the "green" curve) on the BRIDGE and ROBIN datasets.

TABLE 17.2 Mean Average Precision Values for 500 Query Images of BRIDGE and 100 Query Images of ROBIN Dataset Using Different Feature Extraction Techniques.

Feature extraction technique	BRIDGE	ROBIN
VGGNet	0.444	0.741
ResNet	0.440	0.644
InceptionNet	0.348	0.341
SIFT	0.463	0.651
HOG	0.427	0.528
SURF	0.470	0.517
BRISK	0.269	0.328
ORB	0.244	0.334
Ours	**0.603**	0.588

We have also compared the time required for retrieval using different algorithms such as bruteforce, Kd-tree, ball tree for different types of features. Table 17.3 denotes the comparison of time taken in seconds by various retrieval techniques for various features on the two datasets. The bold value in the table denotes the quickest of them, all for a given retrieval method. It shows that our trained model requires less time compared to VGGNet, ResNet, InceptionNet, SIFT, HOG, SURF for all three retrieval algorithms because as seen in Table 17.1, the size of the feature vector obtained by our method is comparatively lower whereas it is higher than those obtained by ORB and BRISK, hence the time required is higher. However, if we compare the time taken by the VGGNet, which is our closest competitor in terms of performance on floor plan retrieval, our timing is very less. This also shows the efficiency of our technique.

17.5 CONCLUSION AND FUTURE SCOPE

This chapter proposes an autoencoder architecture for the image retrieval task that was observed to be better compared to the traditional techniques on the complex BRIDGE dataset. It was seen that the retrieved images had a similar overall shape or layout as that of the query image in most of the test cases. Using the autoencoder also helped in making the retrieval faster, since the reduced number of dimensions of the feature vectors made the distance calculation quicker.

Labeling the BRIDGE dataset partially or completely can help us explore semi-supervised or supervised techniques, respectively. In the future, we are also planning to investigate room semantic-based retrieval, which would address the functional aspects of the floor plan for measuring similarity such as the placement of the various interior components like bedrooms, kitchen.

TABLE 17.3 Time (in Seconds) Required for Retrieval of 500 Query Images of Two Datasets Using Different Feature Extraction Techniques and Different Retrieval Algorithms.

Method	Algorithm					
	Brute force		K-d Tree		Ball tree	
	BRIDGE	ROBIN	BRIDGE	ROBIN	BRIDGE	ROBIN
VGGNet	1.71	0.14	151.45	2.85	121.20	2.03
ResNet	4.68	0.35	627.07	11.56	508.12	8.48
InceptionNet	2.67	0.17	314.94	5.83	243.13	4.39
SIFT	0.80	0.03	49.31	0.89	39.79	0.75
HOG	2.04	0.13	200.14	3.76	161.53	2.80
SURF	0.54	0.03	24.51	0.45	19.41	0.38
BRISK	0.47	0.03	6.68	0.11	6.03	0.09
ORB	0.36	0.02	3.74	0.06	3.10	0.04
Ours	0.52	0.02	24.04	0.45	19.47	0.34

ACKNOWLEDGMENTS

We would like to thank all the reviewers for their insightful comments and suggestions to improve the quality of the paper. The work is partially supported by the project titled "Development of Multimodal Search Framework for Architectural Floor plan" supported by the Science and Engineering Research Board, India under the project id ECR/2016/000953.

KEYWORDS

- **floor plan**
- **retrieval**
- **deep learning**
- **autoencoder**

REFERENCES

1. Ahmed, S.; Liwicki, M.; Weber, M.; Dengel, A. Improved Automatic Analysis of Architectural Floor Plans. In *2011 International Conference on Document Analysis and Recognition*; IEEE, 2011; pp 864–869. https://doi.org/10.1109/ICDAR.2011.177
2. Ahmed, S.; Liwicki, M.; Weber, M.; Dengel, A. Automatic Room Detection and Room Labeling from Architectural Floor Plans. In *2012 10th IAPR International Workshop on Document Analysis Systems*. IEEE, 2012; pp 339–343. https://doi.org/10.1109/DAS.2012.22
3. Aoki, Y.; Shio, A.; Arai, H.; Odaka, K. A Prototype System for Interpreting Hand-Sketched Floor Plans. In *Proceedings of 13th International Conference on Pattern Recognition*, Vol. 3; IEEE, 1996; pp 747–751. https://doi.org/10.1109/ICPR.1996.547268
4. Delalandre, M.; Valveny, E.; Pridmore, T.; Karatzas, D. Generation of Synthetic Documents for Performance Evaluation of Symbol Recognition & Spotting Systems. *International J. Doc. Analys Recogn. (IJDAR)* **2010,** *13* (3), 187–207. https://doi.org/10.1007/s10032–010–0120-x
5. Dosch, P.; Masini, G. Reconstruction of the 3d Structure of a Building from the 2d Drawings of Its Floors. In *Proceedings of the Fifth International Conference on Document Analysis and Recognition. ICDAR'99* (Cat. No. PR00318); IEEE, 1999; pp 487–490. https://doi.org/10.1109/ICDAR.1999.791831
6. Goyal, S.; Bhavsar, S.; Patel, S.; Chattopadhyay, C.; Bhatnagar, G. SUGAMAN: Describing Floor Plans for Visually Impaired by Annotation Learning and Proximity-Based Grammar. *IET Image Processing* **2019,** *13* (13), 2623–2635. https://doi.org/10.1049/iet-ipr.2018.5627
7. Goyal, S.; Mistry, V.; Chattopadhyay, C.; Bhatnagar, G. BRIDGE: Building Plan Repository for Image Description Generation, and Evaluation. In *2019 International Conference on Document Analysis and Recognition (ICDAR)*; IEEE, 2019; pp 1071–1076. https://doi.org/10.1109/ICDAR.2019.00174
8. de las Heras, L. P.; Terrades, O. R.; Robles, S.; Sánchez, G. CVC-FP and SGT: A New Database for Structural Floor Plan Analysis and Its Ground Truthing Tool.

Int. J. Doc. Analys. Recogn. (IJDAR) **2015,** *18* (1), 15–30. http://doi.org/10.1007/s10032-014-0236-5.
9. Jain, M.; Sanyal, A.; Goyal, S.; Chattopadhyay, C.; Bhatnagar, G. Automatic Rendering of Building Floor Plan Images from Textual Descriptions in English, 2018. arXiv preprint arXiv:1811.11938.
10. Lladós, J.; López-Krahe, J.; Martí, E. A System to Understand Hand-Drawn Floor Plans Using Subgraph Isomorphism and Hough Transform. *Mach. Vis. App.* **1997,** *10* (3), 150–158. https://doi.org/10.1007/s001380050068
11. Macé, S.; Locteau, H.; Valveny, E.; Tabbone, S. A System to Detect Rooms in Architectural Floor Plan Images. In *Proceedings of the 9th IAPR International Workshop on Document Analysis Systems*, 2010; pp 167–174. https://doi.org/10.1145/1815330.1815352
12. Qiu, H.; Hancock, E. R. Graph Matching and Clustering Using Spectral Partitions. *Pattern Recogn.* **2006,** *39* (1), 22–34. https://doi.org/10.1016/j.patcog.2005.06.014
13. Rusiñol, M.; Borràs, A.; Lladós, J. Relational Indexing of Vectorial Primitives for Symbol Spotting in Line-Drawing Images. *Pattern Recogn. Lett.* **2010,** *31* (3), 188–201. https://doi.org/10.1016/j.patrec.2009.10.002
14. Sharma, D.; Chattopadhyay, C. High-Level Feature Aggregation for Fine-Grained Architectural Floor Plan Retrieval. *IET Computer Vision* **2018,** *12* (5), 702–709. https://doi.org/10.1049/iet-cvi.2017.0581
15. Sharma, D.; Chattopadhyay, C.; Harit, G. A Unified Framework for Semantic Matching of Architectural Floorplans. In *2016 23rd International Conference on Pattern Recognition (ICPR)*; IEEE, 2016; pp 2422–2427. https://doi.org/10.1109/ICPR.2016.7899999
16. Sharma, D.; Gupta, N.; Chattopadhyay, C.; Mehta, S. DANIEL: A Deep Architecture for Automatic Analysis and Retrieval of Building Floor Plans. In *2017 14th IAPR International Conference on Document Analysis and Recognition (ICDAR)*, Vol. 1; IEEE, 2017; pp 420–425. https://doi.org/10.1109/ICDAR.2017.76
17. Khade, R.; Jariwala, K.; Chattopadhyay, C.; Pal, U. A Rotation and Scale Invariant Approach for Multi-Oriented Floor Plan Image Retrieval. *Pattern Recogn. Lett* **2021**. https://doi.org/10.1016/j.patrec.2021.01.020
18. Weber, M.; Liwicki, M.; Dengel, A. A. Scatch-a Sketch-Based Retrieval for Architectural Floor Plans. In *2010 12th International Conference on Frontiers in Handwriting Recognition*; IEEE, 2010; pp 289–294. https://doi.org/10.1109/ICFHR.2010.122

CHAPTER 18

SMART CARD-BASED PRIVACY PRESERVING LIGHT-WEIGHT AUTHENTICATION PROTOCOL FOR E-PAYMENT SYSTEMS

N. SASIKALADEVI

Department of CSE, School of Computing, SASTRA Deemed University, Thanjavur, Tamil Nadu, India

ABSTRACT

Traditional transactional techniques have been gradually supplanted by electronic transactions in recent years. To protect transactional details and to ensure secure electronic transactions, various e-payment mechanisms have been introduced. However, we discovered that earlier electronic payment mechanisms did not need non-repudiation from the customer and thus had several weaknesses. Hence, an authentication protocol satisfying the user's requirements without having vulnerabilities should be designed. To enhance the security and robustness in protocol, biometric-based authentication is required. Biometrics has been widely preferred as a third authenticating factor in password and smartcard-based user authentication protocol. Hence, mutual authentication protocol using Biometrics along with password and smartcard has been designed. To strengthen it, an absolute light-weight protocol has been designed based on ECC that provides low computational cost with high security, high speed and makes it suitable for practical application. We preferred to utilize AVISPA and SPAN animator tool for

Advanced Computer Science Applications: Recent Trends in AI, Machine Learning, and Network Security. Karan Singh, PhD, Latha Banda, PhD & Manisha Manjul, PhD (Eds.)

protocol validation. Informal analysis is also done for protocol verification and validation.

18.1 INTRODUCTION

As the world is moving forward with technology development, the use of electronic payment systems as a payment processing device has also been increasing. As the usage of online shopping increases, payment systems are also provided with secured transactions and reduced cash transactions. One of the most popular forms of e-payment methods are credit and debit cards. There are some alternative payment methods such as net banking, electronic wallet, smart card, bitcoin wallet. Security is an indispensable candidate for any transactions that take place over public network. Following are some of the essential requirements of any e-payment transactions: confidentiality, integrity, availability, authenticity, nonreputability and encryption. To enhance the security, encryption can be done in a very efficient and realistic way to preserve the information being communicated over public network, that is, the sender encrypts the information with a secret code so that only the corresponding receiver can decrypt it using the same code. Likewise, to make any purchase on Internet, the online user needs to share some secret information through public links. If these details are transferred as a plain text, then there is a chance of eavesdropping. This is because any person listening to network can gain access to such sensitive data like card numbers and type and also the complete details of card holder. Denial of service attacks, impersonation attacks, replay attacks, and spoofing attacks are among the additional dangers. The customer may lose faith if the system compromises on his privacy issues. To improve the security of e-payment system, to resist these types of attacks a secured authentication protocol for a payment system should be designed.

18.2 RELATED WORK

A light-weight authentication scheme is used in any applications, because simple symmetric cryptography primitives are used which requires less computations but produces improved results than others. In RFID (Radio Frequency Identification) systems, RFID tags are used for automatic identification. But it gives space for some security issues like easy access of RFID tags by an adversary leading to privacy and forgery problems. In addition, the computational capacity of RFID tags is also limited. To overcome such problems, "Gope and Hwang"[1] introduced a new realistic lightweight authentication scheme for the RFID system which protects security features such as untraceability, forward security, and anonymity in RFID tags. To handle large databases in the environment of Internet of Things (IOT), Cloud Computing has been emerged as a key technology. User authentication in Wireless Sensor Network (WSN) is a severe security issue, since sensor node has limited storage and computing capabilities. Thus, designing a lightweight authentication protocol is very difficult. Gope et al.[3] developed for WSN a realistic lightweight authentication system that enables user anonymity, forward/backward secrecy, complete forward secrecy, and untraceability. Ruhul et al.[2] introduced a novel "lightweight user authentication and key agreement scheme" for WSN which provides "session key agreement" and "mutual authentication property." Since security of data is very crucial for the RFID system, an ultra-lightweight validation convention called Succinct and Lightweight Authentication Protocol (SLAP) has been created in.[4] This protocol only consists of operations like XOR, left rotation, and conversion where these are very easy to implement in tags. Li et al.[5] created an authentication system based on smart cards. Kumari et al.[6] presented a biometric-based authentication system. Jiang et al.[7] proposed a three-factor authentication protocol.

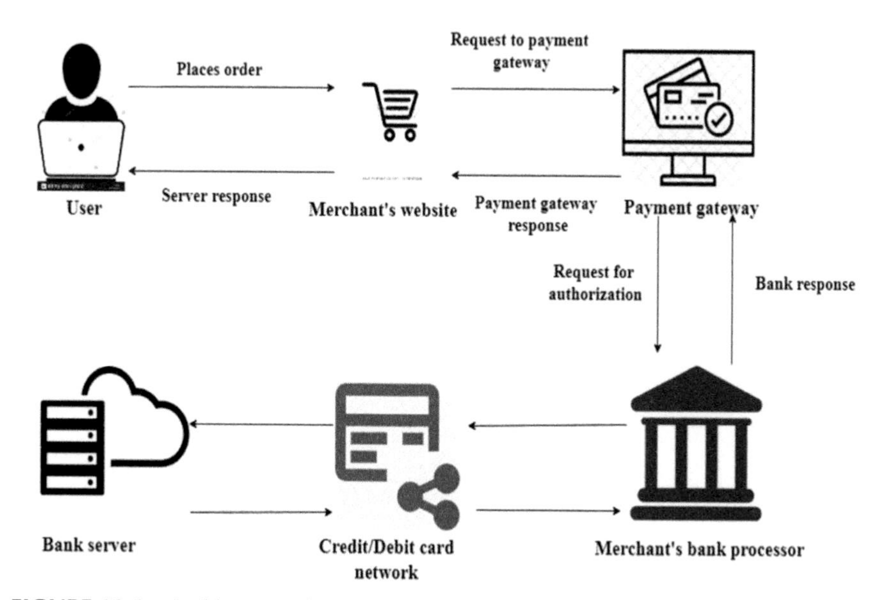

FIGURE 18.1 Architecture of payment system.

18.3 PROPOSED SCHEME

Our proposed authentication scheme for an electronic payment system consists of three phases, namely, user registration phase, payment gateway registration phase, and mutual authentication and session key distribution phase.

18.3.1 USER REGISTRATION PHASE

In this phase, the user registers its details such as its user id, password, and biometrics with the bank server. The bank server encrypts the user credentials and saves it in the smart card and then sends it back to the user.

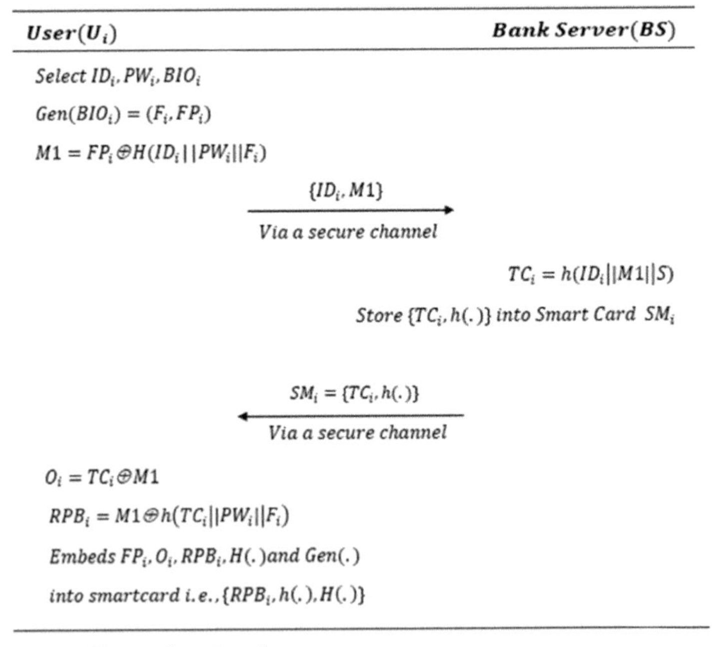

FIGURE 18.2 User registration phase.

18.3.2 PAYMENT GATEWAY REGISTRATION PHASE

Then the payment gateway should be registered with the bank to make the payment securely via the payment gateway. In this phase initially the payment gateway registers its id with the bank server. The bank server encrypts it and keeps it secret between the bank server and the payment gateway.

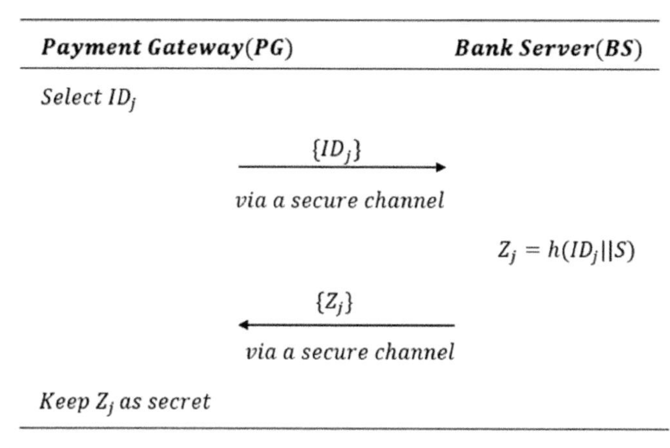

FIGURE 18.3 Payment gateway registration phase.

18.3.3 MUTUAL AUTHENTICATION AND SESSION KEY DISTRIBUTION PHASE

In this phase, the user encrypts its credentials and sends it to the payment gateway for logging into the bank server. Now payment gateway encrypts its credentials and sends both user credentials and its credentials to the bank server. Finally, the bank server verifies both the credentials and then sends the authentication message to the payment gateway and the user.

User(U_i) — Payment gateway(PG)

U: Input ID_i', PW_i', BIO_i'

computes $F_i' = Rep(BIO_i', FP_i)$

$M1' = TC_i \oplus O_i$

$RPB_i' = M1' \oplus H(TC_i||PW_i'||F_i')$

Check if $RPB_i' = RPB_i$

Generate $Ee \in Z_n$

$Eia = Ee.P$

$DID_i = ID_i \oplus h(E_{ia})$

$Qq = h(ID_i||E_{ia}||O_i)$

$Msg1 = \{DID_i, E_{ia}, Qq, M1\}$ → (Public channel)

FIGURE 18.4 Login phase.

FIGURE 18.5 Mutual authentication and distribution of session key.

18.4 INFORMAL SECURITY ANALYSIS

18.4.1 ANONYMITY

U_i's identity is included in $DID_i = ID_i \oplus h(E_{ia})$ and $AID_i = ID_i \oplus h(F_{ja})$ where $E_{ia} = Ee.P, F_{ja} = F_f.P$ in our scheme. If adversary wants user's real identity, he needs to compute E_{ia} from (Ee,P) and F_{ja} from (F_f,P) which means that he needs to compute the ECDH problem, else it is impossible to get

U_i's identity. Likewise, *PG*'s identity is included in $DID_j = ID_j \oplus h(F_{ja})$ and $BID_j = IDj \oplus h(E_{ia})$ where $E_{ia} = Ee.P, F_{ja} = F_f.P$ in our scheme. If the adversary wants payment gateway's real identity, he needs to compute F_{ja} from (F_f,P) and E_{ia} from *Ee,P*. It implies that he will have to do some more math. ECDH problem, else it is impossible to get *PG*'s identity.

18.4.2 MUTUAL AUTHENTICATION

Bank server (*BS*) can authenticate U_i and *PG*'s by verifying (*Qq*, *Xx*) respectively. If both of them are correct then *BS* generates the authentication code (*Yy*, *Ww*) and transfers it to the U_i and *PG* for future verification. With the help of Bank Server (*BS*) *PG* could authenticate the *BS* and U_i by checking the validity of *Yy*. And U_i can be able to authenticate *BS* and *PG* by checking the validity of *Ww.*

18.4.3 FULL FORWARD SECRECY

In the mutual authentication phase, both U_i and *PG* should agree with shared session key $KA_{ij} = h(ID_i \| ID_j \| Ee.F_{ja}) = h(ID_i \| ID_j \| Ee.Ff.P) = h(ID_i \| ID_j \| Ff.E_{ia}) = KA_{ji}$. Even if an adversary intends to compute session key, he requires real identities of U_i and *PG*. As well as he has to compute *Ee,Ff,P* from $E_{ia} = Ee.P$ and $F_{ja} = Ff.P$. As we understood before he needs to solve the ECDH problem to get those parameters. Otherwise, he or she knows the long-term secret parameters, she cannot deliver a substantial meeting key.

18.4.4 RESISTANCE TO PRIVILEGED USER ATTACK

In this phase, U_i submits *M*1 to *BS* instead of PW_i. Then the privileged user of *BS* cannot compute PW_i from $M1 = FP_i H(ID_i \| PW_i \| F_i)$ because of the uniqueness of F_i and secure one-way bio-hashing. By the way, our scheme resists privileged user attack.

18.4.5 RESISTANCE TO MASQUERADE ATTACK

During mutual authentication, the user initially generates a legal request $Msg1 = \{DID_i, E_{ia}, Qq, M1\}$. To masquerade user, an adversary needs to compute $Qq = h(ID_i \| E_{ia} \| O_i)$. We have already shown that the adversary cannot get correct Oi and Qq even adversary has any two of three factors such as smartcard, password, or biometric information.

18.4.6 RESISTANCE TO REPLAY ATTACK

Let us assume that an adversary intercepts the message $Msg1$ = $\{DID_i, E_{ia}, Qq, M1\}$ in an intention to impersonate U_i and replay back to PG. But PG will be aware of this replay attack, while verifying $Zz = h(BID_j \| Ww \| F_{ja} \| Tt \| ID_i \| ID_j \| KA_{ij})$. So given that the adversary has no idea about the nonce $\{Ee, Ff\}$, the adversary cannot be able to generate the valid session key $KA_{ij} = h(ID_i \| ID_j \| EeFfP)$

18.4.7 RESISTANCE TO PAYMENT GATEWAY SPOOFING ATTACK

To impersonate PG, an adversary needs $Z_j = h(ID_j \| S)$ to generate verification challenge $Xx = h(DID_i \| E_{ia} \| Qq \| F_{ja} \| ID_j \| Z_j)$. Here, $h(\cdot)$ is a secure one-way hash function and S is kept secret by BS, spoofing the payment gateway is challenging.

18.4.8 RESISTANCE AGAINST DOS ATTACK

In DOS attack, the attacker may prevent to establish the secure communication in between user and payment gateway, payment gateway and bank server. Our proposed protocol resists against Dos attack, since we are using an EAP-IKEv2 (which is Extensible Authentication Protocol—Internet Key Exchange Version 2) which provides mutual authentication and establishes session key.

18.5 SIMULATION RESULTS USING AVISPA

The simulation result of the suggested method in AVISPA is provided in this section. AVISPA is a robust security verification tool that may be used to ensure that an authentication mechanism is secure.

```
% OFMC
% Version of 2006/02/13
SUMMARY
  SAFE
DETAILS
  BOUNDED_NUMBER_OF_SESSIONS
PROTOCOL
  /home/span/span/testsuite/results/fully.if
GOAL
  as_specified
BACKEND
  OFMC
COMMENTS
STATISTICS
  parseTime: 0.00s
```

18.6 CONCLUSIONS

Our proposed three-factor authentication scheme is designed using ECC which gives low computation and communication costs. In addition, AVISPA simulation results show that the proposed system can survive both passive and aggressive assaults. In addition, informal security analyses were performed to prove that the proposed scheme counters various malicious attacks like masquerade attack, privileged user attack, and replay attack. In addition to this, our proposed work ensures user nameless, mutual authentication and secrecy in full forward. In the future, we will minimize the computational complexity of our protocol without compromising security properties.

ACKNOWLEDGMENTS

This part of this research work is supported by Department of Science and Technology (DST), Science and Engineering Board (SERB), Government of India under the ECR grant (ECR/2017/000679/ES).

KEYWORDS

- **mutual authentication**
- **biometrics**
- **smart card**
- **elliptic curve cryptosystem (ECC)**
- **AVISPA**

REFERENCES

1. Gope, P.; Hwang, T. A Realistic Lightweight Authentication Protocol Preserving Strong Anonymity for Securing RFID System. *Comput. Secur.* **2015,** *55*, 271–280.
2. Amin, R.; Biswas, G. P. A Secure Light Weight Scheme for User Authentication and Key Agreement in Multi-Gateway Based Wireless Sensor Networks. *Ad Hoc Netw.* **2016,** *36*, 58–80.
3. Gope, P.; Hwang, T. A Realistic Lightweight Anonymous Authentication Protocol for Securing Real-Time Application Data Access in Wireless Sensor Networks. *IEEE Trans. Ind. Electr.* **2016,** *63* (11), 7124–7132.
4. Luo, H.; Wen, G.; Su, J.; Huang, Z. SLAP: Succinct and Lightweight Authentication Protocol for Low-Cost RFID System. *Wireless Netw.* **2018,** *24* (1), 69–78.
5. Li, H.; Li, F.; Song, C.; Yan, Y. Towards Smart Card Based Mutual Authentication Schemes in Cloud Computing. *KSII Trans. Internet Inf. Syst. (TIIS)* **2015**, *9* (7), 2719–2735.
6. Kumari, S.; Li, X.; Wu, F.; Das, A. K.; Choo, K. K. R.; Shen, J. Design of a Provably Secure Biometrics-Based Multi-Cloud-Server Authentication Scheme. *Fut. Gen. Comput. Syst.* **2017,** *68*, 320–330.
7. Jiang, Q.; Zeadally, S.; Ma, J.; He, D. Lightweight Three-Factor Authentication and Key Agreement Protocol for Internet-Integrated Wireless Sensor Networks. *IEEE Access* **2017,** *5*, 3376–3392.

CHAPTER 19

DEC-GA: GENETIC ALGORITHM-BASED DETERMINISTIC ENERGY-EFFICIENT CLUSTERING PROTOCOL FOR IOT TRANSACTIONS IN CUSTOMER MANAGEMENT TOOLS

INDU DOHARE and KARAN SINGH

School of Computer and Systems Sciences, Jawaharlal Nehru University, New Delhi, India

ABSTRACT

Deterministic energy-efficient clustering protocol (DEC) is a dynamic, distributive, and self-organized method to reduce the energy consumption in the network. The DEC protocol promises a better and fixed number of Cluster Heads (CH) selections based on the remaining energy in its every round. DEC has various demerits like non-consideration of intra-cluster distance, disregard of the degree of the node. To address these issues of DEC, this chapter proposes Genetic Algorithm (GA)-based Deterministic Energy-Efficient Clustering Protocol, DEC-GA. In this work, K-means clustering is used for initial clustering on the homogeneous network and hybrid DEC-GA is used for CH selection. MATLAB-based recreation output shows that DEC-GA outwits on conventional DEC and improves the lifetime of the network. The performance matrices are evaluated and the calculated results indicate a significant improvement over the conventional protocols.

Advanced Computer Science Applications: Recent Trends in AI, Machine Learning, and Network Security. Karan Singh, PhD, Latha Banda, PhD & Manisha Manjul, PhD (Eds.)

19.1 INTRODUCTION

At the base degree of IoT design, all devices by and large structure a remote sensor network that detects the atmosphere and gathers the information from it. Sensor hubs can accomplish correspondence among them or to unceasing base stations. These sensors play out the detection of information, handling of information and likewise accomplish steering of the information.[1] By and large, sensor hubs are sent in a unattended and assorted antagonistic network that makes the force stockpiling basic and makes it extremely challenging to re-energize them for reuse.[2] In this manner, for supporting the power of sensor organization, elongation of organization lifespan, better organization correspondence, preservation of organization's energy, and saving the adaptability of hubs are required.[3]

The network can be divided into leveled or flat design wherein progressive construction prompts an alternate utilization of hubs, for example, hubs having high energy are utilized to measure and transmission of the message while hub having little energy used to detect the objective region in-line. A portion of the various hierarchal routing protocols are DEC,[4] LEACH,[5] SEP.[6] Cluster-based energy effective procedure ended up being a capable strategy for a major size organization. Figure 19.1 portrays clustered design of sensor network; it remembers sensor gadgets for tremendous number which are coordinated in grouped organization. Every source node transmits data to CH in all clusters and CH transmits this data to the sink which is established within the radio range.

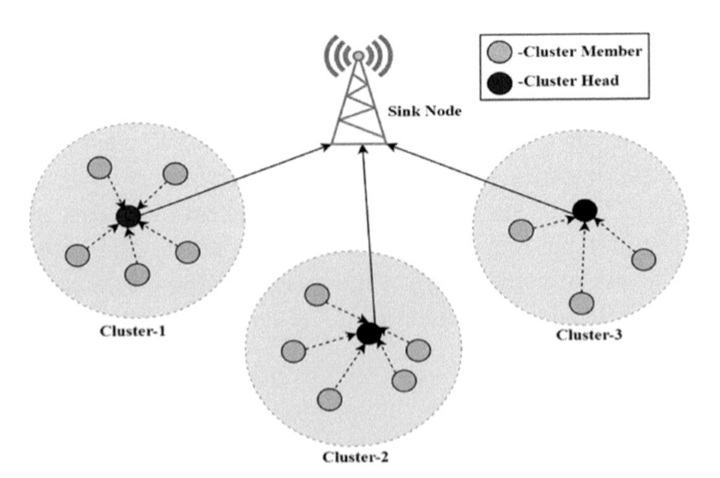

FIGURE 19.1 Clustered construction of sensor-assisted IoT.

The enormous amount of energy is used for the communication in WSN by each node and nodes consume energy in forwarding the information from the sender node to the sink node in terms of battery power. In clustered routing protocols, each cluster has some of the nodes grouped together and each cluster has one of the head nodes identified as a CH. CH collects the data from neighbor nodes and transfers these data packets with the minimum cost of transmission to the sink node or base station.[3]

A few methods have been produced for power proficient organization in that grouping-based protocol is ended up being an effective procedure for a major size organization. Sensor network climate can be of progressive or in a level construction in which progressive design prompts an alternate utilization of hubs, for example, nodes with high battery power are helpful in transmitting the data when hub consuming lower energy used to detect the objective region data. A portion of the various leveled and homogeneous directing calculations are LEACH,[5] PEGASIS,[7] and HEED.[8] In a heterogeneous organization structure, SEP[6] utilizes typical and advanced two sorts of hubs, in which energy circulation is different as per the hubs.

Genetic Algorithm (GA) is an optimization method that solves the problem on the basis of natural selection process with a large number of solutions.[9] GA randomly generates the population as chromosomes and length of each chromosome is equal. This protocol viability is in tackling NP issues. In this chapter, we proposed an upgraded capacity of DEC by GA. This chapter initially partitioned the organization by the utilization of k-implied grouping and then after DEC-GA it is used for group head choice. It ensures a steady number of CH in each cycle.

The rest of the chapter is divided into six sections: Section 19.2 analyzes the literature of WSN directing conventions and foundation and summary about GA. Section 19.3 gives a concise portrayal of the framework model depiction. Section 19.4 discusses the proposed model. Recreation results and relative investigation are outlined in Section 19.5. At last, our work and future not really set in stone in Section 19.6.

19.2 LITERATURE REVIEW AND BACKGROUND

Different clustering and routing strategies are set up in WSN and these methodologies demonstrate the exhibition of WSN yet at the same time confronting a portion of the disadvantages. Dispersed clustering-based steering convention LEACH[5] utilizes randomized pivot for the determination of CH that turns in all hubs. CH in each bunch make and give TDMA to all its part hubs for adjusted power utilization and to stay away from the crash. LEACH does not discuss about the outstanding energy in the determination of CH and group, likely any hub can be CH. The SEP[6] protocol is the extension of the LEACH algorithm in terms of energy distribution among nodes. SEP divides the energy into two levels of hierarchy. Advanced nodes more often have probabilities to become CH as they have high energy than normal nodes. The central version of LEACH is (LEACH-C),[11] in this, all node sends their local statistics like location, energy status to the base station which classified these nodes for the clusters and cluster head. This protocol produces better results in energy consumption and packet delivery rate than LEACH. Another extended version of LEACH which uses iteration for the formation of the cluster is Hybrid Energy-Efficiency Distributed Clustering (HEED).[8] It is a clustering protocol that uses the residual energy of every node for CH selection.

DEC[4] is a deterministic clustering algorithm that utilizes the leftover power for the choice of CH. This calculation limits the doubts during the time spent CH determination. In this convention, the arrangement stage is improved yet the consistent state stage has been set like that of LEACH convention. The nodes having the base excess power are chosen as a CH.

GA[9] is proved to be superior than the existing optimization problem and also shows the stability in solving the combinatorial optimization problems. GA is used in WSN to solve many problems, in solving the energy problem of sensor network it proved to be efficient. GA initializes the population as the same length of the chromosomes and evaluates the fitness value for every chromosome. Chromosome with a better objective value is closer to the optimal solution. To attain a better result mutation is applied which randomly selects the chromosomes to evaluate a better solution. The next population is generated with the help of crossover and mutation. The new generation always selects some better solution from the previous solution to confirm that new solution is at least as fit as previous. The entire process is repeated till the stopping criteria do not meet.

Customarily in WSN, GA plays an extraordinary part as it is utilized in bunching, ideal hub organization, hub situating, and information accumulation. A significant number of the creators use GA for grouping and bunch head determination. In this setup, in LEACH-GA,[10] is a hybrid technique of genetic algorithm and LEACH. In this, GA is used to predict the optimal value of probability effectively. In[11] the author proposes a clustering technique for heterogeneous network based on the genetic algorithm. The performance of the algorithm has been evaluated on the basis of first node die and last node die. In Ref. [12], author proposes LEACH like genetic algorithm based clustering method for the lifetime maximization of the network. The algorithm has two phases: set-up stage and steady phase. The set-up phase is initial cluster phase and in steady phase all the clusters remain the same, only cluster head rotation takes place.

Motivated by these research strategies, we present a methodology that can create a further developed life expectancy and guarantees that the best reasonable group head will get chosen. This methodology conveys a more perfect way out for power exhaustion in nodes empowered IoT. The article can be outlined in the accompanying focuses:

(1) Cluster head is chosen based on the cost capacity of every node and each step is self-deciding of the ensuing round like DEC.
(2) DEC-GA guarantees a stable *Nopt* CH is selected.
(3) The k-means is used to create a group of nodes in the first round of simulation and these groups are steady in different rounds, so the proposed model decreases the overhead expense of making clustering in further rounds.
(4) Genetic algorithm is applied in various steps of the DEC-GA model to select the cluster head by the use of crossover and mutation operators.

19.3 SYSTEM MODEL

This segment portrays the energy model and the network model which are used in the simulation.

19.3.1 NETWORK SCENARIO

The elements of the simulation region are taken as $100 \times 100m^2$ and N nodes are randomly organized in this region. The sink node is deployed at (75,150) in the network. We have taken homogeneous network in which initial energy of the nodes is $0.5J$.

19.3.2 ENERGY MODEL

We deliberated a similar energy model for power consumption and information conglomeration energy as utilized in a portion of the prior investigations.[4] The scheme is planned at the physical layer of WSN that utilized ascertain energy loss of all nodes at the hour of correspondence of nodes. Over the distance(*d*), the proliferation method is applied among two stations, the intensification energy (*E*) is disseminated for free-space and over the multipath; it is multi-hop communication used for transferring a pack. So, the power ingesting can be considered for transferring (*b*) bit packet over distance *dis* as

$$(b, dis) = E_{Tx-elec}(k) + E_{Tx-elec}(b, dis)$$
$$\begin{cases} bE_{elec} + bE_{fs} dis^2 & if\ dis < d_0; \\ bE_{elec} + bE_{fs} dis^4 & if\ dis \geq d_0 \end{cases} \quad (19.1)$$

All the parameters that are curtailed in the equation are portrayed in Table 19.1. Here $d_0 = \sqrt{\frac{E_{fs}}{E_{amp}}}$. The power disbursed on the communication for getting a packet is

$$E_{Rx}(b) = E_{elec} b \quad (19.2)$$

For transporting the $b - bit$ data among nodes we expected a symmetric communication channel.

19.4 DEC-GA SCHEME

This part illustrates a definite clarification of our recommended model. The model is separated into three phases, in the first phase of the DEC-GA model, N_{Opt} numbers of nodes are randomly selected from all nodes so that selected nodes are not coming communication range of previously selected nodes. That indicates that DEC-GA fixes the number of CHs. The second phase is K-means clustering phase; in this phase whole sensor nodes are clustered around the nodes which are selected in phase 1 of the model. The third phase of the model is the main DEC-GA algorithm for the cluster head selection.

Each phase of the model is shown in Figure 19.2 and the description of the model is given in subsections.

19.4.1 N_{Opt} SELECTION PHASE

We named this phase as N_{Opt} selections phase because of the assurance of the specific number of N_{Opt} CH. N_{Opt} are fixed number of CHs. These N_{Opt} are not under correspondence scope of one another. For N_{Opt} choice first hub N_{Opt} haphazardly elected from developed hub as cluster head and for the next CH the hub that isn't coming in the scope of right off the bat chosen CH appoint as second bunch head. This cycle rehashed until N_{Opt} CH selected. This cycle functions distinctly for the first cycle of DEC-GA calculation. These CHs develop driven for the subsequent stage to group part choice.

19.4.2 K-MEANS CLUSTERING PHASE

K-Means is utilized to cluster the nodes of network area. In this phase N_{Opt} nodes are already identified from phase first of the proposed model; these nodes act as the initial cluster centers. Further, when $N - N_{Opt}$ nodes discover any nodes from N_{Opt} at least possible Euclidean distance $\sqrt{\left((x1-x2)^2+(y1-y2)^2\right)}$ it becomes the member of that node.

19.4.3 DEC-GA PHASE

Phase 3 is the main phase of the proposed model; in this phase we offered a complete DEC-GA algorithm 1 for CH election. The algorithm begins with the introduction stage, in this stage first and foremost work out the fitness value of every chromosome (eq 19.3). In the event that halting rules are fulfilled stop the emphasis, in any case repeat for $t + 1$. In every round the node that gave the best global result is selected as the CH for the cluster that is having a place. The proposed algorithm relies upon the accompanying fitness function:

$$\textit{Minimize}: \quad fit = c1 \times f1 + c2 \times f2 + c3 \times f3 \tag{19.3}$$

where $c1 + c2 + c3 = 1$, $f1$ is the residual power (eq 19.4) of node, $f2$ is the distance among cluster members, and $f3$ is the node degree. Every parameter can be determined by subsequent conditions:

$$f1 = \sum_{i=1}^{K} \frac{1}{E_{ni}} \tag{19.4}$$

$$f2 = \sum_{i=1}^{K} |CM_i| \tag{19.5}$$

The node degree (eq 19.5) demarcated as the summation of all the nodes is in its radio communication range.

$$f3 = \sum_{i=1}^{K} Dis(N_i, CM_i) \tag{19.6}$$

Intra-cluster distance (eq 19.6) can be demarcated as the average distance among node and its CMs. The objective value determined in the premise of these boundaries and this calculation runs in a few cycles and when a CH discovered lesser power than other, GA's factors, the whole process runs thus and CH is created.

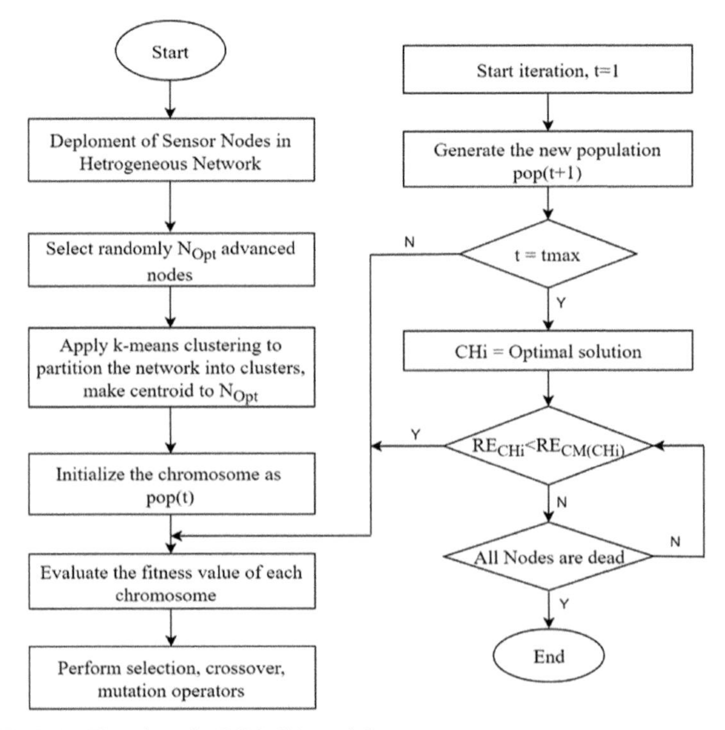

FIGURE 19.2 Flowchart for DEC-GA model.

Selection, Crossover, and Mutation operator of genetic algorithm plays a vital role in optimal solution generation. In selection operation combine Roulette wheel election method has been used to elect the chromosome with high fitness value as a parent. Chromosomes with best fitness value are also secured for the next generation. To generate new individual crossover is used in such a way so that it can cross neighboring individuals. If cluster head is affected during crossover a new cluster head selection takes place from cluster members only. The mutation digits are 0 to 1 and 1 to 0 in our work. These all process has been repeated till the maximum limit has been reached. Finally, the best chromosome with best fitness value has been selected for the CH.

Algorithm 1: Propose DEC-GA Algorithm for Clustering

Input:
Set of nodes $S = \{S_1, S_2, ..., S_N\}$
and N
is total number of Chromosomes
The measurement of atoms d
Output: N_{Opt}
number of CH.

Step 1) *Initialize chromosomes, pop(t)* $p_i, \forall i, j, 1 \leq i \leq N, 1 \leq j \leq d$
and fixed the N_{Opt}
number of CH.

Step 2) *Initially; randomly elect* $N_{Optn} = |N_{Opt}|$
nodes from the set of nodes that are not in the communication range of one another.
Step 3) *Apply K-means to isolate all nodes in clusters by consolidating* N_{Optn}
nodes from step 2.
Step 4) *For i* = 1 *to N*

Calculate objective function of each chromosome by eq (19.3)
End
Step 5) *Perform Selection, crossover and mutation operators.*
Step 6) *update population pop(t+1)*
Step 7) *for t* = 1 *to tmax*

Repeat step 5 and 6.
Endfor
Step 8) *For i* = 1 *to* N_{Optn}

For $j = 1$ *to* $|CM|_j$

If ($RE_{CHi} < RE_{CMij}$

)

Recap steps 4–7.

Endif

Endfor

Endfor

Step 9) *Repeat step (8) until all energy of all nodes finished.*

19.5 SIMULATION RESULTS AND PERFORMANCE ANALYSIS

This section assesses DEC-GA algorithm exhibition as far as network lifetime and remaining energy. This part likewise portrays the exhibition lattices in a word. At long last, a far-reaching result investigation of the presented scheme model with near outcomes to standard calculations is illustrated. Table 19.1 depicts the significant boundaries utilized in the reproduction of DEC-GA model. The organization is planned and reproduced on MATLAB.

The suggested scheme contrasted the schemes from literature DEC[4] and LEACH-GA.[10] The exhibition frameworks are utilized; network lifetime and leftover energy of the hubs. We extend network lifespan as far as Half Node Die (HND) and First Node Die (FND). The outcomes accomplished by the implementation of the suggested DEC-GA calculation are portrayed in Figures 19.3 and 19.4. Table 19.1 discusses the simulation parameters description.

TABLE 19.1 Simulation Parameter.

Parameter	Value
Network area	100*100 m
Number of nodes	100 nodes
Size of population	100
Chromosome length	100
Selection type	Roulette wheel
Crossover rate	0.7
Mutation rate	0.1
Sink location	(75,150)
E_{ele}	50 nJ/bit>

TABLE 19.1 *(Continued)*

Parameter	Value
E_{amp}	100 pJ/bit/m^2
b	4000 bits
D_{th}	30 m

19.5.1 PERFORMANCE EVALUATION IN TERMS OF NETWORK LIFESPAN

In our suggested model lifespan is estimated in half node die and first node die that are illustrated in Figure 19.3a and b. This shows that DEC-GA achieves a better lifetime than LEACH-GA and DEC as the first node and half of the nodes are dying earlier in these algorithms. The figures delineate the effectiveness of the proposed convention in improving the network lifetime. This is because the ensuing target work doesn't just uses remaining energy yet, but utilizes inter-cluster distance and the node degree.

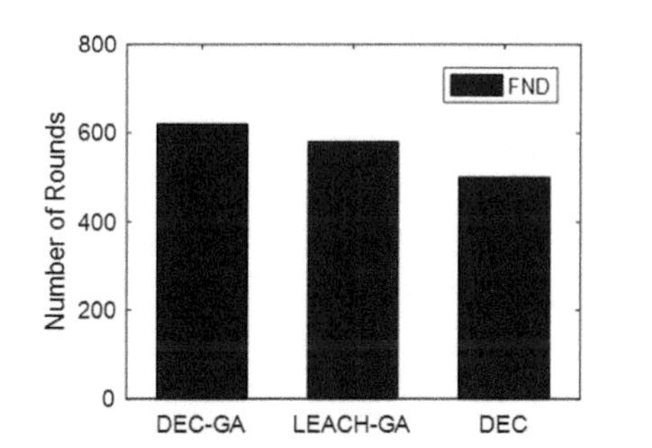

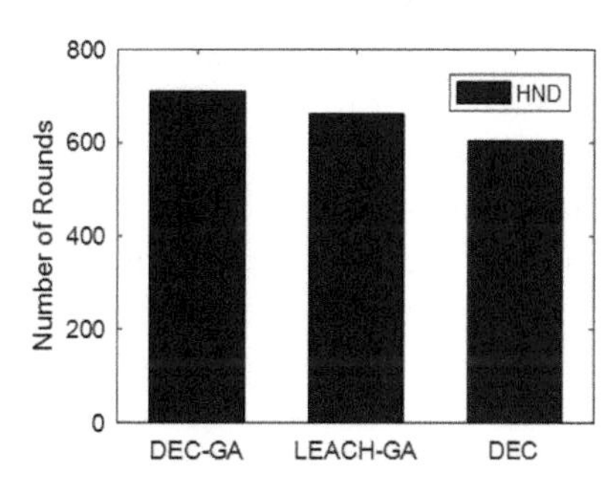

FIGURE 19.3 Network lifespan in terms of (a) FND, (b) HND.

19.5.2 PERFORMANCE EVALUATION IN TERMS OF REMAINING ENERGY

We evaluate the outcomes of the simulation in terms of the remaining energy. The assessment of DEC-GA is performed with DEC and LEACH-GA. It is seen that DEC-GA beats these current estimations in Figure 19.4. This is a direct result of the way that DEC-GA searches for a higher extra energy center to each pack in every round.

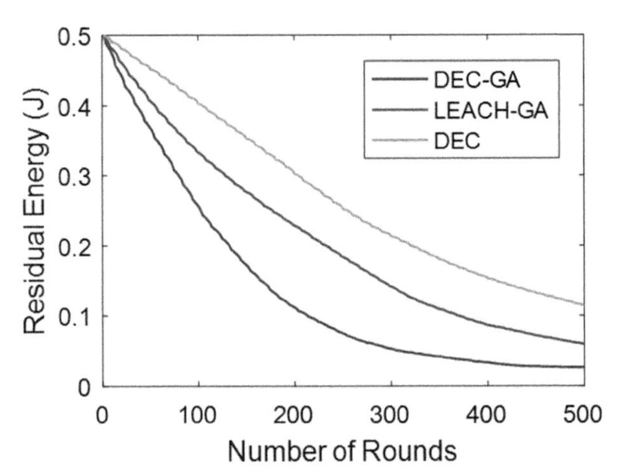

FIGURE 19.4 Performance evaluation in terms of remaining energy.

19.6 CONCLUSIONS

In this article, we present the DEC-GA model for cluster head selection to enhance the lifetime of the network. This is further developed; scheme has to get ridded of a few insufficiencies of the regular convention by captivating different factors in objective function separated from remaining energy. The baseline protocols utilize residual energy for the determination of CHs; our proposed DEC-GA algorithm upgrades the method of choice for cluster head by utilizing an objective function that relies upon the inter-cluster distance, residual energy of the nodes, and the node degree. The simulation result of propose algorithm DEC-GA demonstrates that energy depletion for the first node is achieved at around 605 rounds, while in the conventional (DEC) algorithm it is at around 450 rounds, so our algorithm performs better. Finally, the proposed model effectively expands the network lifespan against the DEC algorithm.

KEYWORDS

- **IoT**
- **wireless sensor networks (WSN)**
- **genetic algorithm**
- **deterministic energy-efficient clustering protocol (DEC)**

REFERENCES

1. Pantazis, N. A.; Nikolidakidirs, S. A.; Vergados, D. D. Energy Efficient Routing Protocols in Wireless Sensor Networks: A Survey. *IEEE Commun. Surveys Tuts.* **2013,** *15* (2), 551–591.
2. Bedi, G. Review of Internet of Things (IoT) in Electric Power and Energy Systems. *IEEE Internet Things J.* **2018,** *5* (2), 847–870.
3. Heinzelman, W.; Chandrakasan, A.; Balakrishnan, H. Energy Efficient Communication Protocol for Wireless Microsensor Networks. In *IEEE Proceedings of the 33rd Hawaii International Conference on System Sciences (HICSS '00)*, 2000.
4. Derohunmu, F. A. A.; Deng, J. D.; Purvis, M. K. A Deterministic Energy-Efficient Clustering Protocol for Wireless Sensor Networks. In *Proceedings of Seventh IEEE International Conference on Intelligent Sensors Sensor Networks and Information Processing*, 2011; pp 341–346.
5. Heinzelman, W.; Chandrakasan, A.; Balakrishnan, H. Energy Efficient Communication Protocol for Wireless Sensor Networks. In *Proceedings of Hawaii International Conference on System Sciences*; IEEE Computer Society, 2000; pp 3005–3014.
6. Smaragdakis, G.; Matta, I.; Bestavros, A. SEP: A Stable Election Protocol for Clustered Heterogeneous Wireless Sensor Networks. In *Second International Workshop on Sensor and Actor Network Protocols and Applications*, 2004.
7. Lindsey, S. PEGASIS: Power-Efficient Gathering in Sensor Information Systems. *IEEE Aerospace Conf. Proc.* **2002,** *3* (9–16), 1125–1130,
8. Younis, O. HEED: A Hybrid Energy-Efficient Distributed Clustering Approach for Ad Hoc Sensor Networks. *IEEE Trans. Mob. Comput.* **2004,** *3* (4), 660–669.
9. Yuan, X. A Genetic Algorithm-Based, Dynamic Clustering Method Towards Improved WSN Longevity. *J. Netw. Syst. Manage.* **2017,** *25* (1), 21–46.
10. Pal, V.; Singh, G.; Yadav, R. P. Cluster Head Selection Optimization Based on Genetic Algorithm to Prolong Lifetime of Wireless Sensor Networks. *Procedia Comput. Sci.* **2015,** *57*, 1417–1423.
11. Elhoseny, M. Balancing Energy Consumption in Heterogeneous Wireless Sensor Networks Using Genetic Algorithm. *IEEE Commun. Lett.* **2014,** *19*, (12), 2194–2197.
12. Bayraklı, S.; Erdogan, S. Z. Genetic Algorithm Based Energy Efficient Clusters (Gabeec) in Wireless Sensor Networks. *Procedia Comput. Sci.* **2012,** *10*, 247–254.

CHAPTER 20

HASHTAG RECOMMENDATION SYSTEM TO TRACK STREAMING NEWS AND CAPTURE DYNAMIC EVOLUTION TO ENSURE HIGH COVERAGE (CADENCE)

K. GEETHA, N. SASIKALADEVI, DEEBIKA J., and VINDHYA GURU RAO

School of Computing, SASTRA Deemed University, Thanjavur, Tamil Nadu, India

ABSTRACT

Social media is becoming an increasingly important tool for journalists to find news content and to distribute the stories to their audience. Social media often releases news about current affairs that are trending, which the mainstream media might not have been aware of. The newsrooms hence rely on social media feeds to broadcast and publish the news. For the media to track news, hashtags are used. The term "Hashtag" allows users to group messages from the Internet and extract information that pertained to a specific theme of context. The group of words can be combined and prefixed by a hashtag. Hashtags contextualize the entire story in minimum words, which helps in keeping track of the news. In this work, we propose a model to find the most suitable hashtag for news articles and tweets using a probabilistic approach. The proposed probabilistic neural network (PNN) model provides

Advanced Computer Science Applications: Recent Trends in AI, Machine Learning, and Network Security. Karan Singh, PhD, Latha Banda, PhD & Manisha Manjul, PhD (Eds.)

the probability metric for each hashtag that can be appropriate for a specific event. As a result, for data segmentation, this model can integrate news and social-stream in real time and further processing. The proposed model can deliver recommendations and narrate the incidents that happened before and after the current hot event. This can aid the media industry to streamline its delivery. The recommended hashtags can be used by the journalist for associating the news stream with the social stream. This strong association can address challenging real-time issues and can capture the dynamic evolution of news. Our model is implemented using news articles from RSS feeds and tweets from Twitter Streaming API.

20.1 INTRODUCTION

Microblogging plays an ardent role in promoting and universalizing news content. People broadcast their views as microblogs.[1] Twitter is one of the most prevalent social media platforms for information distribution and gathering. A recent study says that around 336 million users are active on Twitter, and 76 percent of these people use Twitter for news. Due to the vast amount of users Twitter has helped predict the trends.[4] For media such as Newsrooms, it is very important to keep track of the current events and trends. Various newsrooms are using Twitter as a method for obtaining information.[2] Almost every story on Twitter is associated with a hashtag. These hashtags can be regarded as the description of the story in a word or two prefixed with a hashtag. For example, #MeToo. In literature, hashtags have been used for advocacy and story tracking.[5–7] In this work, we propose a hybrid recommender system that recommends the best hashtags to search with to get the most suitable tweets for the current news article. Hence using these recommended hashtags, the entire story can be tracked.

20.2 RELATED WORK

Many of the researchers doing work have gone into developing a hashtag suggestion system for Twitter. In,[8] a low-rank weighted matrix factorization-based method has been introduced to suggest hashtags purely depend on a history of the users rather than the context of the tweets. A ranking method called the HF-IFU, which is a variant of the TF-IDF, has been proposed to overcome the challenges such as data sparseness and relevance.[9] Few other works[10–12] regard hashtags as topics and map the other content to them using content similarity. Many works in literature use LDA (Latent Dirichlet Allocation) and LSI (Latent Semantic Indexing) to capture the topics. Alvari[3] proposes a matrix factorization-based approach for Twitter hashtag recommendation.

Twitter hashtags have been mapped with real-time streaming news with excellent precision and coverage.[7,13] A Learning To Rank (L2R) model has been proposed in[13] which recommends the best set of hashtags for news articles. Sixteen classification algorithms have been compared in[7] and it is found that the pointwise Random Forest approach outperforms other pair-wise and list-wise approaches. In this work, a set of relevant hashtags are identified using the Random Forest-based L2R model and the probability of the most suitable hashtag among the set of relevant hashtags is identified using a probabilistic neural network.

20.3 PROPOSED METHODOLOGY

We propose a hybrid recommender system to recommend the best possible hashtags to relate news and tweets.

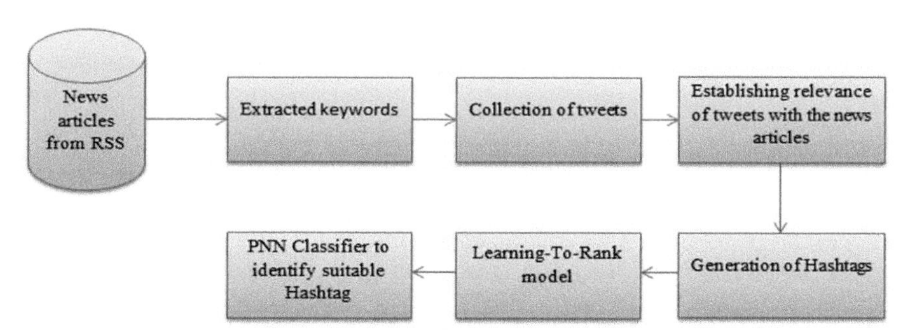

FIGURE 20.1 System architecture.

20.3.1 COLLECTING NEWS ARTICLES FROM RSS FEEDS

RSS (Rich Site Summary) is a web feed that can be used to retrieve updates of online content in a computer-readable format. RSS feeds have data in a semi-structured format, that is, XML. Six reputable newspaper feeds are considered for data collection. The feeds are collected using python packages such as feedparser and Newspaper.

20.3.2 EXTRACTION OF THE FEATURES FROM FEEDS

Query generation mechanism is very important to boost the number of relevant tweets for the news article. Shi[7] presented four alternative query-generating techniques, namely, (i) TF-IDF +POS, (ii) TF-IDF+POS+NER, (iii) Alchemy API, and (iv) URL. Alchemy API is a commercial tool that consumes a lot of time. We consider the TF-IDF +POS tagging approach in this work as it is suggested to give good results next to Alchemy API.

20.3.3 TF-IDF

Processing unstructured data is complicated as it involves additional work of looking for a methodology to represent the unstructured in a structured format. TF-IDF is a weighing technique that provides weight for the words in the document based on their frequency and occurrence throughout the corpus. Term Frequency compares a word's frequency in a single text to its frequency throughout the whole corpus. Considering only frequency may not be accurate because words like "is", "the" occur in more frequency than important words. To weigh down such words IDF (Inverse Document Frequency) is used. IDF is the size of the no. of documents to the no. of words containing that document. Hence this provides the rareness of the word. By combining both TF and IDF, we arrive at a suitable weight for the words. More the weight, more significant is the word.

$$TF\ (t,d) = \left(\frac{f_{(t,d)}}{no.of\ words\ in\, d} \right) \quad (20.1)$$

$$IDF\ (t)\ = ln\left(\frac{n_{documents}}{n_{documents\ containing\ t}} \right) \quad (20.2)$$

$$TF\text{-}IDF = TF\ (word)\ *\ IDF(word). \quad (20.3)$$

20.3.4 PART OF SPEECH (POS)

The technique of identifying a word in a text as belonging to a certain part of speech is known as POS tagging. This methodology selects the top nouns/ phrases in the document by giving precedence to noun phrases, proper nouns, and other nouns. The chosen nouns are ranked based on the TF-IDF scores.

A comparison of TF-IDF and POS+TF-IDF methods is shown in the table. POS+TF-IDF is found to be good outcomes.

The top words from these articles are used as hashtags and are later recommended as suitable keywords for story tracking.

20.3.5 TWEETS COLLECTION

Twitter provides various API to extract data. We used the streaming API of Twitter to extract relevant tweets based on the keywords extracted with the previous method. With streaming API each article is allocated a certain amount of time to extract data. The tweet data collected are preprocessed using various methodologies as mentioned in Wisdom.[14]

20.3.6 ESTABLISHING RELEVANCE

The similarity measure is a distance measure that reflects the degree of closeness between two target objects. There are several similarity metrics. Choosing a suitable similarity metric is very important for our application as not a single similarity metric is suitable for all the works. Different similarity measure techniques are discussed in Huang.[16] According to the literature cosine and Jaccard coefficients are said to be the best approaches. We have considered the cosine and Jaccard similarity approach in this work. Different results were observed with both methods which are discussed in the next section of this work.

20.3.7 COSINE SIMILARITY

By calculating the cosine of the angle the two vectors create in respective dot product space, cosine similarity returns the similarity between them. If the angle is zero, then it is most similar, the larger the angle, the least the similarity. This method is independent of the vector length as two vectors can be of any length.[15] The formula for the cosine similarity coefficient is given by Eq. (20.4)

$$cos(\Theta) = \frac{A.B}{|A||B|} = \frac{\sum_{i=1}^{n} A_i B_i}{\sqrt{\sum_{i=1}^{n} A_i^2}\sqrt{\sum_{i=1}^{n} B_i^2}} \tag{20.4}$$

where "Ai and Bi are the components of vector A and B" respectively.

20.3.8 JACCARD SIMILARITY

The Jaccard coefficient, sometimes called the "Tanimoto coefficient," calculates the similarity value as the ratio of object intersection to object union. The Jaccard index is referred to as the Jaccard similarity coefficient and intersection over union is a measure for calculating the likeness and difference of sample sets. This can be calculated by given formula (20.5).

$$J_{(A,B)} = \frac{P \cap Q}{P \cup Q} \tag{20.5}$$

J(P, Q) refers to the Jaccard similarity among any two P and Q documents.

The Jaccard distance estimates the uniqueness between test sets, corresponding to the Jaccard coefficient, and is obtained by removing the Jaccard index value from 1. This is shown in (20.6).

$$d_j = 1 - J_{(P,Q)} \tag{20.6}$$

where d_j is the distance and J(P, Q) refers to the Jaccard similarity between any two documents like P and Q.

The factor relevance is used as a characteristic for Learning to Rank Model. The relevance here is established through the similarity coefficient. The similarity matrix is constructed between both news articles and each tweet of that news article. The value is considered for the feature label of relevance class based on the threshold value.

20.3.9 LEARNING TO RANK MODEL

Learning To Rank (L2R), which is also known as "Machine Learned Ranking," is used in the construction of ranking models for information retrieval systems. The significance of this method is that it can be reused several times after training with few manually labeled data. There are different types of L2R models namely listwise, pairwise, and pointwise.

Listwise approach: This approach is the most complicated compared to the other two types. The algorithms in this approach directly look into the list of documents and come up with an optimal ordering.

Pairwise approach: This works with pair and cost function. The algorithms look at each pair of documents with the cost function at every instance and come up with the optimal pair of rank and compare it with the ground truth.

Pointwise approach: The pointwise approach essentially takes a single document and trains the classifier on it and predicts the relevance for the current query. Hence, in this approach score of each document is independent of other documents of the query.

According to the literature, the Pairwise L2R approach is found to give better results compared to the rest two approaches. For real-time streaming data, the pairwise approach is found to give the best result due to less training data.[7] Since it is real-time, the pointwise approach seems more relevant as the ranking is based on the current query. We have implemented the pairwise RandomForest L2R method in this approach to get the set of relevant words that can be suggested for story tracking.

Ensemble learning is a methodology that combines various types of algorithms or the same algorithm more times to arrive at suitable predictions. Random Forest is one such ensemble technique for supervised learning. We use the inbuilt sklearn python package for implementation as it is found to give better results than rankLib according to Shi.[7]

20.3.10 PNN CLASSIFIER

The simplest type of artificial neural network is a feed-forward neural network (FFNN), in which connections between nodes do not form a cycle. PNN is a FFNN, developed by D. F. Specht.[17] It is a realization of the statistical algorithm Kernal discriminate analysis, which has all operations that are categorized into a four-layer "multi-layered feed-forward network." The significant meaning of this neural organization based classifier incorporates: fast training method, an intrinsically parallel structure, ensured to join to an ideal classifier as the size of the preparation set increments, and extra preparing tests can be embedded or taken out without vast retraining.[18] The basic architecture of PNN consists of four layers, an i/p layer, pattern layer, summation layer, and o/p layer. Figure 20.2 depicts the architecture of a PNN where ×1, ×2, ×3, ×4 represent the inputs.

Input layer (i/p): The input layer nodes are passive, that is they do not amend changes in the data. They generate multiple values at the output from a single input value. Each neuron is completely connected to the neurons of the next layer.

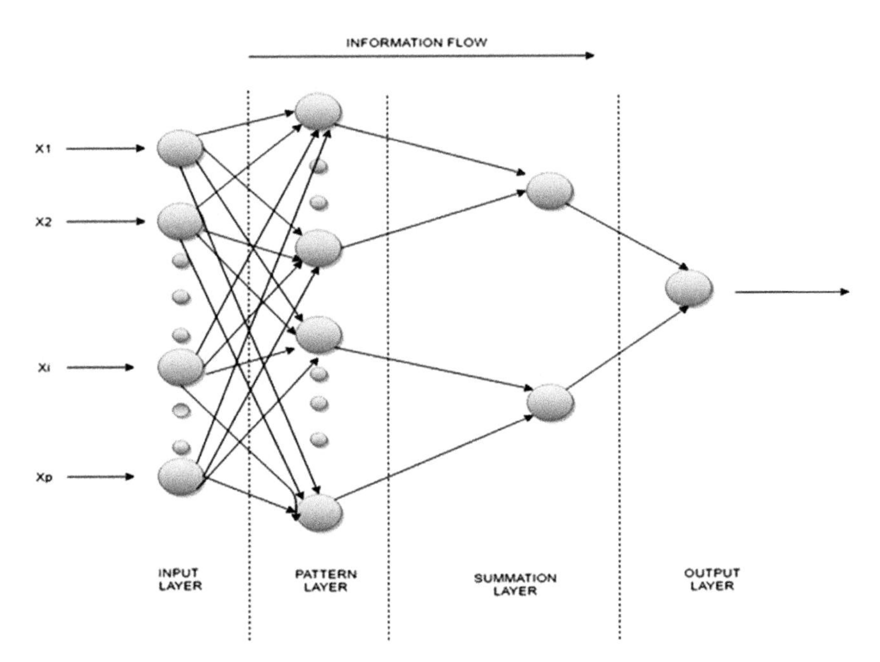

FIGURE 20.2 PNN architecture.

Pattern layer: This is a hidden layer that contains the pattern units. The number of patterns in this layer equals the classes of the training set. Each pattern node forms a product of the input pattern vector x with the weight vector W_i, $Z_i = xW_i$, which are given as an input for an activation function such as sigmoid given by $S(x) = \frac{1}{(1+e^{(-x)})}$

Summation layer: For each given class, the output of the pattern layer is given as input in this layer.[19] It merely sums the inputs from the pattern layer, which correspond to the class from which the pattern for training was selected. It provides the probability for input vector x to belong to class C_i.

Output layer: One neuron in the output layer makes a classification judgement for the input using the Bayes classification algorithm. The output is calculated by prob for class

$$x = C(x) = \arg\left[\max\ j = 1..m \left\{ \frac{1}{(2\pi)^{d/2}\sigma^d} \frac{1}{N} \sum_{j=1}^{N_j} \exp\left(-\frac{\left(x - x_j^i\right)^t \left(x - x_j^i\right)}{2\sigma^2} \right) \right\}\right. \quad (20.7)$$

where m is the number of classes.

The most vital advantage of PNN compared to other classifiers is that training is instantaneous. The probabilistic neural network proposed in Specht[17] can be used to estimate the posterior probabilities of each class. The activation function used can be of linear or nonlinear type. Nonlinear activation functions are the most used, as the linear activation functions are not confined to a range. The range of nonlinear functions is from ∞ to $+\infty$. For each class label, we want the posterior probability of each cluster. The sigmoid function is suitable for models where we have probability as an output. Sigmoid exists between 0 and 1, and probability also ranges from 0 to 1. Hence sigmoid is the suitable transfer function.

The set of important words extracted using the query generation method are considered class labels and they are trained with the suitable news data. Hence for every tweet, the model gives out the most suitable keyword that can be used for story tracking.

20.3.11 ALGORITHM

Input: News articles collected from RSS feeds.
Output: list of keywords for each news article.
for each news article.

(a) *Generate important word list using tf-idf + pos tagging scheme, say w=[w1,w2,..wn]*
(b) *Using streaming api, collect tweets using w for 40 seconds*
(c) *Calculate the similarity value between the news article and each tweet collected.*
(d) *Input the similarity column for random forest classifier to classify the set of words as w-relevant or w-irrelevant*
(e) *Train w-relevant with a news article in PNN*
(f) *Input the tweets to PNN to figure out the most suitable keywords using the probability*
end for.

20.4 EXPERIMENTAL ANALYSIS

a. Data Collection

News channels and newspaper proprietors that update news online allow users and applications to track news through RSS feeds. APIs are available today for easy access to news articles. We have considered few reputed news articles mentioned in Table 20.1.

TABLE 20.1 Dataset.

Newspapers	Rss feed links	Date
NDTV	"http://feeds.feedburner.com/NDTV-LatestNews"	18/10/2018
Times of India	"https://timesofindia.indiatimes.com/rssfeedstopsto-ries.cms?x=1"	12.03 PM
The Hindu	"https://www.thehindu.com/news/national/?service=rss"	
Deccan Chronicle	"https://www.deccanchronicle.com/rss_feed/"	
India Today	"https://www.indiatoday.in/rss/1206578"	
TheIndian Express	"https://indianexpress.com/feed"	

The important words resulting through TF-IDF and POS-Tagging are shown in Table 20.2.

TABLE 20.2 Resulting Important Words.

News article	Words
"MJ Akbar is the most high-profile person to exit his job in the wake of the #MeToo campaign.\n\nMJ Akbar, who is accused of sexual harassment, has filed a defamation case against journalist Priya Ramani, the first woman to name him in the growing #MeToo movement in India."	"Akbar," "Ramani," "MJ," "MeToo," "sexual," "name," "Ms," "capacity"
"Neena Gupta, AyushmannKhurrana and the cast of Badhaai Ho (Courtesy badhaaihofilm)\n\nCast: AyushmannKhurrana, Sanya Malhotra, Neena Gupta, Gajra Rao, SurekhaSikri\n\nDirector: Amit Ravindernath Sharma\n\nRating: 3.5 stars (out of 5)\n\nDirector Amit Ravindernath Sharma\'s Badhaai Ho deserves unstinted kudos."	"Badhaai" "Ho" "mother" "Khurrana"
In India, cybercrime is on the rise and Mumbai Police wants you to stay alert.\n\nRiding on the topical wave yet again, Mumbai Police has won netizens over with their latest tweet on Dussehra. A graphic shared by them has started a conversation about the "new age Ravana" in the cyber world. Urging Mumbaikars to stay alert, the tweet points out that the demons of today\'s times are "cyberbullying"	"Mumbai" "alert" "evil" "happy" "Dussehra" "strength"

Table 20.3 shows the comparison between two query generation mechanisms. It can be seen that TF-IDF + POS Tagging outperforms TF-IDF in terms of outcomes. The TF-IDF mechanism involves words such as "with," "her" which are not keywords. On adding POS Tagging we get only nouns. We have considered only "NN," "NNP," "NNP," "NNPS," "JJ," "JJR," "JJS" tags in POS Tagging.

TABLE 20.3 Results of Preprocessing.

Article	TF-IDF	TF-IDF+POS tagging
Neena Gupta, AyushmannKhurrana and the cast of Badhaai Ho (Courtesy badhaaihofilm)\n\nCast: AyushmannKhurrana, Sanya Malhotra, Neena Gupta, Gajra Rao, SurekhaSikri\n\nDirector: Amit Ravindernath Sharma\n\nRating: 3.5 stars (out of 5)\n\n"Director Amit Ravindernath Sharma\'s Badhaai Ho deserves unstinted kudos.	Badhaai Ho With Her That From Mother In She Khurrana	"Badhaai" "Ho" "mother" "Khurrana"

To collect Twitter content, Twitter's streaming API was used and tweets were collected using the important words as a filter. The article was given a 40-second window to collect as many tweets (in live streaming) containing at least one of the important words of the news article. A different number of tweets has been collected for each article as shown in Table 20.4.

TABLE 20.4 Articles and Tweets Collected.

News article	The topic of the article	Number of tweets collected in a 40s time window
1	Badhaai Ho release	199
2	MeToo	693
3	Navami	156
4	Mumbai cybercrime	808
5	Rise in petrol price	176
6	Water tanker strike	333
7	Sabarimala case about women	164

TABLE 20.4 *(Continued)*

News article	The topic of the article	Number of tweets collected in a 40s time window
8	Urban Naxals	72
9	Ashish Pandey surrender	676
10	Ayodhya Ram temple	339
11	Xbox collaboration with Microsoft	1725
12	Twitter about deleted tweets	1501
13	Facebook about spam ads	620

20.4.1 RELEVANCE ESTABLISHMENT

To further filter out more relevant tweets out of the collected ones, similarity indices were used. Relevance was established between the news article and each tweet extracted through the keywords. We have considered two similarity indices. Detailed results of the cosine and Jaccard similarity coefficients are shown (See Tables 20.5 and 20.6).

For the same article and the same set of tweets, both cosine and Jaccard similarity coefficients have been implemented.

TABLE 20.5 Relevance of the Tweet, an Observation for Jaccard Similarity.

Maximum value	Tweet corresponding	Jaccard value for the same tweet
0.46984843351399486	RT @IYC: IYC is honored to be able to represent the voice of women in India.\nThe road ahead is difficult, but we will journey on in the fi…	0.051470588235294115

TABLE 20.6 Relevance of the Tweet, an Observation for Cosine Similarity.

Maximum value	Tweet corresponding	Cosine value for the same tweet
0.08396946564885496	"RT @KPadmaRani1: BJP's Union Min @mjakbar has been accused of "sexual harassment and assault" by as many as 20 women, who have absolutely n…"	0.2957887112070247

It can be observed that the magnitude of cosine is quite greater than the Jaccard coefficient. According to literature, this is due to the ignorance of the same words by the Jaccard index. A threshold was set to filter out more relevant tweets and the tweets resulting in higher similarity are shown in Table 20.6 with important words found again through TF-IDF. These are the possible words for hashtags recommendation (See Table 20.7).

20.4.2 RANDOM FOREST

These words were run through the Learning-To Rank model. Random Forest Classifier was implemented on these words using the similarity value found of the tweet as the feature vector. The test set was classified correctly and the confusion matrix on the test set is shown in Table 20.8.

20.4.3 PNN

After establishing relevance and finding the most suitable words for the hashtag recommendation, PNN classification was done to assert which word out of them was more suitable as a hashtag. The random forest technique could only assert if a word is appropriate as a hashtag or not. The result of PNN is shown in Table 20.9.

TABLE 20.7 Recommended Hashtags.

TWEET 7	TWEET 13	TWEET 15
Even	ramprasad_c	Atheist_Krishna
aartic02	Still	Everyone
Hats	Employ	Except
Brave	Harassment	Father
Women	Only	Brother
Intimidated	Media	Molesters
97	Lower	Feminists
Lawyers	Standards	Supporting
Testify	Politicians	Movement
Court	While	
		TWEET 31
TWEET 27	TWEET 30	ANI

Harassment
Workplace
t.co/cdUGD88MIE
Sexual
https
TWEET 38
Shattered
Glass
t.co/WwUc7G2pAJ
Thanks
CBCNB
GrippedMagazine
Mentosibo
Metoo
Maga
Latest

TheRestlessQuil
More
t.co/cq0FoKmMes
Jatin
Das
MeTooIndia
MeToo

TWEET 154
PhilMcCrackin44
Dianne
Wants
Misconduct
Investigation
Reopened
Democrats
Take
Control
Senate

Gone
Shahnawaz
Hussain
Minister
Modi
govt...
t.co/XjJnjZyasU
MeTooIndia

TWEET 47
desh_bhkt
Yuck
Plant
Seed
Matki
t.co/zRGwalhVHb
Let
Jatin
Das
MeTooIndia

TABLE 20.8 Confusion Matrix.

	size of test set: 169	
	size of training set	
	Confusion Matrix	
Predicted->	Relevant	Irrelevant
Actual		
Relevant	164	0
Irrelevant	0	5

TABLE 20.9 PNN Results.

Tweet	#MeToo	#MS	#capacity	#Sexual Harrasement	#Akbar	#name	#MJ	#ramani
1	0.7158	0.3882	0.2748	0.1686	0.1552	0.1216	0.1105	0.0199
2	0.5077	0.3873	0.4529	0.1580	0.0854	0.1260	0.1500	0.0317

TABLE 20.9 *(Continued)*

Tweet	#MeToo	#MS	#capacity	#Sexual Harrasement	#Akbar	#name	#MJ	#ramani
3	0.6867	0.3243	0.4122	0.0944	0.1098	0.3470	0.2398	0.0169
4	0.7663	0.2716	0.2078	0.1149	0.0935	0.1123	0.2792	0.0076
5	0.4851	0.2087	0.1556	0.2267	0.2418	0.1441	0.1137	0.0350
6	0.4851	0.2087	0.1556	0.2267	0.2418	0.1441	0.1137	0.0350
7	0.4851	0.2087	0.1556	0.2267	0.2418	0.1441	0.1137	0.0350
8	0.4851	0.2087	0.1556	0.2267	0.2418	0.1441	0.1137	0.0350

A graph on the possible usage of a hashtag for the tweet considered is shown in Figure 20.3.

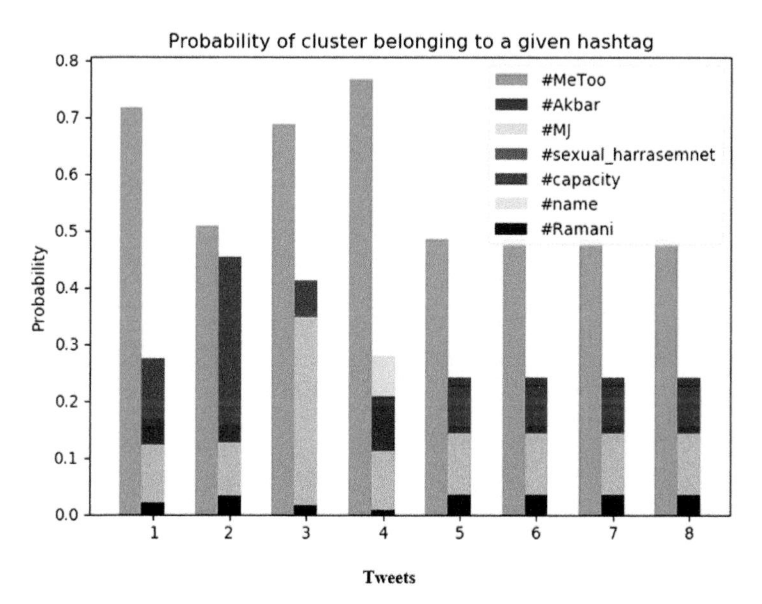

FIGURE 20.3 PNN statistics.

20.4.4 TWEETS

Figure 20.3 displays the probability of a hashtag that can suit well for the considered tweet. For the want of space, the number of tweets considered for portrayal has been reduced to eight. Journalists can easily infer the hot event to be flushed from the maximum value attained by the hashtags. This can promote their activities in tracking the details for that hashtag. It is obvious from Figure 20.3 that #MeToo has got significant reach and hence can be used for further tracking of the news related to that.

20.5 CONCLUSIONS

In this work, we presented a hybrid system that recommends suitable keywords as hashtags for tracking a story. We also presented the contrasting results of two similarity metrics with precise results. Further, the implementation of the L2R approach followed by the PNN classifier resulted in suitable words. This system has a limitation for the size of data. The processing time is huge for even a single article.

At present Hashtags can be found on even YouTube channels. Also, many websites add tags at the end of their pages to improve their results during Google searches. Science journals have publications that contain keywords about the focus of research in that paper. Hence, an author writing some content can get recommended with similar articles and journals. An efficient tracking approach to these kinds of content on the websites must be available so that more data analytics, inferences from results, and recommendations based on the content can take place. For a huge data size, the time taken for processing grows enormously, and hence Hadoop platform can be recommended for efficient processing.

KEYWORDS

- **hashtags**
- **PNN**
- **recommendation system**
- **tweets**
- **news**
- **random forest**

REFERENCES

1. Zhao, D.; Rosson, M. B. How and Why People Twitter: The Role That Micro-Blogging Plays in Informal Communication at Work. In *Proceedings of the ACM 2009 International Conference on Supporting Group Work*; ACM, 2009; pp 243–252.

2. Moon, S. J.; Hadley, P. Routinizing a New Technology in the Newsroom: Twitter as a News Source in Mainstream Media. *J. Broadcast. Electr. Media* **2014,** *58* (2), 289–305.
3. Alvari, H. Twitter Hashtag Recommendation Using Matrix Factorization, 2017. arXiv preprint arXiv:1705.10453.
4. Lu, R.; Yang, Q. Trend Analysis of News Topics on Twitter. *Int. J. Mach. Learn. Comput.* **2012,** *2* (3), p.327.
5. Saxton, G. D.; Niyirora, J. N.; Guo, C.; Waters, R. D. # Advocating For Change: The Strategic Use of Hashtags in Social Media Advocacy. *Adv. Soc. Work* **2015,** *16* (1), 154–169.
6. Kywe, S. M.; Hoang, T. A.; Lim, E. P.; Zhu, F. December. On Recommending Hashtags in Twitter Networks. In *International Conference on Social Informatics*. Springer: Berlin, Heidelberg, 2012; pp 337–350.
7. Shi, B.; Poghosyan, G.; Ifrim, G.; Hurley, N. Hashtagger+: Efficient High-Coverage Social Tagging of Streaming News. *IEEE Trans. Knowl. Data Eng.* **2018,** *30* (1), 43–58.
8. Alvari H. Twitter Hashtag Recommendation using Matrix Factorization. arXiv preprint arXiv:1705.10453. 2017 May 30.
9. Otsuka, E.; Wallace, S. A.; Chiu, D. A Hashtag Recommendation System for Twitter Data Streams. *Comput. Soc. Netw.* **2016,** *3* (1), 3.
10. Ma, Z.; Sun, A.; Yuan, Q.; Cong, G. Tagging Your Tweets: A Probabilistic Modeling of Hashtag Annotation in Twitter. In *Proceedings of the 23rd ACM International Conference on Conference on Information and Knowledge Management*; ACM, 2014; pp 999–1008.
11. Hoang-Vu, T. A.; Bessa, A.; Barbosa, L.; Freire, J. Bridging Vocabularies to Link Tweets and News. In *International Workshop on the Web and Databases*; WebDB, 2014.
12. Gong, Y.; Zhang, Q.; Huang, X. Hashtag Recommendation Using Dirichlet Process Mixture Models Incorporating Types of Hashtags. In *Proceedings of the 2015 Conference on Empirical Methods in Natural Language Processing*; 2015; pp 401–410.
13. Shi, B.; Ifrim, G.; Hurley, N. Learning-to-Rank for Real-Time High-Precision Hashtag Recommendation for Streaming News. In *Proceedings of the 25th International Conference on World Wide Web*. International World Wide Web Conferences Steering Committee, 2016; pp 1191–1202.
14. Wisdom, V. An Introduction to Twitter Data Analysis in Python, 2016. DOI: 10.13140/RG.2.2.12803.30243.
15. Zahrotun, L. Comparison Jaccard Similarity, Cosine Similarity and Combined Both of the Data Clustering with Shared Nearest Neighbor Method. *Comput. Eng. App. J.* **2016,** *5* (1), 11–18.
16. Huang, A. Similarity Measures for Text Document Clustering. In *Proceedings of the Sixth New Zealand Computer Science Research Student Conference (NZCSRSC2008)*; Christchurch, New Zealand, 2008; pp 49–56.
17. Specht, D. F. Probabilistic Neural Networks (a One-Pass Learning Method) and Potential Applications. WESCON'89, 1989; pp 780–785.
18. El Emary, I. M.; Ramakrishnan, S. On the Application of Various Probabilistic Neural Networks in Solving Different Pattern Classification Problems. *World Appl. Sci. J.* **2008,** *4* (6), 772–780.
19. Rao, P. N.; Devi, T.; Kaladhar, D.; Sridhar, G.; Rao, A. A. A Probabilistic Neural Network Approach for Protein Superfamily Classification. *J. Theor. Appl. Inf. Technol* **2009,** *6* (1).

PART III
Network Security Applications

CHAPTER 21

MITIGATION OF RPL STATELESS ADDRESS AUTO-CONFIGURATION IPV6 SPOOFING ATTACK IN IOT

SAURABH YADAV, SHISHIR GANGWAR, PRATEEK KEMBHAVI, B. R. CHANDAVARKAR, and SNEHA KAMBLE

Department of Computer Science and Engineering, National Institute of Technology Karnataka, Surathkal, Karnataka, India

ABSTRACT

As wireless network usage is expanding, attempts are being made to disrupt the smooth functioning of the network. Due to power and memory constraints inherent in the Internet of Things (IoT) devices, implementing security measures becomes an even more significant challenge. There are many ways in which confidential information can be extracted from networks. IP address spoofing is one such attack. IP address spoofing is an illegal, unauthorized change of IP address. Malicious elements in the network impersonate as a genuine client by copying their IP address. As a result, attackers can intercept network communication between genuine clients. A false IP address makes it possible for the attacker to bypass existing security mechanisms, which creates severe security and confidentiality problems. This chapter proposes a method to resolve the issue by ensuring that an incoming node in a network always gets a unique IP address, thus making it difficult for the attacker to disguise itself as a genuine client and carry out the attack.

Advanced Computer Science Applications: Recent Trends in AI, Machine Learning, and Network Security. Karan Singh, PhD, Latha Banda, PhD & Manisha Manjul, PhD (Eds.)

21.1 INTRODUCTION

The Internet of Things (IoT) is a network of devices used in daily life like home appliances, gadgets, vehicles, etc. These devices are embedded with sensors, software, and connectivity, making it possible for them to communicate with each other. It is also possible to activate and control such devices remotely. Such technology has abundant applications in different fields like home automation, security, fashion, sports, and fitness.[1–3] IoT devices come with specific inherent challenges, and hence, working with them is more complicated than working with standard devices like laptops and desktops. These devices are constrained with lower memory, processing, communication, and energy capabilities, while offering high mobility and portability. Due to their power and memory constraints and high mobility, a separate protocol, called the Routing Protocol for Low-power and Lossy Networks (RPL), is used for network topology formation and routing.

RPL uses Stateless Address Auto-Configuration (SLAAC) for configuring the devices. Whenever a new node joins the network, it captures the network prefix through the Router Advertisement (RA) and attaches interface address to it in order to obtain its IPv6 address. However, this feature can be exploited if a third-party node knows the interface address of the actual client. It can easily configure itself with the IPv6 address of the original node and join the network as an imposter compromising its security. The proposed solution aims to alleviate this issue and is demonstrated in Contiki OS. Contiki[4] was one of the first operating systems to provide IP connectivity for sensor networks. It is an open-source operating system which is implemented in C. The application programs can be dynamically loaded and unloaded on an event-driven kernel at run-time.

The chapter is organized as follows: In Section 21.2, additional literature that examines the problem of IP spoofing is compiled. Section 21.3 elucidates the working of the RPL protocol. Section 21.4 explains how SLAAC works in Contiki. Section 21.5 describes how the attack is carried out. Section 21.6 describes the simulation setup carried out for demonstrating the attack. Section 21.7 puts forth the proposed solution to the problem of IP spoofing. Section 21.8 concludes the chapter and mentions the possible work that can be carried out in the future.

21.2 EXPERIMENTAL METHODS AND MATERIALS

IPv6 over Low-Power Wireless Personal Area Networks (6LoWPAN) has made it possible for IoT devices to communicate using IPv6. It enables the IPv6 packets to be sent and received over a network, by encapsulation and compression technique.[7] A lot of research work is carried out on security, associated with 6LoWPAN, which is very intricate due to IPv6. Demonstration on IP spoofing attack is also available.

The topology in RPL is built using Destination Oriented Directed Acyclic Graph (DODAG). An algorithm to carry out attacks using spoofed DODAG Information Solicitation (DIS) message is described in (6). As defined in the algorithm, with the help of the DODAG Information Object (DIO) packet, the attacker first learns the IPv6 address of nodes in the network. Section 21.3 provides details about DIS and DIO messages. Further, it selects one of the nodes and sends a probe packet to check whether the node is in sleep or wake-up mode. Depending on the state of the victim node, the attacker will send a spoofed DIS message to the border router if the node is in sleep mode else it waits. The border router gets the IPv6 and Media Access Control (MAC) address that is used to make changes to its routing table using spoofed RPL message. The attacker switches an entry with the victim node in the routing table, which results in sending data to the attacker node.

Yong-Joon Lee et al.[7] discuss different detection methods using spoofed IP address against Distributed Denial of Service (DDoS) attacks. The method includes the Bogon filtering, Unicast Reverse Path Forwarding (uRPF), Transmission Control Protocol (TCP) intercept, and Hop Count Filtering (HCF) Techniques. The method suggested in the paper discusses detecting spoofed IP addresses for IoT devices. It distinguishes reliable TCP packets from regular traffic flowing into the DDoS shelter. Hence when the attack occurs, the DDoS shelter identifies whether it is counterfeit or not by identifying the packets from the traffic and by analyzing the data about the Operating System (OS). If the node is found as counterfeit, the relevant IP traffic is blocked by the shelter.

Mitigation of Denial of Service (DoS) attacks on wireless sensor networks (WSN) with IPv6 is examined in Oliveira.[8] The authors have proposed a countermeasure system that reduces the DoS and DDoS attacks triggered remotely. The approach is built on 6LoWPAN neighbor discovery, which works only at edge routers. The mechanism protects the Wireless Sensor Network (WSN) from the attacks like DoS and DDoS.

21.3 RPL PROTOCOL

RPL is an IPv6 routing protocol used for low power and lossy networks. It is based on distance vectors and operates on IEEE 802.15.4; it uses multi-hop and many-to-one and one-to-one communications. Figure 21.1 shows a sample DODAG that is formed as a result of the RPL protocol.[9]

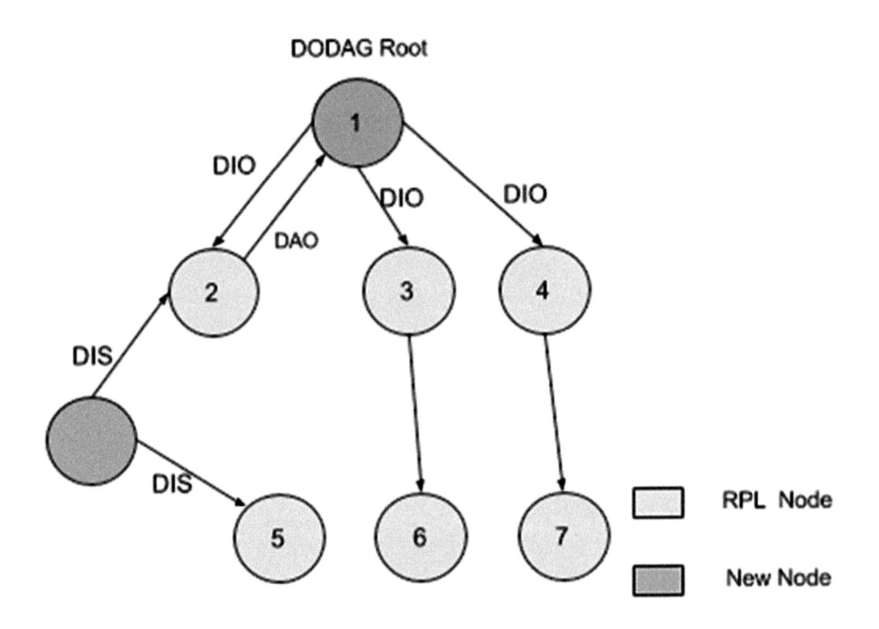

FIGURE 21.1 RPL DODAG formation.

Network topology is created according to the Directed Acyclic Graph (DAG), which segregates to numerous DODAGs, and each DODAG has one sink or router or root. In case if more roots are present in a DODAG, they are aligned in common backbone link.

To create a DODAG, exchange of three types of messages takes place:-

- DODAG Information Solicitation (DIS)—This message is used to request information from nearby networks, each network defined by a DODAG.
- DODAG Information Object (DIO)—This message is sent in response to the DIS messages from the DAG. It periodically updates

the information of the nodes of the network. This message is periodically multi-casted to neighbor nodes.

- DODAG Advertisement Object (DAO)—This message is used to propagate the destination information in the direction of DODAG.

Since RPL runs in power and memory-constrained nodes, it is a reactive protocol. The routes are created when they are needed, rather than maintaining the routing tables over time.

The nodes are connected to the root devices through a multi-hop path. These root devices (also called sinks) are responsible for data collection and coordination duties. For each of them, a DODAG is created, taking into consideration the link cost, status information, and objective function.

21.4 STATELESS ADDRESS AUTO-CONFIGURATION

IPv6 defines mechanisms for the assignment of both stateful and stateless address auto-configuration. In stateful address auto-configuration, the server dynamically assigns the addresses to devices, drawing it from a pool of available addresses. This approach is similar to the Dynamic Host Configuration Protocol (DHCP) used in IPv6.

The stateless autoconfiguration allows various devices to connect to the network without any help from the server like DHCP. Using the DIO messages, the DODAG root sends the network prefix via the Prefix Information Option. The RPL nodes use this option for the purpose of Stateless Address Auto-Configuration (SLAAC) from a prefix advertised by a parent and advertise its own address. DIO messages are sent at regular intervals and in response to device solicitation messages. By conjoining the interface of the node with the prefix given in the DIO messages, a node configures its global IPv6 address.

The inherent vulnerability of this mechanism is discussed in Section 21.5. There have been works optimizing the security feature of SLAAC. Further analysis of SLAAC in Low power and Lossy Networks (LLN) is provided in Montavont.[10]

21.5 IP SPOOFING ATTACK

IP spoofing is a procedure for altering the IP address of any device which connects to a network. The IP address of a node is generated by the operating system and should not be changed. But, there are many tools that can enable attackers to change the IP address. As a result of IP spoofing, the attacker can circumvent general authentication procedures and masquerade as an authorized user thereby gaining access to the network elements. The attacker can now launch all kinds of attacks like Denial of Service (DoS) and Man in the middle attack disrupting the normal network operations.

The DIO messages are disseminated throughout the network at regular intervals of time. Whenever a new node joins the network, with the help of SLAAC, the node captures the network prefix through the DIO messages. An IP address is generated by combining the 64-bit network prefix supplied by the router and the 48-bit MAC address which is unique for every device. However, this feature can be exploited if a third-party node knows the interface address of the actual client. It can easily configure itself with the IPv6 address of the original node and join the network as an imposter compromising its security.

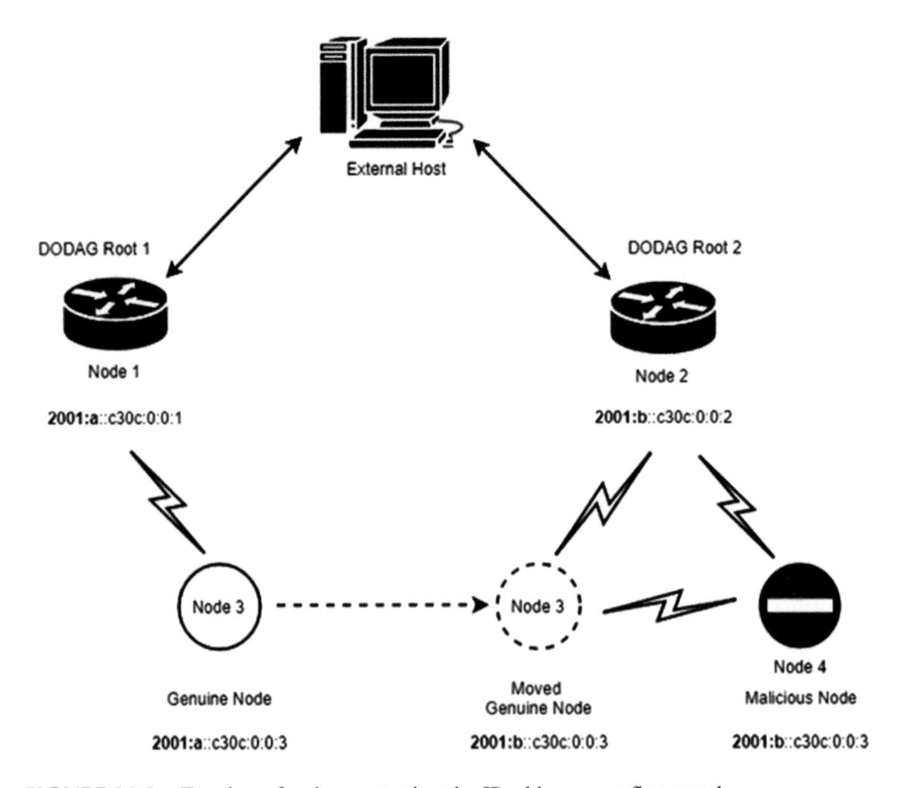

FIGURE 21.2 Topology for demonstrating the IP address spoofing attack.

21.6 SIMULATION SETUP

The topology shown in Figure 21.2 will be referred for the purpose of this demonstration. In this topology, there are four nodes. The first two nodes act as the routers. These routers form their own independent networks and have their own prefixes. The routers are connected to a common external host, which means that the networks formed by both the routers can be accessible from the external host. In fact, what these routers create is a form of interface between the external host and their own respective networks. Nodes 3 and 4 are clients that are connected to or are attempting to connect to these two routers. Node 3 is a genuine client. Node 4 is a malicious client. The malicious client knows the IP address of the genuine client. Now, the attacker, who is in possession of information about the genuine node, changes (or spoofs) its MAC address to that of the genuine node. At this instant, both the nodes, node 3 (genuine node) and node 4 (malicious node), have the same MAC address. IP address is formed by combining the prefix with the MAC address. As was mentioned in the previous section, first 64-bit of network prefix (which are sent by the router to the node) and the 48-bit of the node's individual MAC address are combined to obtain the 128-bit Global IPv6 address of the node.

At present, node 3 is in the network of router 1 and node 4 is in the network of router 2. Since both the clients are in separate networks, they both receive different prefixes from their respective routers, but their MAC addresses are the same. Now consider the scenario where node 3 tries to move to the coverage of router 2. This movement of node 3 is depicted by a dotted line in Figure 21.2. When both nodes 3 and 4 are in the coverage of router 2, they receive the same prefix that is supplied by router 2. Since IP address is formed by the combination of the prefix and MAC address, both nodes 3 and 4 will have the same IP address owing to the fact that both have the same pair of prefixes and MAC address. This creates a conflict in the network.

Contiki allows the user to check the neighbors of the router via the browser. Figure 21.3 shows that there is just one neighbor, where in fact there are two neighbors but with the same IP address.

To attempt to understand which nodes can communicate with the network, a ping is initiated from the external host. The result of the ping can be verified from the built-in pcap packet analyzer in Contiki. Figure 21.4 is a snapshot of the same.

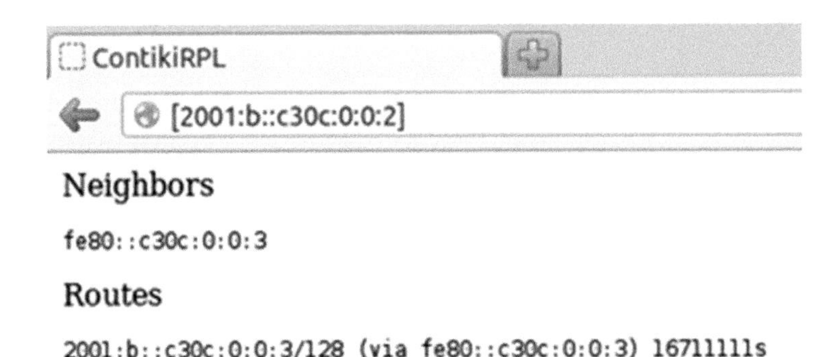

FIGURE 21.3 Browser Window listing the neighbors and routes from router 2.

In Figure 21.4, in the area highlighted by the topmost rectangle, it can be seen that router 2 sends an ECHO REQ (echo request) to node 4 and receives an ECHO RPLY (echo reply) after a few messages are exchanged. Further, after the exchange of a few more messages, in the area highlighted by the bottom-most rectangle, it is seen that router 2 sends an ECHO REQ to node 3 as well. After the exchange of a few messages, router 2 receives an ECHO RPLY from node 3. Thus, we can infer that both nodes 3 and 4 can be pinged from the external host via router 2. This means that router 2 is communicating with both the nodes as is evident from Figure 21.4. Therefore, whatever information is sent from the router to the actual node is also being sent to the malicious node. As a result, in scenarios where confidential information is being passed from the router to the client, a malicious client having spoofed its MAC address will be able to obtain the confidential information. This endangers the confidentiality of information.

Time	From	To	Data
03:59.772	2	4	5: 15.4 A
04:00.202	3	2	97: 15.4 D C1:0C:00:00:00:00:00:03 0xFFFF|IPHC|IPv6|ICMPv6 RPL DIO|2001000B 00000000 C30C0000 00000002 040E0008 0C0A0700 01000001 00FFFFFF 081E4040 000...
04:00.219	3	2,4	97: 15.4 D C1:0C:00:00:00:00:00:03 0xFFFF|IPHC|IPv6|ICMPv6 RPL DIO|2001000B 00000000 C30C0000 00000002 040E0008 0C0A0700 01000001 00FFFFFF 081E4040 000...
04:00.223	3	2	97: 15.4 D C1:0C:00:00:00:00:00:03 0xFFFF|IPHC|IPv6|ICMPv6 RPL DIO|2001000B 00000000 C30C0000 00000002 040E0008 0C0A0700 01000001 00FFFFFF 081E4040 000...
04:00.339	2	-	127: 15.4 D C1:0C:00:00:00:00:00:02 C1:0C:00:00:00:00:00:03|FRAG1|IPHC|IPv6|ICMPv6 ECHO REQ 0|20130016 D6F9A85C 57840B00 08090A0B 0C0D0E0F 10111213 1415...
04:00.344	2	4	127: 15.4 D C1:0C:00:00:00:00:00:02 C1:0C:00:00:00:00:00:03|FRAG1|IPHC|IPv6|ICMPv6 ECHO REQ 0|20130016 D6F9A85C 57840B00 08090A0B 0C0D0E0F 10111213 1415...
04:00.348	4	2	5: 15.4 A
04:00.350	2	4	36: 15.4 D C1:0C:00:00:00:00:00:02 C1:0C:00:00:00:00:00:03|FRAGN|30313233 34353637
04:00.352	4	2	5: 15.4 A
04:00.363	4	2	126: 15.4 D C1:0C:00:00:00:00:00:03 C1:0C:00:00:00:00:00:02|FRAG1|IPHC|IPv6|ICMPv6 ECHO RPLY 0|20130016 D6F9A85C 57840B00 08090A0B 0C0D0E0F 10111213 141...
04:00.367	2	4	5: 15.4 A
04:00.369	4	2	36: 15.4 D C1:0C:00:00:00:00:00:03 C1:0C:00:00:00:00:00:02|FRAGN|30313233 34353637
04:00.371	2	4	5: 15.4 A
04:01.069	3	2	76: 15.4 D C1:0C:00:00:00:00:00:03 C1:0C:00:00:00:00:00:03|IPHC|IPv6|ICMPv6 RPL DAO|1E4000F7 2001000B 00000000 C30C0000 00000002 05120080 2001000B 0000...
04:01.093	3	2,4	76: 15.4 D C1:0C:00:00:00:00:00:03 C1:0C:00:00:00:00:00:03|IPHC|IPv6|ICMPv6 RPL DAO|1E4000F7 2001000B 00000000 C30C0000 00000002 05120080 2001000B 0000...
04:01.095	4	2,3	5: 15.4 A
04:01.106	4	2	76: 15.4 D C1:0C:00:00:00:00:00:03 C1:0C:00:00:00:00:00:02|IPHC|IPv6|ICMPv6 RPL DAO|1E4000F7 2001000B 00000000 C30C0000 00000002 05120080 2001000B 0000...
04:01.109	2	4	5: 15.4 A
04:04.500	2	-	127: 15.4 D C1:0C:00:00:00:00:00:02 C1:0C:00:00:00:00:00:03|FRAG1|IPHC|IPv6|ICMPv6 ECHO REQ 0|20130017 D7F9A85C 3F8B0B00 08090A0B 0C0D0E0F 10111213 1415...
04:04.520	2	3	127: 15.4 D C1:0C:00:00:00:00:00:02 C1:0C:00:00:00:00:00:03|FRAG1|IPHC|IPv6|ICMPv6 ECHO REQ 0|20130017 D7F9A85C 3F8B0B00 08090A0B 0C0D0E0F 10111213 1415...
04:04.524	3	2	5: 15.4 A
04:04.526	2	3	36: 15.4 D C1:0C:00:00:00:00:00:02 C1:0C:00:00:00:00:00:03|FRAGN|30313233 34353637
04:04.527	3	2	5: 15.4 A
04:04.539	3	2	126: 15.4 D C1:0C:00:00:00:00:00:03 C1:0C:00:00:00:00:00:03|FRAG1|IPHC|IPv6|ICMPv6 ECHO RPLY 0|20130017 D7F9A85C 3F8B0B00 08090A0B 0C0D0E0F 10111213 141...
04:04.593	3	2,4	126: 15.4 D C1:0C:00:00:00:00:00:03 C1:0C:00:00:00:00:00:03|FRAG1|IPHC|IPv6|ICMPv6 ECHO RPLY 0|20130017 D7F9A85C 3F8B0B00 08090A0B 0C0D0E0F 10111213 141...
04:04.597	4	2,3	5: 15.4 A
04:04.600	3	2,4	36: 15.4 D C1:0C:00:00:00:00:00:03 C1:0C:00:00:00:00:00:03|FRAGN|30313233 34353637

FIGURE 21.4 Packet capture analysis of IP address spoofing attack.

21.7 PROPOSED METHOD TO PREVENT IP SPOOFING

IP spoofing is an attack that needs to be prevented to ensure the safety of the network. The attacker is able to spoof its IP address because it can easily predict the IP address of the genuine client. Since IP address is a combination of the prefix and the MAC address, and since the MAC address of the genuine client is already known to the attacker, it is very easy to predict the IP address of the client.

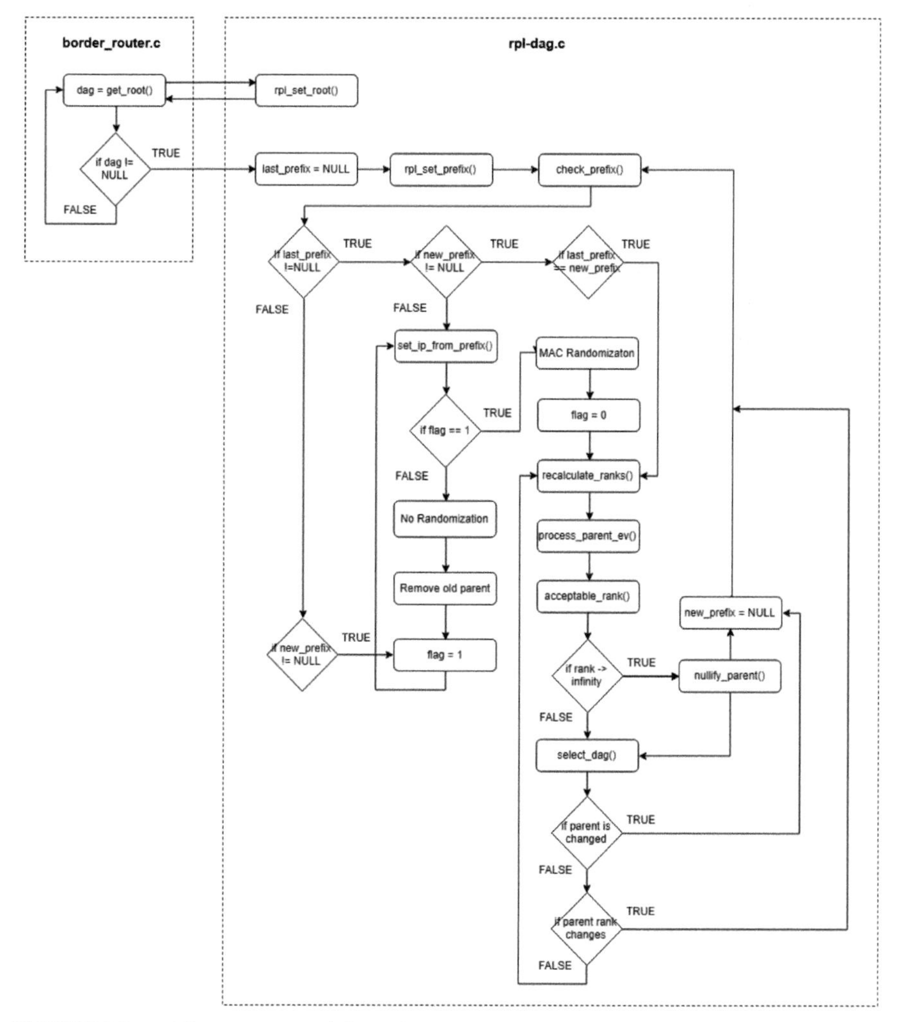

FIGURE 21.5 File structure and flow chart.

To make the MAC address unpredictable, a system to randomize the MAC address is proposed. Randomizing the MAC address will make it impossible for an attacker to predict the MAC address and in turn the IP address of the actual node. Whenever a node is moving from network to network, the IP prefix will keep changing according to the network, but the MAC address is constant. To prevent attackers from copying the true client's MAC address, the MAC address needs to be randomized whenever a node moves from network to network. In this way, the attacker will not be able to spoof its MAC address because it is impossible for him to predict what the true client's MAC address will be.

This system for randomizing the MAC address will have to be integrated with the existing RPL protocol. Since the RPL protocol handles the process of stateless address auto-configuration, it automatically assigns the nodes their IP address based on the prefix which is received from the router and the nodes' in-built MAC address. In effect, the proposed system amends the RPL protocol to prevent MAC address spoofing attack. Figure 21.5 explains where and when to implement the system of randomization of MAC address in Contiki.

The procedures to set the IP address from the prefix and the MAC address are defined in the file *rpl-dag.c*, as can be seen from Figure 21.5. To implement the RPL protocol, control is transferred from *border_router.c* to *rpl-dag.c* . To decide when to randomize the MAC address, a *flag* variable is used. This variable is set to one when the conditions are favorable to randomize the node's MAC address. When there is no need to randomize the MAC address, this *flag* variable is set to zero. To track changes in the movement as well as to check the stability of the network, two fields, *last_prefix* and *new_prefix*, are used. The *last_prefix* field indicates the most recent prefix that was obtained before the node joined this router. The *new_prefix* field indicates the new prefix that is to be adopted by the node. As a result of having a randomized MAC address, the IP address of the node would be difficult to predict for an attacker. The *flag* is set to zero once the randomization process is completed. Further, at times, because of the movement of a node and due to network fluctuations, the rank of the router (parent to the node) changes. These changes are tracked by recalculating the ranks periodically. During the recalculation of ranks, the rank of the parent is checked based on the distance between the node and the router and the time it takes to reach the router from the node. If the rank is different from the previous rank, it is noted. If the rank change is drastic and the node is out of the coverage of the parent (router), this implies that the node has moved (the rank of the parent is infinity). Accordingly, the parent is removed and a new parent is chosen by selecting a new DODAG. Since the parent is changed, the *new_prefix* field is set to zero. Further, while checking prefixes, after noting that the *new_prefix* field has been set to zero, the node understands that the old prefix must be purged.

After purging the old parent, the *flag* variable is set to one, because a new IP will be assigned to the node depending on the router it joins. Since the *flag* is set to one, a random MAC is assigned to the node and a new IP address is generated from this MAC address.

If the rank change during rank recalculation is not drastic, that is, the node has just moved within the network the new rank is noted (not infinity). After any rank change, the prefixes are checked again but now both the fields are nonzero and equal. When this happens, no change takes place.

In Figure 21.2 while using the randomization of the MAC address, if node 3 moves to the coverage of router 2, there will not be any network conflicts because the IP address of node 3 will be difficult to predict on account of the randomization that happens when it moves from one network to the other.

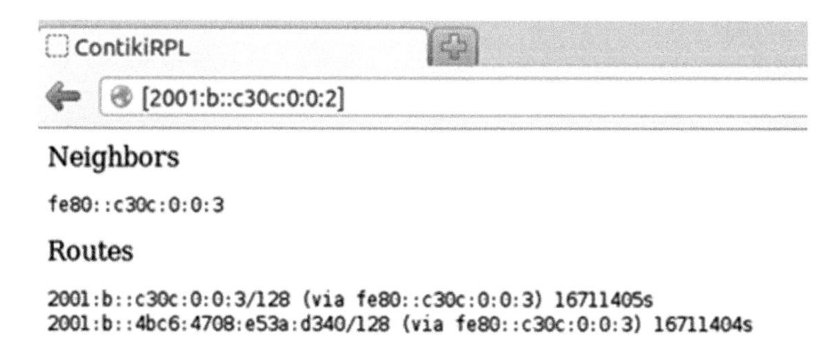

FIGURE 21.6 Browser Window listing the neighbors and routes from Router 2 after implementing the solution.

Contiki enables users to obtain information about the routes of the routers via the browser. On typing the router address on the browser address bar, we get the result as shown in Figure 21.6.

As is evident from Figure 21.6, there are now two routes listed from the router instead of just one. This means that the genuine client was able to successfully make a connection with the router. Thus, the attack orchestrated by the malicious client did not affect the ability of the genuine client to make a connection with the router.

21.8 CONCLUSIONS

Despite the convenience of RPL's stateless address auto-configuration, we are faced with issues in some cases. A vulnerability in RPL's stateless address auto-configuration makes it possible for attackers to impersonate other nodes; this attack is called the IP spoofing attack. IP spoofing, therefore, leads to loss of security and confidentiality of data. We put forth a solution in terms of our proposed system. The proposed system prevents IP spoofing attack by randomizing the MAC address as and when required, mainly when it leaves from one network to join another network. Preventing IP spoofing ensures that nodes will be able to move freely from network to network without the threat of an attacker spoofing their IP address.

Hashing of MAC address, as opposed to just randomization of MAC address, will be more efficient. Hashing the MAC address by any SHA algorithm will make it tougher for an attacker to predict the IP address of any node. Implementation of the hashing function will have to be done, while making sure that it does not violate the lightweight property of the RPL protocol for IoT devices. IoT devices are supposed to be lightweight; hence, care should be taken to ensure that the implementation of hashing does not overload the IoT devices. A proper lightweight hashing function is something that will make it fundamentally impossible for attackers to predict IP addresses of nodes.

KEYWORDS

- **IP spoofing**
- **genuine client**
- **malicious client**

REFERENCES

1. Ebling, M. R. IoT: From Sports to Fashion and Everything In-Between. *IEEE Pervasive Computing* **Oct–Dec 2016,** *15*, 4, 2–4.
2. Al-Kuwari, M.; Ramadan, A.; Ismael, Y.; Al-Sughair, L.; Gastli, A.; Benammar, M. Smart-Home Automation Using IoT-Based Sensing and Monitoring Platform. In *Proceedings of the IEEE 12th International Conference on Compatibility, Power Electronics and Power Engineering (CPE-POWERENG 2018)*; Doha, 2018; pp 1–6.

3. Wang, Y.; Chen, M.; Wang, X.; Chan, R. H. M.; Li, W. J. IoT for Next-Generation Racket Sports Training. *IEEE Internet of Things J.* **Dec. 2018,** *5* (6), 4558–4566.
4. Dunkels, A.; Gronvall, B.; Voigt, T. Contiki—A Lightweight and Flexible Operating System for Tiny Networked Sensors. In *Proceedings of the 29th Annual IEEE International Conference on Local Computer Networks (LCN '04)*; IEEE Computer Society: Washington, DC, 2004; pp 455–462.
5. Kushalnagar, N.; Montenegro, G. Network Working Group, Request for Comments: **4919,** August 2007.
6. Mavani, M.; Asawa, K. Experimental Study of IP Spoofing Attack in 6LoWPAN Network. In *Proceedings of the 7th International Conference on Cloud Computing, Data Science & Engineering—Confluence*, Noida, 2017; pp 445–449.
7. Lee, Y-J.; Baik, N-K.; Kim, C.; Yang, C-N.; Study of Detection Method for Spoofed IP Against DDoS Attacks. *Personal Ubiquitous Comput.* Feb **2018,** *22* (1), 35–44.
8. Oliveira, L. M. L.; Rodrigues, J. J. P. C.; de Sousa, A. F. Jaime Lloret; Denial of Service Mitigation Approach for IPv6 Enabled Smart Object Networks. *Concurrency Comput.* Jan **2013,** *25* (1), 129–142.
9. Winter, T.; Thubert, P.; Brandt, A. RPL: IPv6 Routing Protocol for Low-Power and Lossy Networks. 2012 Internet Engineering Task Force (IETF) Request for Comments: **6550,** ISSN: 2070–1721.
10. Montavont, J.; Cobârzan, C.; Noël, T. Theoretical Analysis of IPv6 Stateless Address Autoconfiguration in Low-Power and Lossy Wireless Networks; Proceedings of the IEEE RIVF International Conference on Computing & Communication Technologies - Research, Innovation, and Vision for Future (RIVF), *Can Tho* **2015,** 198–203.

CHAPTER 22

SECURE AND EARLY DETECTION FRAMEWORK FOR COVID-19: STANDARDIZATION OF CLINICAL PROCESS

VINAY PATHAK[1], ACHINTYA KUMAR PANDEY[2], and RAJAT KISHOR VARSHNEY[2]

[1]*School of Computer and Systems Sciences, Jawaharlal Nehru University, New Delhi, India*

[2]*Department of CSE, Greater Noida Institute of Technology, Greater Noida, India*

ABSTRACT

Coronavirus (COVID-19) is the major outbreak in the whole country or we can say across the globe. Till now, no vaccine has been discovered to prevent this epidemic. Quarantine and self-isolation is the only preventive option from this epidemic/pandemic. COVID-19 is spreading through human touch or by touching the things that came in contact with an infected person. This situation is very critical for those people who are staying alone, that is, elder people without their kids, working youth in different city than their home town, and quarantine people due to their travel history. Real-time monitoring of these types of people is very important. Because in this pandemic situation nobody knows that when and how anyone can get infected (due to touching the daily necessary food supply items and many more reasons are possible) and those newly infected people need medical help on an emergency basis; also authorities (medical/WHO/police) should be aware of the medical situation of quarantine people due to their travel history.

Advanced Computer Science Applications: Recent Trends in AI, Machine Learning, and Network Security. Karan Singh, PhD, Latha Banda, PhD & Manisha Manjul, PhD (Eds.)

In this chapter, we proposed a model that will monitor real-time activity of people who are either living alone or quarantine. If any unusual health situation occurs and it is captured into online monitoring video like continuous sneezing, throat infection, and person laying down in the same position more than 3–4 h (symptoms of COVID-19) and they also send help-needing symbol like Victory sign by fingers then automatically our system will send an alert message to registered numbers, that is, relatives, neighbors, and nearest hospital or family doctor for medical help. For privacy reason, image processing will be applied on real-time monitoring video and for security and confidentiality purpose, alert message will be encrypted by security algorithm so that no unauthorized person changes the value of alert message. To detect the health condition, our algorithm uses a two-way approach. Initially, the sensor collects information regarding symptoms such as Roll, Pitch and Net acceleration of body, Pitch and Roll; it helps to identify health status. Secondly, to identify the fingers' sign, image processing has been used. Here, victory sign made by fingers is used to identify when someone wants to tell that they need help. Thus, our work in this chapter (Extended Algorithm for fall detection using IoT and Image Processing(EAFDI^2) works on all possible situations and gives 97% correct results.

22.1 INTRODUCTION

Since the existence of earth many kinds of virus and bacteria were associated with earth. Few of them (virus/bacteria) were considered pandemic/epidemic and created a disaster in human life and nature. It destroys the families and also raises the economical problem throughout the world. To control such kinds of pandemic, primarily focus of scientist and doctors are cleanness of human body surrounding. Thereafter, they invent the vaccine of the particular disease after a long experiment on living hoods.

The coronavirus is also such type of pandemic which has again created such type of problem. This virus was first discovered by group of virologists (J D Almeida, D M Berry, C H Cunningham, D Hamre, M S Hofstad, L Mallucci, K McIntosh, and D A J Tyrrell) in 1968 in a bat, and they reported that there is no communication between bat and human bodies; therefore, no vaccination was required at that time.[1] It was published in Nature journal in 1968. The name of corona is taken from the word crown because its structure appears like a crown.

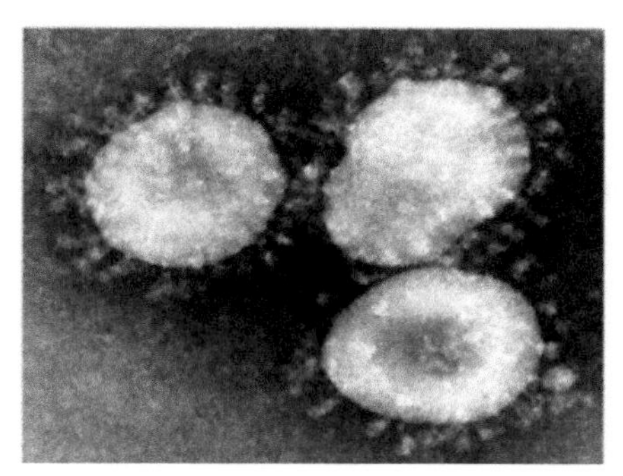

FIGURE 22.1 Virions of coronaviruses, Sun's corona appeared at the time of an eclipse.

In 2019 as per available records and data, first identified in the city of Wuhan, the novel coronavirus disease, or COVID-19, is a member of the family of coronaviruses which can cause mild conditions like the common cold to potentially lethal ones like the severe acute respiratory syndrome or SARS. The COVID-19 strain found in Wuhan has been found to be closely linked to those found in bats, and could have initially spread in the Wuhan seafood wholesale market, where live animals are sold and slaughtered.

However, according to China this infection is spread through human to human contact, similar to seasonal influenza, and thus its fast spread around the world. This is all the more so given the way that the episode occurred during the Lunar New Year season, when 3 billion outings are relied upon to be made. As communicable property of this disease it spread out in the Wuhan city, China and later on this virus touched the boundary of other countries and created pandemic situation within those countries.[2] COVID-19 virus changes its mutation that is more problematic for researchers and scientists involved in innovation of vaccine.

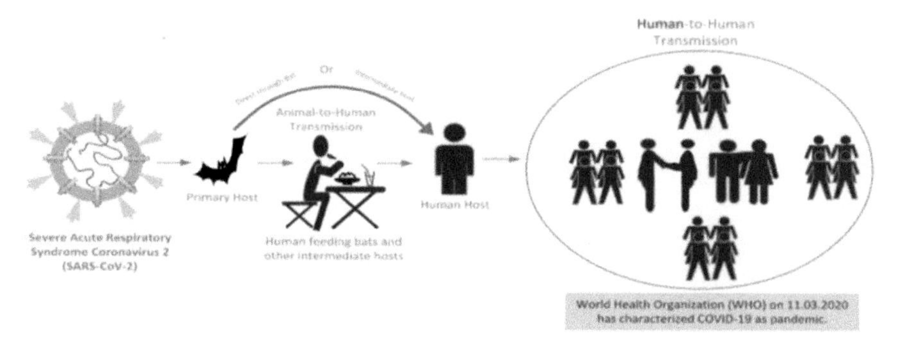

FIGURE 22.2 Spread of coronavirus.

A wide research is continuously going on to get rid of from this virus. In this view, scientists and doctors are working day and night to make an effective vaccine, to avoid a huge economic and life loss but still no success has been found. If we look into the data available in world meter website then total numbers of active cases are approx 2 million in the world. In total 3 million cases have been reported and death cases are more than 200 k as per data available in the website.[3]

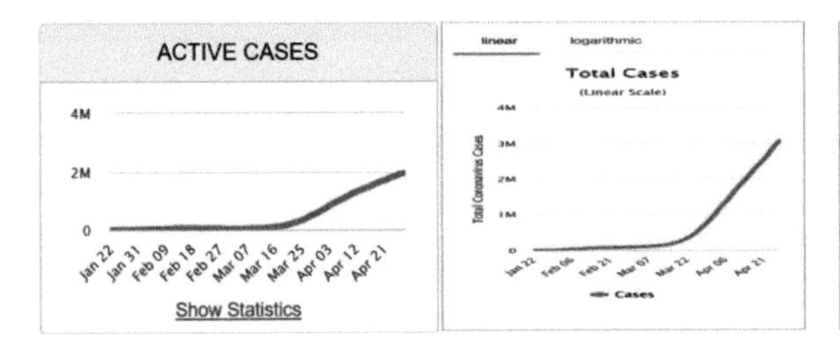

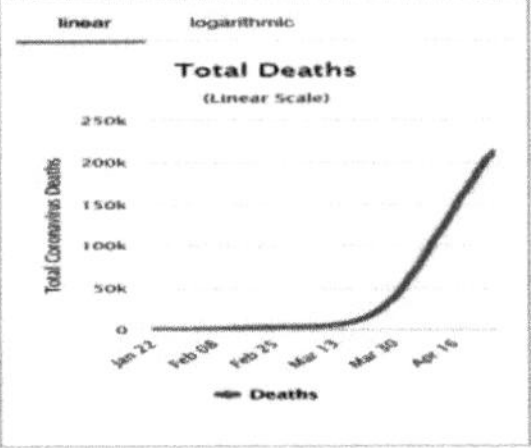

FIGURE 22.3 Active cases, total cases, and total deaths.

As per the data available on the same website total number of reported cases is 31 k, active cases in India are more than 21 k, and total deaths are more than 900 till date.

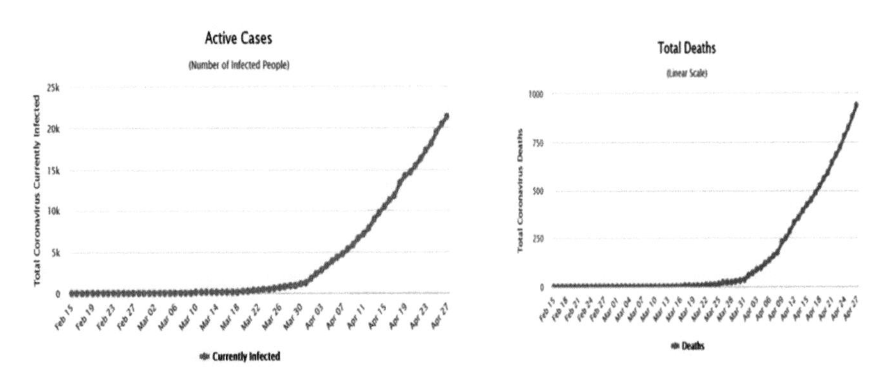

FIGURE 22.4 Total number of active cases and deaths in India.

No vaccine has been developed till now to get rid of this coronavirus. And due to its transmissible nature, it is spreading very fast within the human beings in all over the globe. Lots of research is going on to predict the number of possible cases in next few months so that governing bodies can arrange the required facilities and can take necessary steps to run the country in smooth functioning. Otherwise, in the lack of food and medical facility mob can create disturbance in everyone's life. Until no effective vaccine is to be invented till that time quarantine and self-isolation is the only way to stop this virus from affecting human beings. Due to lack of medical equipment as compared to population of India, corona cases are still in control and very less as compared to other developed countries because of complete lockdown imposed on whole country. Quarantine and self-isolation is playing a vital role in stopping spread of coronavirus from infected peoples to normal peoples. Actual numbers of infected peoples are still not known to the government because it is next to impossible for government to scan each and every person in India. Due to lack of complete scanning again sources of spreading virus are unknown. People can get infected by street vendors, grocery suppliers, and many more sources are possible to spread virus to those people who are home quarantine.

The state of the art of this work is that the given model will observe real-time activity of self home quarantine people and if any symptoms of COVID-19 will be found in quarantine person then it will send an alert message to medical bodies for providing medical facility to infected person. Second, those people who are staying alone and if they feel that some medical help is required and they are not in position to contact anyone then they can show victory sign image by fingers to the camera and the moment camera will capture this sign immediately it will send an emergency medical help message to nearest hospital or registered mobile numbers.

22.2 KEY CONTRIBUTION

The key contribution of this chapter is finding the COVID-19 symptoms within a quarantine person. First of all, sensors will collect the data whether the quarantine person is in a normal position or not; if any abnormal situation occurs like continuous sneezing, cough, and abrupt changes in body parts. For training and testing purpose 2000 sample data were used and collected from 15 people.

Second, if quarantine person needs medical help then he or she can use victory sign made by fingers. For this purpose image processing is used and once these types of images will be identified then immediately alert/ help message will be sent to registered mobile numbers for help. Also, for security purpose end-to-end encryption algorithm is used so that an intruder cannot alter the actual message.

22.3 LITERATURE SURVEY

In Chaolin Huang's paper[4] features of corona patients were described on the basis of 41 people admitted in Wuhan, China. Mostly people were men (30 out of 41), their median age was 49 years. In that one family cluster was also found. From there it was observed that it is community spread. In their medical reports few common symptoms were identified like fever (40 out of 41), fatigue or myalgia (18 out of 41), and cough (31 out of 41).

In Anuradha Tomar's paper[2] prediction of upcoming infected cases in India was given on the basis of data-driven approach based on long short-term memory concept. They predicted the number of cases in next 30 days and numbers were quite correct. For prediction they used previous data and applied their mathematical-based model to get the numbers. It was very helpful for government and health officials to take preventive steps before the time.

In Arti M.K.'s paper,[5] a mathematical model is used to predict number of possible corona cases in India. A tree-based model was used to predict the coming cases. In paper the author assumes that lockdown is working effectively to stop virus into community spread. Results are quite satisfactory when we consider only cases reported by governing bodies, but situation is more critical reason being cases not detected due to lack of random testing into all areas.

The Indian Council of Medical Research (ICMR) is also working on a mathematical model[6] which will help to tell the number of infected people in next few months. According to the chief of ICMR isolation is the only way to stop this infection into community. Their study will use testing and isolation as a defence mechanism for ending the virus spreading.

Experts of ICMR used optimistic and pessimistic concepts in mathematical modeling and found good results.[7] If isolation and quarantine will be followed strictly then up to 62% spread of coronavirus can be reduced. Its limitation is that results will be 62% only when scenario will be optimistic (Valve of Ro is 1.5); otherwise, this percentage will be reduced that comes under pessimistic scenario.

In paper [12], the researcher used a device to collect data and used transmission medium to send data for analysis. Information was classified into different categories, but the main problem with the system is that the whole algorithm works for known classification of data. The moment any information is processed which does not belong to defined classification, it starts giving false alarm. Also if any new human wears the device then again it generates false alarm. The whole system needs lot of work to prepare a complete system.

In paper [8], the researcher used two segment feature extraction method. First, it extracted features through an online tool and then applied a machine learning approach. As per paper, results are correct more than 99.9% mathematically. The system needs to check against the real-time results. Researchers' main focus was to save power consumption and they achieved. The limitation of the proposed model is that it will work on the same frequency, that is, heart rate of human should be fixed then only power consumption will be achieved. If the heart rate increases then power consumption may increase depending upon the heart rate.

In paper [9], the researcher proposed a model based on thermal sensors to detect the fall of elderly persons at home. It is good in the sense that privacy will be maintained. They used three algorithms to implement the system, that is, GRU, LSTM, and Bi-LSTM and results were accurate up to 93% in ideal environment but in real-time observation there are many obstacles like the perfect implementation place so that it can detect all the data without any interrupt, temperature, etc. It is good but still needs to be implemented and tested in real-time environment.

In paper [13] the author use Microsoft Kinect camera to capture human movement based on skeleton. The author proposed a model in which instead of object movement, skeleton movement will be captured.

In paper [10] the author used Microsoft Kinect to collect all data based on 3D skeleton from elderly at home. It is good because elderly do not need to wear any device and this whole system protects their privacy too. In the proposed model, they train all types of possible scenarios and the system will generate alarm if any such type of condition will be recorded. For the authenticity of the proposed model, available data sets are used and results are quite good but still real-time application is not developed or not tested.

In paper [14] the researcher uses a wireless video sensor network for monitoring the patients. Their main focus was on saving power consumption, memory, and fast transmission. The researcher uses stationary frames for sensing the environment and the proposed algorithm is able to achieve the desired up to three times better than previous algorithms used in other papers. They have used test videos as input. Real-time tracking of humans is still not implemented and it is in their future work.

In paper [11] the researcher uses four feature vector combinations to collect data from the patient. They used an unsupervised learning approach to train the system and the Gaussian mixture model was used to create the cluster of different situations. In this approach, the threshold value was used to differentiate between fall state and non-fall state. Wristband was used to collect all the data from the patient and it is quite effective. False fall detection was still a big problem.

These researchers are using different approaches, but the existing problem is precision and trust with the system. So, in this chapter, we have used an accelerometer to detect the quarantine person's symptoms and whenever any person used victory sign from fingers then that image will also be sent on registered mobile numbers to stop continuous watching of quarantine person's video.

22.4 RESEARCH GAP

ICMR is emphasizing on isolation and quarantine of corona-infected people; also the government is using complete lockdown to reduce the virus spread percentage. Many people are staying alone, that is, elder people, working people in different cities, and self or forced quarantine people. Specially, when nobody is there to take care of them then how governing bodies will monitor them. Till now, no solution is proposed yet. In this chapter, a model is given which will send alert messages on real time to registered mobile numbers or through email with the help of image processing and IoT so that medical help can be provided in emergency cases. The used algorithm is strong enough to identify the quarantine people's medical situation and also to secure the alert message from intruders.

22.5 PROPOSED ALGORITHM

Extended Algorithm for fall detection using IoT and Image Processing (EAFDI^2).

22.6 FLOW CHART

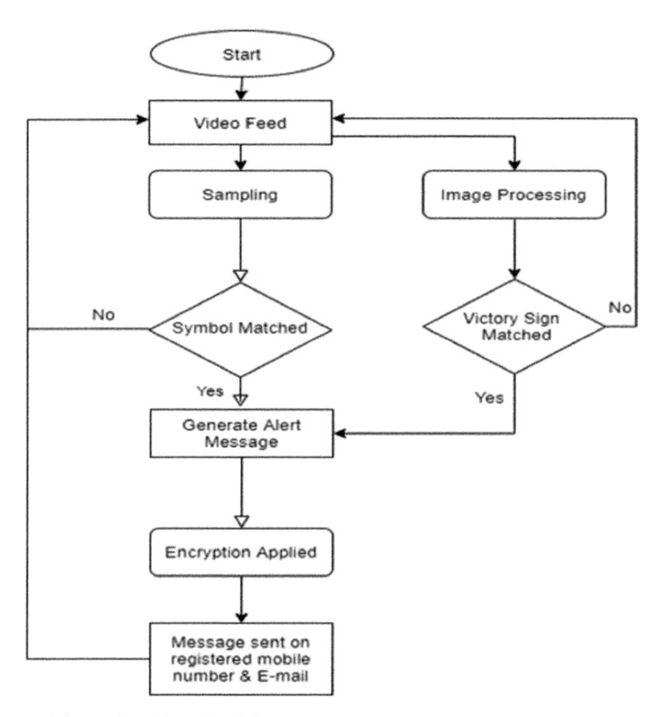

FIGURE 22.5 Flow chart for algorithm.

22.7 EXPERIMENTAL SETUP

Hardware: Arduino UNO:

Arduino UNO is an open-source microcontroller created by Arduino.cc. It is widely used microcontroller in this segment (Arduino range) because of its simplicity, ease of use, and its capacity of being interfaced to different development sheets. It is created based on ATmega328P computer chip.

The board comprises both advanced and simple pins and a programmable Arduino IDE. The force supply can be given through a USB cable or an external 9 V battery. It can accept the voltages between 7 and 20 V. Some other specification includes a 32 kB of flash memory with nearly 16 MHz clock speed.

Accelerometer:

ADXL335

It is a three-axis accelerometer. We have used it to find the angular position and net acceleration. Some of the specifications are

- Three-axis sensing
- Small and compact packaging
- Works on low power
- Works on supply (1.8–3.6v)
- 10,000 g shock survival
- Good temperature resistance

Bluetooth module:

HC-05

It is used to transmit the data collected by the sensor to PC or device. It is easy to use and encapsulated. It can work with any USB Bluetooth adapter. It works on very low power supply 3.3 V. It is also known for being compatible with master mode, slave mode, and both master–slave mode.

Piezo-buzzer:

It uses piezoelectricity to generate sound and the frequency can be given by the user in IDE.

22.8 RESULT AND ANALYSIS

The proposed scheme provides instantaneous response that will increase the attention of doctor on patients and be helpful in reducing happy hypoxia death of COVID-19 people. Image processing and capturing is going on continuously so that if any symptom matched with predefined database algorithm start working. In the proposed algorithm we will try to automate the entire process to reduce the burden and for improving overall efficiency of the entire system involved in COVID-19 management.

22.9 CONCLUSION AND FUTURE SCOPE

The proposed model can identify sneezing and cough symptoms by analyzing body movements. And also it works when a quarantine person seeks for immediate help by showing victory sign by fingers onto camera. This is a quite trustable model for medical help bodies as well as for family members. Limitation of this model is that if somehow we can feel the body temperature of a quarantine person then it will be working as a human who is observing the health situation of a quarantine person continuously. Also it needs network connectivity all the time. In the absence of network it will not be able to send alert messages on registered mobile numbers.

KEYWORDS

- **COVID-19**
- **encryption**
- **IoT**

REFERENCES

1. https://www.cebm.net/covid-19/coronaviruses-a-general-introduction/
2. Anuradha, T.; Neeraj, G. Prediction for the Spread of COVID-19 in India and Effectiveness of Preventive Measures. *Sci. Total Environ.* **2020,** 728.
3. https://www.worldometers.info/coronavirus/
4. Huang, C.; Wang, Y. et al. Clinical Features of Patients Infected with 2019 Novel Coronavirus in Wuhan, China. Published Online January 24, 2020. https://doi.org/10.1016/ S0140-6736(20)30183-5
5. Arti, M. K. Modeling and Predictions for COVID 19 Spread in India. DOI: 10.13140/RG.2.2.11427.81444. https://www.researchgate.net/publication/340362418_Modeling_and_Predictions_for_COVID_19_Spread_in_India
6. https://economictimes.indiatimes.com/industry/healthcare/biotech/healthcare/icmr-initiates-study-to-predict-the-rate-of-covid-19-infections-in-india/articleshow/74768015.cms?utm_source=contentofinterest&utm_medium=text&utm_campaign=cppst
7. https://www.indiatoday.in/science/story/coronavirus-in-india-social-distancing-quarantines-reduce-covid19-cases-icmr-study-1659140-2020-03-24
8. Cucchiara, R.; Prati, A.; Vezzani, R. A Multi-Camera Vision System for Fall Detection and Alarm Generation. *Expert Syst. J.* **2007,** *24* (5), 334–345. CrossRefGoogle Scholar.
9. Kerdjidj, O.; Ramzan, N.; Ghanem, K.; Amira, A.; Chouireb, F. Fall Detection and Human Activity Classification Using Wearable Sensor and Compressed Sensing. *J. Ambient Intell. Humanized Comput.* January **2019**.
10. Nart-Charif, H.; McKenna, S. J. Activity Summarisation and Fall Detection in a Supportive Home Environment; ICPR 2004, 2014.
11. Nho, Y.; Lim, G. J.; Kim, D.; Kwon, D. User-Adaptive Fall Detection for Patients Using Wristband. In *IEEE/RSJ International Conference on Intelligent Robots and Systems (IROS)*, 2016.
12. Saleh, M.; Jeannes, B. L. R. Elderly Fall Detection Using Wearable Sensors: A Low Cost Highly Accurate Algorithm. *IEEE Sens. J.* DOI: 10.1109/JSEN.2019.2891128
13. Taramasco, C.; Rodenas, T.; Martinez, F.; Fuentes, P.; Munoz, R.; Olivares, R.; Albuquerque, V.; Demongeot, J. A Novel Monitoring System for Fall Detection in Older People. In *IEEE Access* **2016,** 4. DOI: 10.1109/ACCESS.2018.2861331
14. Veeraputhiran, A.; Sankararajan, R. Feature Based Fall Detection System for Elders Using Compressed Sensing in WVSN. *Wireless Netw.* **2017**. DOI: 10.1007/s11276-017-1557-3.

CHAPTER 23

PROBABILISTIC IMAGE ENCRYPTION-BASED SECURE SURVEILLANCE FRAMEWORK FOR AN IOT ENVIRONMENT

MOHD SHARIQ and KARAN SINGH

School of Computer and Systems Sciences, Jawaharlal Nehru University, New Delhi, India

ABSTRACT

With the advancement in Internet and Communication Technologies, the sensitive information of the transmitted contents can be easily leaked via insecure communication from across the globe. In the digital era, different multimedia technologies have emerged across the globe. Moreover, a large amount of data such as images and audio-visual documents (such as video, speech, and music) is digitized at a low cost and can be stored over a public network. Images have been widely used for searching, sharing, and uploading at various platforms, which make them insecure against adversaries. To address the security issues, image encryption is becoming more prominent from secure image communication aspect. In the last decades, a wide range of chaotic map-based image encryption schemes have been introduced which suffer from high computation overhead and low-key space. This chapter puts forward a probabilistic image encryption-based secure surveillance framework for the IoT environment. The proposed scheme has been analyzed under some security test measures such as entropy, NPCR, UACI, and correlation coefficient. The obtained results confirm that our scheme is resistant to various known security

Advanced Computer Science Applications: Recent Trends in AI, Machine Learning, and Network Security. Karan Singh, PhD, Latha Banda, PhD & Manisha Manjul, PhD (Eds.)

attacks. Furthermore, the proposed scheme consumes comparatively low computation overhead.

23.1 INTRODUCTION

Nowadays, a broad range of the digital images is popularly generated, transferred, and stored over the communication network. In addition, the images have been widely shared, searched, and uploaded in each organization, which can be attacked by some third party or can be an attacker, thus an image is vulnerable to various attacks.[1,2] Therefore, image encryption is becoming more prominent in the area of secure image communication. As the sensitive data of the images is being widely used in a variety of multimedia applications, image security is becoming more crucial from storage and communication aspects. To mitigate security flaws in the storage and communication of sensitive data, many cryptographic techniques have been proposed for ensuring secure communication that must satisfy the integrity, authentication, and data confidentiality properties.[3,4] The transmitted data related to encryption can be converted in the form of unpredictable format using some secret keys as shown in Figure 23.1. However, it can be vulnerable to an adversarial attack targeted at decrypting the data by employing state-of-the-arts cryptographic techniques and also performing various other security attacks.

Over the last decades, several encryption methods have been introduced. Encryption is considered to be efficient and classic method to solve such problems. Moreover, the conventional cryptographic methods including AES and DES typically have been used for the textual information. Due to high correlation and large data, these methods are infeasible for multimedia information. The major goal of the image encryption method is to provide privacy like in video surveillance systems.[5] The main aim is to protect the information and contents of the image with higher efficiency which can be attained by employing modern techniques rather than conventional cryptographic methods. With regards to privacy protection, Region of Interest (RoI) describes encrypting the entire video that can be used to encrypt the limited privacy-sensitive areas due to resource-constrained IoT devices. Besides, RoI contains the sensitive information of the image. Many cryptographic methods can be applied in privacy protection such as RSA and AES but the major drawback of these methods is, they cannot efficiently meet the real-time features of the video surveillance systems.

23.1.1 PROBLEM DEFINITION AND MOTIVATION

In the digital era, images are being widely used on every social media platform. The wireless communication channel is vulnerable to various attacks performed by adversaries, thus presenting the data privacy and security concerns. In this context, the attacker or adversary can perform several known security attacks and can obtain the sensitive content of the useful images. To address such security flaws, image encryption approaches have been taken into account. Therefore, we have put forward a probabilistic image encryption-based secure surveillance framework for the IoT environment in this chapter. The proposed scheme is efficient and robust and also prevents various security attacks as compared with the previous existing schemes.

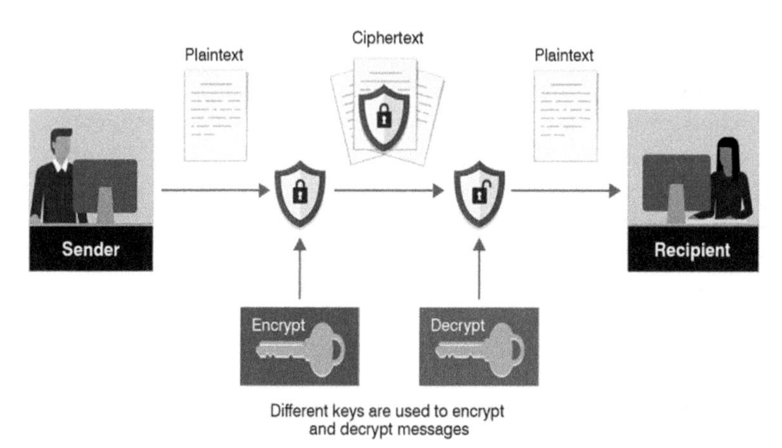

FIGURE 23.1 Process of image encryption.

23.1.2 CONTRIBUTION

- This chapter puts forward a probabilistic image encryption-based secure surveillance framework for IoT environments.
- We used a video summarization approach issued to extract the contained information of frames with the help of visual sensors processing capabilities.
- In case, whenever an abnormal event is being detected by keyframes, then subsequently an alert signal has been sent to the concerning authorities.
- The experimental study confirms the effectiveness of our scheme for execution time and robustness.
- Our scheme shows superiority in terms of security features as to other image encryption algorithms.

23.1.3 ORGANIZATION

The remaining part of this chapter is structured as follows. A brief study of existing related work with their techniques, advantages, and pitfalls is presented in Section

23.2. The preliminaries and notations used are presented in Section 23.3. Next, the proposed probabilistic image encryption-based secure surveillance framework for IoT environment is presented in Section 23.4. Further, Section 23.5 evaluates the performance analysis along with experimental setup and result discussion of our scheme. Finally, Section 23.6 presents the concluding remark.

23.2 RELATED WORK

This section mainly focuses on several image encryption approaches proposed by many authors.[6–12] The various classes of encryption techniques that are employing different kinds of 2D chaotic maps are as follows:

Mondal and Mandal[13] proposed a secure and lightweight chaos and DNA computing-based image encryption scheme. Their scheme produces two "pseudo-random number" sequences with the help of PRNG by employed chaotic logistic map. Thereafter, the first and second PRN sequences are used for permutation of the plain image and generate random DNA sequences, respectively. The authors ensure that their scheme resists known security attacks.

Hua and Zhou[14] introduced a "2D Logistic Adjusted Sine map-based image encryption scheme" called (LAS-IES). The used map has better unpredictability and ergodicity. The properties of confusion and diffusion have been used in their work which efficiently encrypts various types of images.

Hamza and Titouna[15] proposed a novel Zaslavsky chaotic (ZC) map-based encryption scheme for securing digital images. The used ZC map utilizes the pseudo-random generator which generates the key encryption. The permutation-diffusion processes have been adopted for ensuring the confusion and diffusion features for the obtained encrypted images. The authors have shown that their scheme can withstand various security attacks. However, the secret keys have been selected based on the real numbers, which is the main issue of their scheme.

Rafik Hamza[16] proposed a novel Chen chaotic system-based algorithm for PRN sequence generators for image-cryptographic applications. The algorithm generates cryptographic keys and PRNG solves the issue related to the non-uniform probability distribution by using Chen chaotic system. Their algorithm meets a variety of security features including robustness against differential and statistical attacks, key sensitivity, and large key space.

Preishuber et al.[17] proposed an empirical security analysis and the main motivation for employing various chaos-based image encryption as compared to conventional cryptographic encryption. The authors have shown various security measures tests experimentally for chaos-based encryption schemes.

Wu et al.[18] proposed DNA and 2D Henon-Sine Map-based image encryption called 2D-HSM. This map has better pseudo randomness, ergodicity as compared with other existing chaotic maps. The DNA XOR operation and DNA random encoding have been used for improving the efficiency of permutation-diffusion images. In this scheme, the authors have shown their scheme can withstand several security attacks including differential, exhaustive, and statistical attacks. However, Chen et al.[19] claimed that their scheme is not secure. To fix the shortcomings, they subsequently introduced the substitution boxes to improve the effect of XOR operations and DNA random coding.

Kaur et al.[20] proposed "a nondominated sorting genetic algorithm" based color image scheme and 5D chaotic map-based local chaotic search for image encryption. The used input image can be split into various sub-bands with the help of a "dual-tree complex wavelet transforms (DTCWT)." Besides, such sub-bands can be diffused with the help of used secret keys, which are computed from the optimizing 5D chaotic map. Furthermore, the DTCWT inverse has been employed to compute the ciphered image.

23.3 PROPOSED SCHEME

The proposed scheme shows a pictorial representation by a flow graph given in Figure 23.2.

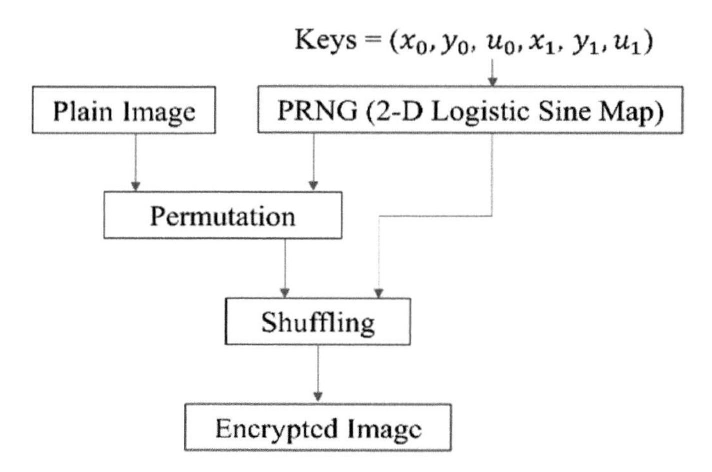

FIGURE 23.2 Pictorial representation of our scheme.

23.3.1 2D-LOGISTIC MAP

It can be defined as a "discrete dynamic system," which shows a chaotic behavior of the evolution of attractors and orbits.[21] Compared to 1D logistic map,[22] the used map is having an extraordinary complex random behavior, which can be denoted by

$$\begin{cases} x_{i+1} = r(3y_i + 1)x_i(1 - x_i) \\ y_{i+1} = r(3x_{i+1} + 1)y_i(1 - y_i) \end{cases}$$

where r denotes the system's parameter and the value of r for the 2D-Logistic system will vary in the range between 1.1 and 1.19[21] because of its chaotic behavior. The value of (x_i, y_i) in the above equation denotes the point of the loop.

Algorithm 1. Chaotic Sequences generation Using LASM.

Input: (x_0, y_0, u, P)

Output: Sequence.

Process:

1: $[a, b, c] \leftarrow size\ (P)$

2: Summation = $\sum_i \sum_j P$

3: ***if*** Summation = 0

$S \leftarrow 0$;

else

$S0 = 2 + abs\ (\log 10\ (\text{sum}^{-1}))$

$S = e^{(S0)} \times Sum^{-1g}$

end

4: $x = x_0 + S$; $y = y_0 + S$; $u = u_0 + S$

5: Sequence ← zeros $(a \times b \times c, 1)$

6: for $i = 1$ *to* $\lfloor (a \times b \times c)/2) \rfloor$

$x_{i+1} = r(3y_i + 1)\, x_i\, (1 - x_i)$

$y_{i+1} = r(3x_i + 1) y_i\, (1 - y_i)$

Sequence $(2i)$ = floor $\lceil 10^{14} \times x_{i+1} \rceil$ mod 256

Sequence $(2i + 1)$ = floor $\lceil 10^{14} \times y_{i+1} \rceil$ mod 256

end

23.3.2 ENCRYPTION METHOD

***Phase* 1:** We assume that the keyframe I of size $[a \times b \times 3]$. Algorithm 1 produces the chaotic sequences of numbers, which are denoted by P_1 as follows:

$$P_1 = PRNG(x_0, y_0, u_{0'0})$$

Phase 2: In this phase, the initial processing is performed of our scheme as follows:

- $[I_R I_G I_B] \leftarrow I$.
- $N_R = LSB(I_R) \oplus I_G \oplus I_B$
- $N_G = N_R \oplus I_B$.
- $N_B = N_R \oplus I_G$.
- $N_1 \leftarrow [N_R N_G N_B]$, reshape the three matrices $(N_R N_G N_B)$ into a 1-Dimension vector N_1.
- $N_{initial} = C_1 \oplus P_1$.

Phase 3: This phase produces two chaotic sequences of numbers P_2 and P_3 by XORing P_1 with $PRNG(x_0, y_0, u_0, N_{initial})$ are as follows:

- Produce $P_2 = PRNG(x_1, y_1, u_1, N_{initial})$.
- Produce $P_3 = PRNG(x_0, y_0, u_0, N_{initial}) \oplus P_1$.

Phase 4: To compute the indices sequences of π and π', this phase sorts the computed chaotic sequences of numbers P_2 and P_3 in increasing order. Therefore, the permutation matrices of generated sequences are as follows:

- $Sort(P_1) = P_1' = [\pi_1, \pi_2, \pi_3, \ldots, \pi_{a \times b \times 3}]$.
- $Sort(P_2) = P_2' = [\pi'_1, \pi'_2, \pi'_3, \ldots, \pi'_{a \times b \times 3}]$.

Phase 5: Performed shuffling N using sort index of new computed sequences. Then we employed P-box of P'_2 and P-box of P'_3, respectively.

23.3.3 DECRYPTION METHOD

The main goal of the decoding process is to retrieve the original keyframe by applying the inverse encryption method. The exact secret keys have been used to retrieve the original keyframe from encrypted one. The following steps are used in this process as given below:

Phase 1: Reading the encrypted keyframe $N_{initial}$.

Phase 2: Reshaping the image matrices into a single matrix.

Phase 3: Applying Algorithm 1, the chaotic sequences P_1, P_2, and P_3 are produces as given below:

- Produce $P_1 = PRNG(x_0, y_0, u_0, 0)$.
- Produce $P_2 = PRNG(x_1, y_1, u_1, N_{initial})$.
- Produce $P_3 = PRNG(x_0, y_0, u_0, N_{initial}) \oplus P_1$.

Phase 4: For restoring the original pixel's position, the bijection property of P_2 and P_3 permutation matrices are employed. Then, the inverse permutation-box of P_3 and P is employed.

Phase 5: Repeating *Phase 4* by changing P-box order, that is, we are using the inverse P-box of P_2 first, followed by using the inverse P-box of P. Then we can compute a matrix that is represented as N_4.

Phase 6: The steps of the final processing are as given below:

$N_{Final} = N_4 \oplus P_1$, reshaping the computed matrix into three $N'_R N'_G N'_B$ matrices for the following RGB matrices as follows:

- Reshape $N_R \oplus N'_R \oplus N'_G \oplus N'_B$.
- Reshape $N_G \leftarrow N'_G \oplus N'_R$.
- Reshape $N_G \leftarrow N'_G \oplus N'_R$.

Phase 7: The computed matrix is denoted by "*N*" which consists of N_R, N_G, and N_B matrices indicate the decrypted keyframes.

23.4 EXPERIMENTAL SETUP

This section demonstrates that the performance evaluation has been performed for simulation and analysis with different security assessment metrics. The experimental setup needed MATLAB R2016a in Windows 10 professional, an *i*7-4790 CPU of 3.60 GigaHertz, and 16 GigaByte of RAM. Figure 23.3. (*a*, *i*), (*b*, *i*), and (*c*, *i*) (where, *i* = 1,2,3) shows that the resulting keyframes. Furthermore, Table 23.1 illustrates the various security test measures.

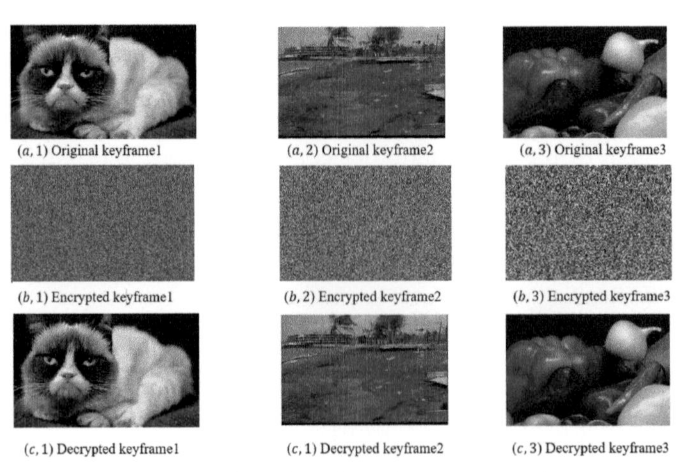

FIGURE 23.3 (*a*, 1), (*a*, 2), and (*a*, 3): original keyframes; (*b*, 1), (*b*, 2), and (*b*, 3); encrypted keyframes, and (*c*, 1), (*c*, 2), and (*c*, 3): decrypted keyframes.

23.4.1 SECURITY ASSESSMENT METRICS

23.4.1.1 KEY SPACE

The key space size can be computed by the total number of different keys used for the proposed image encryption algorithm. To provide resistance from brute force and exhaustive attacks, a good image cryptosystem must ensure a large key space.

23.4.1.2 Key Sensitivity

With regards to a good image cryptosystem, the proposed scheme should ensure the property of sensitivity to the used secret keys. More clearly, a slight modification can be done in the used secret keys that could be the cause of substantial harm in the obtained cipher image.

23.4.1.3 CORRELATION COEFFICIENT (CC)

The CC can be represented by r_{xy}. The value of r_{xy} could be 1 or -1, which indicates a high correlation. The value of r_{xy} is zero, which indicates no correlation.[22] To overcome statistical attacks, the value of r_{xy} must be zero for encrypted key frames or images. This test is perfromed to find out the relation between the same pixels of the cipher and plain images. The computation of CC is done by using the following formula:

$$r_{xy} = \frac{Cov(x, y)}{\sqrt{D(x)}.\sqrt{D(y)}}$$

where $Cov(x, y) = \frac{\sum_{i=1}^{n}(x_i - E(x))(y_i - E(y))}{n}$

$$D(x) = \frac{\sum_{i=1}^{n}(x_i - E(x))^2}{n}$$

$$D(y) = \frac{\sum_{i=1}^{n}(y_i - E(y))^2}{n}$$

$$E(x) = \frac{\sum_{i=1}^{n} x_i}{n}$$

23.4.1.4 DIFFERENTIAL ATTACKS ANALYSIS

NPCR and UACI tests are performed to provide resistance against the differential attack.[23] However, the attackers could slightly modified the plain image for performing this attack, for instance, he/she can modify the one-pixel value to get some useful information between the cipher and plain images. Let C_1 and C_2 are two cipher images of size $M \times N$ and having one-pixel value change between their plain images, the corresponding NPCR and UACI can be calculated by the formulae:

$$NPCR = \frac{\sum_{i,j} D(i,j)}{M \times N}$$

$$UACI = \frac{1}{M \times N} \sum_{i,j} \frac{|C_1(i,j) - C_2(i,j)|}{255} \times 100\%$$

where "D" is representing the number of pixels.

23.4.1.5 INFORMATION ENTROPY

This is considered the most important feature of randomness. This also measures the degree of uncertainties of information content. By Shannon's theory, the computation of the information entropy is calculated by using the following formula:

$$H(I) = -\sum_{i}^{N} P(m_i) \log_2 P(m_i)$$

Where N is the possible gray level values used to measure the "randomness" of an encrypted image and $P(m_i)$ denotes that the occurrence probability of symbol m_i. The high value of entropy $H(I)$ demonstrates that the images are more uniform or more randomness in the image. The entropy is always assumed to be closed to 8 for an ideal random image.

23.4.1.6 HISTOGRAM TEST

The image histogram test demonstrates the characteristics of the distribution of its involved pixel values.[24] In other words, Figure 23.4 shows the regularity of pixel's values.

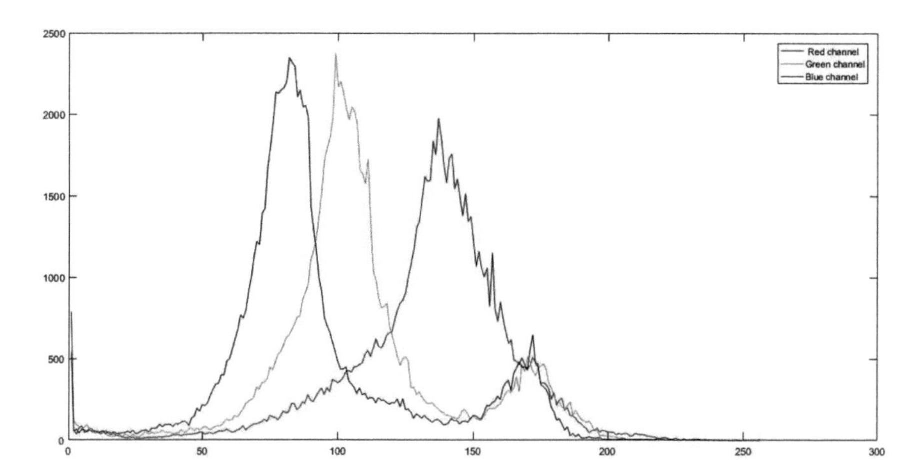

FIGURE 23.4 Histogram.

TABLE 23.1 Various Test Measures of Our Scheme.

Image	Entropy	Correlation	NPCR	UACI
Keyframe1	7.9978	0.9716	99.6149	33.4292
Kayframe2	7.9977	0.9860	99.6035	33.5352
Keyframe3	7.9981	0.9376	99.5698	33.1009

23.5 CONCLUSIONS

This chapter puts forward a probabilistic image encryption-based secure surveillance framework for IoT environments. The 2D logistic sine map is employed to generate the pseudo-random numbers sequence. Then we used an efficient video summarization approach issued to extract the contained information of frames with the help of the processing capabilities of visual sensors. After obtaining and analyzing the experimental results, the proposed scheme ensures various known security test measures such as NPCR, UACI, entropy, and correction coefficient. Furthermore, the obtained results show that our scheme resists various known security attacks with comparatively low computation overhead. In context of processing and storage-constrained devices, the proposed scheme is better suited in such scenarios for IoT systems.

KEYWORDS

- **Internet and Communication Technologies**
- **image encryption**
- **NPCR**
- **UACI**
- **security**

REFERENCES

1. Akhavan, A.; Samsudin, A.; Akhshani, A. A Symmetric Image Encryption Scheme Based on Combination of Nonlinear Chaotic Maps. *J. Franklin Inst.* **2011,** *348*, 1797–1813.
2. Taha, M.; Parra, L.; Garcia, L.; Lloret, J. An Intelligent Handover Process Algorithm in 5G Networks: The Use Case of Mobile Cameras for Environmental Surveillance. *Proc. 2017 IEEE Int. Conf. Commun. Workshops*, **2017,** 840–844.
3. Liu, Y.; Nie, L.; Han, L.; Zhang, L.; Rosenblum, D. S. Action2Activity: Recognizing Complex Activities from Sensor Data. *Proc. Int. Joint Conf. Artif. Intell.* **2015,** 1617–1623.
4. Liu, Y.; Nie, L.; Liu, L.; Rosenblum, D. S. From Action to Activity: Sensor-Based Activity Recognition. *Neurocomputing* **2016,** *181*, 108–115.
5. Muhammad, K.; Sajjad, M.; Mehmood, I.; Rho, S.; Baik, S. W. Image Steganography Using Uncorrelated Color Space and Its Application for Security of Visual Contents in Online Social Networks. *Future Gener. Comput. Syst.* **2016**. DOI: 10.1016/j.future.2016.11.029.
6. Hua, Z.; Zhou, Y. Image Encryption Using 2d Logistic-Adjusted-Sine Map. *Inf. Sci.* **2016,** *339*, 237–253.
7. Huang, X. Image Encryption Algorithm Using Chaotic Chebyshev Generator. *Nonlinear Dyn.* **2012,** *67*, 2411–2417.
8. Xu, L.; Li, Z.; Li, J.; Hua, W. A Novel Bit-Level Image Encryption Algorithm Based on Chaotic Maps. *Opt. Lasers Eng.* **2015,** *78*, 17–25.
9. Wei, X.; Guo, L.; Zhang, Q.; Zhang, J.; Lian, S. A Novel Color Image Encryption Algorithm Based on DNA Sequence Operation and Hyper-Chaotic System. *J. Syst. Softw.* **2012,** *85*, 290–299.
10. Belazi, A.; El-Latif, A. A. A.; Belghith, S. A Novel Image Encryption Scheme Based on Substitution-Permutation Network and Chaos. *Signal Process.* **2016,** *128*, 155–170.

11. Machkour, M.; Saaidi, A.; Benmaati, M. A Novel Image Encryption Algorithm Based on the Two-Dimensional Logistic Map and the Latin Square Image Cipher. *3D Res.* **2015,** *6*, 1–18.
12. Wu, X.; Wang, D.; Kurths, J.; Kan, H. A Novel Lossless Color Image Encryption Scheme Using 2D DWT and 6D Hyperchaotic System. *Inf. Sci.* **2016,** *349*, 137–153.
13. Mondal, B.; Mandal, T. A Light Weight Secure Image Encryption Scheme Based on Chaos & DNA Computing. *J. King Saud Univ.-Comput. Inf. Sci.* **2017,** *29* (4), 499–504.
14. Hua, Z.; Zhou, Y. Image Encryption Using 2D Logistic-Adjusted-Sine Map. *Inf. Sci.* **2016,** *339*, 237–253.
15. Hamza, R.; Titouna, F. A Novel Sensitive Image Encryption Algorithm Based on the Zaslavsky Chaotic Map. *Inf. Secur. J.* **2016,** *25* (4–6), 162–179.
16. Hamza, R. A Novel Pseudo Random Sequence Generator for Image-Cryptographic Applications. *J. Inf. Secur. App.* **2017,** *35*, 119–127.
17. Preishuber, M.; Hütter, T.; Katzenbeisser, S.; Uhl, A. Depreciating Motivation and Empirical Security Analysis of Chaos-Based Image and Video Encryption. *IEEE Trans. Inf. Forensics Secur.* **2018,** *13* (9), 2137–2150.
18. Wu, J.; Liao, X.; Yang, B. Image Encryption Using 2D Hénon-Sine Map and DNA Approach. *Sign. Process.* **2018,** *153*, 11–23.
19. Chen, J.; Chen, L.; Zhou, Y. Cryptanalysis of a DNA-Based Image Encryption Scheme. *Inf. Sci.* **2020,** *520*, 130–141.
20. Kaur, M.; Singh, D.; Sun, K.; Rawat, U. Color Image Encryption Using Non-Dominated Sorting Genetic Algorithm with Local Chaotic Search Based 5D Chaotic Map. *Future Gen. Comput. Syst.* **2020,** *107*, 333–350.
21. Wu, Y.; Noonan, J. P.; Yang, G.; Jin, H. Image Encryption Using the Two-Dimensional Logistic Chaotic Map. *J. Electr. Imag.* **2012,** *21* (1), 013014.
22. Norouzi, B.; Mirzakuchaki, S.; Seyedzadeh, S. M.; Mosavi, M. R. A Simple, Sensitive and Secure Image Encryption Algorithm Based on Hyper-Chaotic System with Only One Round Diffusion Process. *Multimedia Tools App.* **2014,** *71* (3), 1469–1497.
23. Wu, Y.; Noonan, J. P.; Agaian, S. NPCR and UACI Randomness Tests for Image Encryption. *Cyber Journals: Multidisciplinary Journals in Science and Technology, Journal of Selected Areas in Telecommunications (JSAT)* **2011,** *1* (2), 31–38.
24. Wu, X.; Wang, D.; Kurths, J.; Kan, H. A Novel Lossless Color Image Encryption Scheme Using 2D DWT and 6D Hyperchaotic System. *Inf. Sci.* **2016,** *349*, 137–153.

CHAPTER 24

AN ENHANCED APPROACH FOR MULTIMETRIC GEOGRAPHICAL ROUTING IN VANETS USING A FUZZY INTERFACE SYSTEM

AMARPREET SINGH and NAVROOP KAUR

Computer Science Engineering, ACET Amritsar, Punjab, India

ABSTRACT

A Vehicular Ad hoc Network is a characteristic of mobile ad hoc networks where mobile nodes are depicted as vehicles (smart) equipped with network cards, computers, and sensors. In ad hoc network, to exchange traffic information the communication takes place between the vehicles with each other and if there is a need to access internet or request some information then communication takes place between vehicles and base stations placed along the roads. In this kind of networks, the routing is the tedious task to perform. A large number of routing protocols are available. In this study, a novel weight-based approach for data transmission is developed. The fuzzy inference system is applied for evaluating the weight value. Various parameters such as node density, mobility, node status, link lifetime, and PDR are used for evaluating the weight function. The results are analyzed by using the MATLAB simulation platform.

Advanced Computer Science Applications: Recent Trends in AI, Machine Learning, and Network Security. Karan Singh, PhD, Latha Banda, PhD & Manisha Manjul, PhD (Eds.)

24.1 INTRODUCTION

For the duration of the most recent few ages, the particular incidence associated with remote control methods continues to be grown easily, resulting from a lot more extensive convenience as well as rapid powerpoint presentation associated with remote control mobile phone models within various intricate gadgets, by way of example, workstation, PDAs as well as PCs.[1] The considerable variations have already been brought to data methods and advertising devices by the usage of remote control correspondence. A transmitting channel utilized by remote control frameworks, giving extra major adaptability,[6] like this it is actually possible for you the system where cables are definitely troublesome. Companies as well as household industry are finding all these fresh styles of methods in a very vibrant way greatly assist lower operating cost and execution. Wired interchanges may very well be super ceded through remote control correspondences usually.[7] At present a-days, voyaging customers could tactic online on a lot of locations similar to their places of work, properties, as well as on open locations such as events, air flow final, inns, shopping centers, as well as libraries.

The ad hoc system is among the most well-known sensor network primarily based communicating sector.[21] It's a bit totally different from WSN. Around WSN this system adheres to a small system whereas, within ad hoc system, this system was lacking a small system or maybe topology. Such ad hoc system provides a variety of categories for instance Vehicular Ad hoc Networks (VANETs), Mobile ad hoc Networks (MANETs), and also Flying ad hoc Networks. VANETs

Typical Targeted visitors management devices are dependent using a central structure where by the data within the solidity and conditions of the website traffic has been gathered by simply adding cams as well as sensors on the road.[8] This information is shipped to some sort of key model whereby it is actually prepared to create enough decisions. This sort of devices exhibit some sort of relatively big cost of deployment and so are distinguished using a impulse period of the order associated with about a minute with the producing as well as transporting information.[3] In times in which the contract has a big value to get transmitting of info, this kind of hesitate is just not acceptable. Also, the actual occasional and expensive upkeep is actually required for kit that comes with the actual road. Therefore, a large investment in transmission structure as well as sensors is actually required for deploying such a head unit using a larger scale. However, a whole new structures decentralized (or semicentralized) derived from vehicle-to-vehicle communications (V2V, Vehicle to Vehicle) was made recently while using speedy development of mobile transmission systems pursuing devices and data range by simply sensors.[9] An incredible attention with the scientific area, car makers as well as telecom owners have been enticed from this architecture. It really is based on a process distribution, autonomous, and also the cars themselves can cause this kind of structures without help of a restricted structure sending info as well as messages. Some sort of VANET circle is a sign of MANET cpa affiliate networks where by cell nodes usually are stated as cars (smart) equipped with circle cards, PCs, as well as sensors. With ad hoc network, to change website traffic information the actual transmission happens relating to the cars with each other as well as if you have need to have to gain access to internet as well as ask some information and then transmission happens among cars as well as starting programs inserted across the roadways.[14]

Nowadays, there are plenty of effects involving transport for the fiscal along with human assets. Such as, in the USA, in 2009, 30,797 fatalities ended up being brought on a result of the simply reason- traffic crashes, leading to $115 billion expense appeared to be priced to get traffic jams.[12] Consequently, helping the wellbeing along with efficaciousness involving readers are an important process. In intelligent transportation system, adding transport devices along with information technology is usually essentially the most appealing methods, the place VANETs are deemed for a essential component. Intelligent cars or trucks are used to build VANETs, which will are known as exclusive MANETs. Long run setting involving this product is usually visualized mindful about can be smart cars or trucks built with information and facts series equipment (on-board sensors),[13] on-board demonstrate equipment, information, and facts producing equipment (on-board CPU) as well as mobile transmission devices.

24.2 PROBLEM FORMULATION

VANET is a vibrant wireless ad hoc system with regard to communicating in between automobiles with no before-started infrastructure. Creators associated with[1] offered a proficient the navigation process titled AHP-based Multimetric Geographical Routing Protocol. That process essentially utilizes a computed individual considering performance to identify the upcoming hop node inside a explained variety, which could ensure the increased forwarding process. The issue that may be confronted with the work is to explain the extra weight value. It is not easy to explain truly what pounds price will be best to achieve the best results. Though it is drawing very good results in the circumstance they may be concentrating nevertheless it had been a challenging difficulty to locate best pounds price so we have a real need to bring up to date the extra weight price concept.

24.3 PROPOSED WORK

Throughout above area them is determined that this regular direction-finding concept functions based on extra weight cost just about all is afflicted with many concerns including how to pick the ideal pounds cost among the disposable pounds values. Hence the recommended function updates the more common function through changing the technique of pounds cost using Fuzzy game controller based mostly pounds cost examination function. The advantages of this idea are that it does not involve any people assistance, that is, it is not necessary to penetrate extra weight values manually. So in the work, a new unclear based mostly method is implemented to help complete extra weight of each and every node consequently technology-not only pertaining to upcoming hope to complete connection in network. As well in regular function, the selection parameter is usually not having a security issue, as a way an improvement, this node PDR is provided for a selection element as a possible enhancement to the regular work. Throughout regular function, the factors that were useful for examining this CH selection chances are listed below:

Mobility: It refers to the mobility of the node with respect to the speed and Distance.

$$M_{s,i}^{(d)} = P_{D^{D_{avg}}} + P_{A^{A_{s,i}^{(d)}}} + p_{S^{SP_{avg}}} \tag{24.1}$$

Davg Denotes the distance among the vehicles, S*Pavg* denotes the average moving speed $A_{s,i}^{(d)}$ defines the moving angle between lines.

Link Lifetime:

$$L_{x,i} = R - \frac{\sqrt{(x_i - x_s)^2 + (y_i - y_s)^2}}{V_s - V_i} \tag{24.2}$$

Node Status: It defines the current status of the node with respect to the status of the buffer.

$$Q_i^{(t)} = \frac{Q_{max} - Q_i^{(t)}}{Q_{max}} \tag{24.3}$$

Where, Q_{max} refers to the maximum buffer size, $Q_i^{(t)}$ denotes the number of packets in the buffer queue at time t.

Node density:

$$T_i\left(t\right)=\frac{Neighbour\ table\ size()}{R} \tag{24.4}$$

Other than this, the traditional weight value evaluation formulae are as follows:

$$W_{s,i}^{(d)}=P_{M^{M_{s,i}{}^{(d)}}}+P_{L^{L_{s,i}}}+p_{Q^{Q_i^{(t)}}}+p_{T^{T_i}} \tag{24.5}$$

Inside consist of work, the load assessment mechanism may be kept up to date utilizing the fuzzy inference program with regard to calibrating the load functionality and with this a PDR is actually increased as another element with regard to figuring out the selection probability. The next is really a layout of the consist of fuzzy primarily based excess fat range scheme.

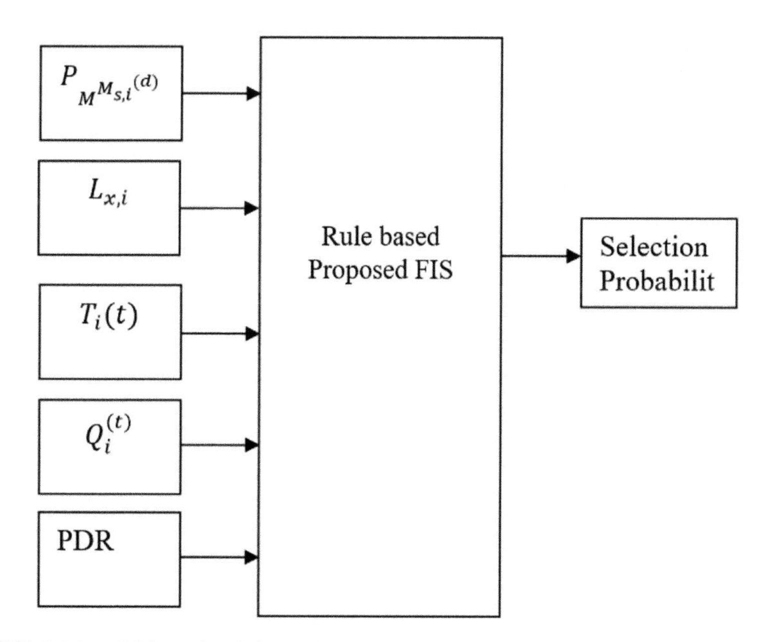

FIGURE 24.1 FIS based weight evaluation framework.

The PDR evaluation is defined as below in eq 24.6.

$$PDR = \sum_{i=1}^{n} \frac{R_i}{S_i} \quad (24.6)$$

Where, *Ri* is used for packets counts that are received at targeted node and *Si* defines the number of packets send by the source node.

The methodology of the proposed work is as follows:

TABLE 24.1 Simulation Setup of Proposed Work.

Parameters	Value
Time for simulation	400
Traffic count	10
Carrier Frequency	5.8 GHz
Length of Data Packet	512 bytes
Physical Layer	IEE802.11p(11Mbps)
Propagation–Model	Two-Ray ground model
Transmission Power	10mW
Traffic Type	UDP

***Step 1**: The first step is to define the initial network parameters such as simulation time, data packet length, carrier frequency, propagation model, traffic type, physical layer etc. The proposed work has the following initial parameters (Table 24.1) as the network setup*

***Step 2**: After defining the initial parameters, the network is deployed. The source node is elected from the deployed nodes in order to initiate the communication process in the network.*

***Step 3**: Implement the next hop selection criteria by using PDR of individual node.*

***Step 4**: Next hop selection is performed for the route creation to transfer the data from a source node to the target node.*

***Step 5**: At last, the data transmission is performed and the performance of proposed work is evaluated.*

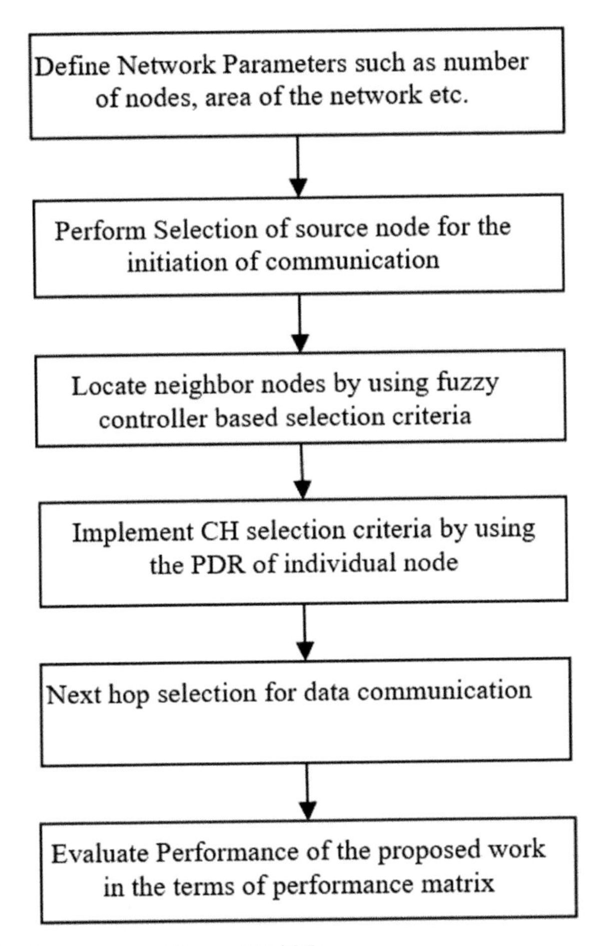

FIGURE 24.2 Framework of Fuzzy AMGRP.

The proposed work implements the Fuzzy-AMGRP routing protocol for VANETs. The fuzzy inference system is implemented for electing the CH nodes and the CH is elected on the basis of the major factors as follows:

- Mobility
- Link Lifetime
- Node Status
- Node Density
- PDR

The MATLAB simulation platform is used for experimental analysis. The execution of the present work is evaluated in the terms of following factors.

Packet Delivery ratio: the packet delivery ratio is a performance evaluation metrics that is specifically used to measure the rate of information bundles conveyed to the objective successfully. It is evaluated as follows:

$$PDR = \sum_{i=1}^{n} \frac{R_i}{S_i} \tag{24.7}$$

Where, n defines the number of source nodes, Ri depicts the number of information parcels got at the objective node, Si is used to define the quantity of information parcels sent by the source hub.

End-to-End Delay: this parameter is utilized to quantify the normal postponement taken by the information bundles to arrive at the objective hub. The end-to-end delay in proposed work is evaluated as follows:

$$End\ to\ End\ delay = \frac{1}{\sum_{i=1}^{n} R_i} \left(\sum_{i=1}^{n} \sum_{j=1}^{R_i} TR_{ij} - TS_{ij} \right) \tag{24.8}$$

Where, the TR denotes the time of receiving the jth data packet that has been transmitted by the ith foundation at the target & $TSij$ denotes transfer time of the jth data packet by the ith source node.

Normalized Routing Overhead: It depicts the proportion of absolute control packets corresponding to the total delivered packets in the network. It is measured by using the following formulation:

$$NRL = \frac{1}{n} \left(\sum_{i=1}^{n} \frac{1}{R_i} \left(\sum_{j=1}^{R_i} \sum_{k=1}^{P_{ij}} C_{ijk} \right) \right) \tag{24.9}$$

The count of control bytes at the kth hop by the jth packet sent at the ith source node is denoted by the variable C_{ijk}.

Average Hop Count: it is an average number of hops required to transmit the data to the base station.

$$AHC = \frac{1}{n} \left(\sum_{i=1}^{n} \frac{1}{R_i} \left(\sum_{j=1}^{R_i} \sum_{k=1}^{P_{ij}} H_{ijk} \right) \right) \tag{24.10}$$

The variable H_{ijk} denotes the number of kth hop traversed by the jth data packet to reach the ith source.

24.4 RESULT ANALYSIS

Figure 24.3 depicts the packet shipping ratio of the presented work. PDR is analyzed dependant upon the quantity of nodes readily available inside network. Back button axis inside data reveals the quantity of nodes and *y* axis calibrates the details for packet shipping ratio and yes it differs amongst 0 and 1. The data symbolizes any time the quantity of nodes inside network is 50 the PDR is analyzed in the area 0.5, while the quantity of nodes is 75, the PDR is 0.566 and many others when the count number with nodes reaches for the 250, as well as acquired PDR is 0.919, respectively. Consequently, it truly is turned out that this offered deliver the results provides the greatest PDR while using varied quantity of nodes inside network.

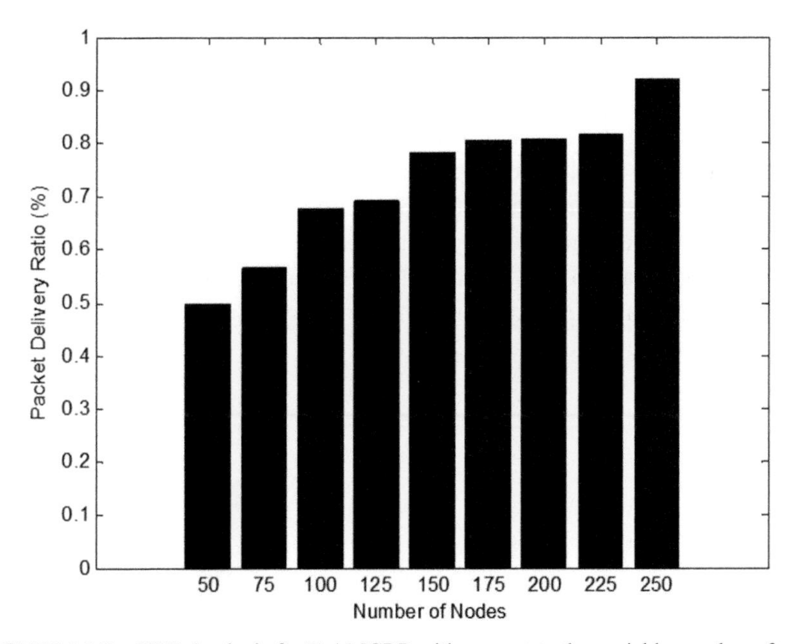

FIGURE 24.3 PDR Analysis for F-AMGRP with respect to the variable number of nodes.

The graph in Figure 24.4 delineates the end-to-end delay for F-AMGRP protocol. The end-to-end delay should be low in order to attain the efficient performance of the network. As per the graph, it is obtained that the end-to-end delay for minimum number of nodes is higher and the end-to-end delay for high number of nodes is lower.

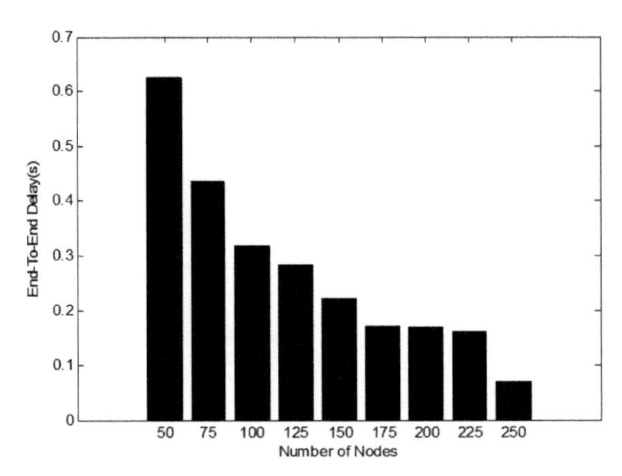

FIGURE 24.4 End to end delay analysis for F-AMGRP with respect to the variable number of nodes.

Similarly, the chart number 5 makes clear the normalized routing overhead obtained by way of using the FAMGRP throughout MATLAB. The actual normalized routing overhead specifies the full quantity of regulate packets with respect to the whole details packets transported to the destination. The actual graph and or chart proves the fact that network having the very best quantity of nodes have got the very best normalized routing overhead, that is, 5.05. The actual graph 6 shows the regular hop add up with regard to consist of work. The average hop add-up with regard to 50 nodes is actually 2 for 250 nodes it can be fewer than 0.2.

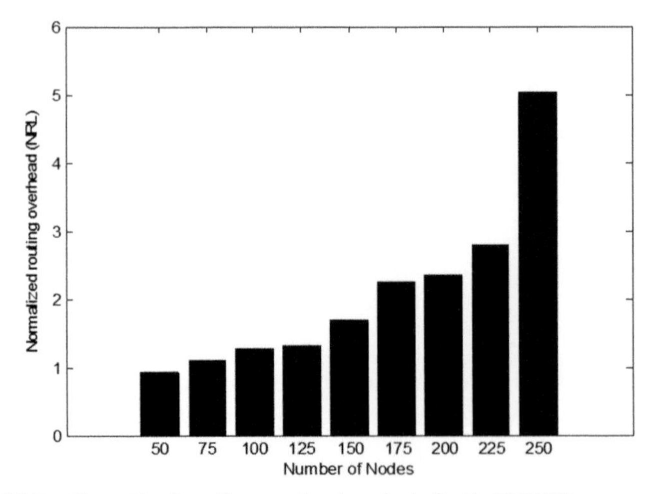

FIGURE 24.5 Normalized routing overhead analysis for F- AMGRP.

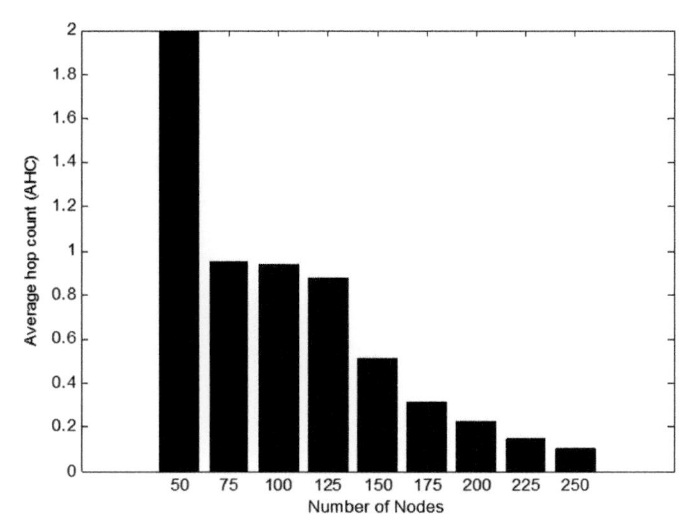

FIGURE 24.6 Average hop count analysis for F-AMGRP with respect to the variable number of nodes.

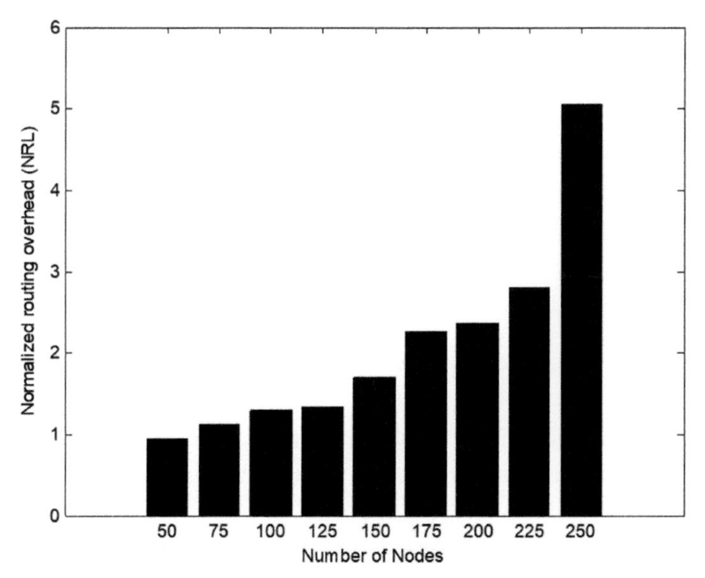

FIGURE 24.7 Normalized ROUTING OVERHEAD ANALYSIS for F- AMGRP.

Similarly, the chart number 5 makes clear the normalized routing overhead obtained by way of using the FAMGRP throughout MATLAB. The actual normalized routing overhead specifies the full quantity of regulate packets with respect to the whole details packets transported to the destination. The actual graph and or chart proves the fact that network having the very best quantity of nodes have got the very best normalized routing overhead, that is, 5.05

The actual graph 6 shows the regular hop add up with regard to consist of work. The average hop add up with regard to 50 nodes is actually 2 for 250 nodes it can be fewer than 0.2

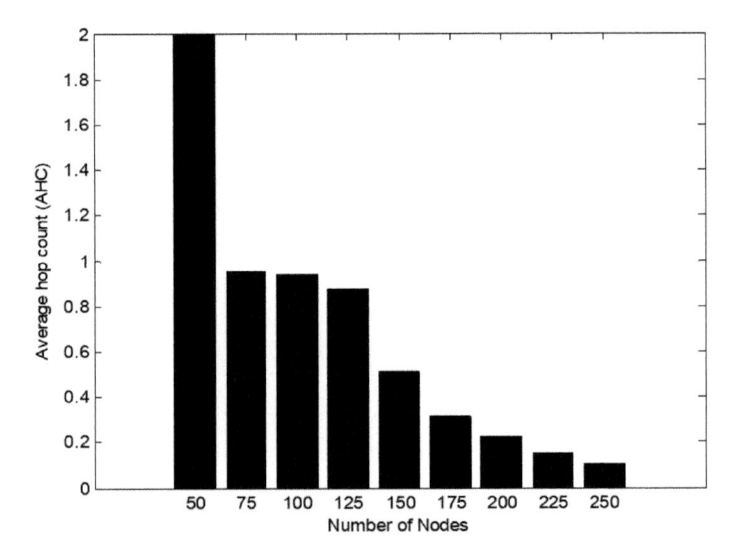

FIGURE 24.8 Average hop count analysis for F-AMGRP with respect to the variable number of nodes.

24.5 CONCLUSIONS

The particular VANETs is a car system as well as vibrant throughout nature. The following dynamicity brings about different routing difficulties inside the network. So that you can prevail over the direction-finding difficulties or node range issues, the concept of weight price seemed to be developed. Judging by the useful weight price the nodes tend to be selected. The original AMGRP approach elects the nodes by the manually assessed weight price in which brings about less overall performance inside the output. Therefore, the current perform offers an enhanced model associated with traditional AMGRP by adding the concept of fuzzy inference system as well as stability to be able to it. The particular wooly is usually employed to measure the weight values. For the purpose of stability aspect, the PDR is usually increased as a possible extra parameter to your weight function. Later on, the current perform may very well be implemented as well as opposed to traditional techniques so that the expertise in the consist of perform may very well be evaluated.

KEYWORDS

- **communication networks**
- **vehicular Ad hoc networks**
- **routing**

REFERENCES

1. Dharani Kumari, N. V.; Shylaja, B. S. *AMGRP: AHP-Based Multimetric Geographical Routing Protocol for Urban environment of VANETs*; Elsevier, 2017; pp 1–12.
2. Kaur, Er. N.; Singh, Er. A.; Singh, Er. P. Routing Protocols in Vanets: A Review. *Int. J. Innov. Eng. Technol.* **2019,** *12* (2).
3. Goudarzi, F.; Asgari, H.; Al-Raweshidy, H. S. Traffic-Aware VANET Routing for City Environments—A Protocol Based on Ant Colony Optimization. *IEEE* **2018,** *99*, 1–11.
4. Asline Celes, A.; Edna Elizabeth, N. *Verification Based Authentication Scheme for Bogus Attacks in VANETs for Secure Communication*; IEEE, 2018.
5. Kiruba Sandou, D.; Jothy, N.; Jayanthi, K. *Secured Routing in VANETs Using Lightweight Authentication and Key Agreement Protocol*; IEEE, 2018.
6. Kanwar, K.; Parveen, S. A Brief Review of Geographical Routing Protocols of VANET. *IJIRCCE.* **2017,** *5* (3), 1–6.
7. Venkatramana, D. K. N.; Srikantaiah, S. B.; Moodabidri, J. SCGRP: SDN-Enabled Connectivity-Aware Geographical Routing Protocol of VANETs for Urban Environment. *IEEE.* **2017,** *6* (5), 102–111.
8. Das, D. *Distributed Algorithm for Geographic Opportunistic Routing in VANETs at Road Intersection*; IEEE, 2017.
9. Qureshi, K. N.; Bashir, F.; Iqbal, S.; Anwar, R. W. *Systematic Study of Geographical Routing Protocols and Routing Challenges for Vehicular Ad hoc Networks*; IEEE, 2017.
10. Agustina, E. R.; Hakim, A. R. *Secure VANET Protocol Using Hierarchical Pseudonyms with Blind Signature*; IEEE, 2017.
11. Yiliang, H.; Xi, L.; Di, J.; Dingyi, F. *Attribute-Based Authenticated Protocol for Secure Communication of VANET*; IEEE, 2017.
12. Chhatani, R.; Quazi, M. A.; Rawat, P. A Hybrid Approach for Secure Communication Over LTE VANET in Urban Road Scenario, 2017.
13. Ali, S.; Nand, P.; Tiwari, S. Secure Message Broadcasting in VANET Over Wormhole Attack by Using Cryptographic Technique, 2017.

14. Akabane, A. T.; Pazzi, R. W.; Madeira, E. R. M.; Villas, L. A. *CARRO: A Context-Awareness Protocol for Data Dissemination in Urban and Highway Scenarios*; IEEE, 2016.
15. Eiza, M. H.; Owens, T.; Ni, Q. Secure and Robust Multi-Constrained QoS Aware Routing Algorithm for VANETs. *IEEE*. **2016,** *13* (1), 32–45.
16. Nema, M.; Stali, S.; Tiwari, R. *RSA Algorithm Based Encryption on Secure Intelligent Traffic System for VANET Using Wi-Fi IEEE 802.11p*; IEEE, 2016.
17. Guo, X-Y.; Chen, C-L.; Gong, C-Q.; Leu, F-Y. *A Secure Official Vehicle Communication Protocol for VANET*; IEEE, 2016.
18. Tripathi, V. K.; Venkaeswari, S. *Secure Communication with Privacy Preservation in VANET- Using Multilingual Translation*; IEEE, 2015.
19. Yang, S.; He, R.; Li, S.; Lin, B.; Wang, Y. *An Improved Geographical Routing Protocol and Its OPNET-Based Simulation in VANETs*; IEEE, 2015.
20. Tavakoli, R.; Nab, M. *TIGeR: A Traffic-Aware Intersection-Based Geographical Routing Protocol for Urban VANETs*; IEEE, 2014.
21. Hanan Saleet, Rami Langar, Kshirasagar Naik, Raouf Boutaba, Amiya Nayak, Nishith Goel; Intersection-Based Geographical Routing Protocol for VANETs: A Proposal and Analysis. *IEEE*. **2020,** *60* (9), 4560–4574.

CHAPTER 25

ENERGY-EFFICIENT PRIVACY PRESERVING VEHICLE REGISTRATION PROTOCOL FOR V2X COMMUNICATION IN VANET

N SASIKALADEVI and MANYAM NANDEESH REDDY

Department of CSE, School of Computing, SASTRA Deemed University, Thanjavur, Tamil Nadu, India

ABSTRACT

Vehicles in Vehicular Ad-Hoc Network (VANET) communicate through Dedicated Short Range Communication protocol. VANET mainly consists of three entities, that is, Roadside Unit (RSU), On-Board Unit (OBU), and Trusted Authority (TA). OBUs embedded in vehicles facilitate communication with other vehicles and RSUs. TA acts as a trusted third party who helps in registration of vehicles and identifying malicious identity if any dispute happens. The different modes of communication in VANET are infrastructure-to-infrastructure (I2I), vehicle-to-infrastructure (V2I), and vehicle-to-vehicle (V2V). VANET serves as an application of intelligent transport systems which is widely adopted domestically and abroad. VANET helps in improving driver's safety by exchanging information related to traffic between vehicles and infrastructures. Here, we propose secured lightweight key distribution scheme for VANET using Elliptic Curve Diffie-Hellman (EC-DH). The proposed scheme prevents eavesdropping of the

Advanced Computer Science Applications: Recent Trends in AI, Machine Learning, and Network Security. Karan Singh, PhD, Latha Banda, PhD & Manisha Manjul, PhD (Eds.)

messages between the TA and the vehicle in a VANET. The informal security analysis on the proposed scheme shows the same. The simulation results in the AVISPA show that the proposed scheme makes it difficult for the intruder to eavesdrop on the messages.

25.1 INTRODUCTION

With the technology improving manifold day by day, the interest in Vehicular in Ad-Hoc Networks (VANETs) is also increasing proportionally. This is mainly due to the wide range of applications it can offer that range from traffic management to infotainment services. Real-time communication between vehicles for a better driving experience and safety can be achieved with the help of VANET.[1,2] This communication includes their present status, weather conditions, and traffic that provide the driver with a more efficient and safer driving experience.

Every vehicle in a VANET contains an On-Board Unit (OBU), which facilitates communication between Vehicle and RSU. Vehicles in a VANET communicate through Dedicated Short Range Communication system. And different modes of communication in VANET include R2V, V2V, and V2R. In R2V, communication takes place between RSUs and a vehicle. While in V2V, communication is between two vehicles, and V2R is the communication between Vehicle and RSU. With the help of communications explained above, drivers can come to a proper conclusion about the driving environment and take the necessary action required. In a VANET, vehicles form the majority of nodes which group together to form networks without any prior knowledge of each other. Hence, vehicles are the most vulnerable part of VANETs which can be easily exploited if no proper security measures are taken.

25.2 RELATED WORK

A number of researchers have proposed many privacy-preserving schemes in the last decade. It includes schemes based on pseudonym, ID-based schemes, group signature-based methods, and symmetric cryptography-based approaches. In Ref.,[9] Hubaux and Raya described the privacy and security requirements of VANET and also proposed the pseudonym-based privacy-preserving algorithm. After that, many researchers have followed the work of Hubaux and Raya to propose several group signatures and pseudonym-based approaches. Public Key Infrastructure (PKI) is used to implement pseudonym-based schemes. Certificates generated by PKI are appended with the message signed by corresponding private key. This certificate holds pseudo-identity and the relation between each certificate and pseudo-identity is known only to certification authority (CA). In order to secure cryptographic parameters stored in On-Board Unit, Raya et al.[2] proposed Temper Proof Device (TPD) or Hardware Security Module. But this approach suffers from communication and storage overhead due to pseudonym-based certificates. Another major drawback of this scheme is Certificate Revocation List (CRL). While revoking particular vehicle, all other certificates issued already to that particular vehicle needs to be revoked. It significantly grows the size of CRL exponentially. Therefore, extra overhead is included in the management of CRL.

Zhang et al.[6] proposed the scheme to ensure conditional anonymity using realistic TPD in the place of ideal TPD. Sun et al. proposed a scheme by introducing a hash chains for reducing the CRL and uses a proxy re-signature method to advance the time needed to update the CRL. Later, conditional privacy preserving model is proposed by Lu et al. A vehicle in the network needs to get pseudonym key from RSU which is valid only for short term. Therefore, this approach necessitates omnipresent deployment of RSUs. But, the great drawback of the approach is that the TA requires updating CRL frequently and distributing it to all RSUs. Later, Rajput et al. introduced a hierarchical pseudonym-based model where CA issues primary pseudonym to each vehicle and RSU issues secondary pseudonyms. An identity-based verification method is proposed by Zhang et al., which generates certificates based on pseudo-identity and its corresponding private key with the help of TPD.

25.3 SYSTEM MODEL

The system model consists of two entities: TA and vehicles.

- Trusted Authority (TA): It is responsible for generating key pair, that is, private and public key for each participating vehicle and issuing secret certificates. TA is a trusted third-party agent.
- Vehicles: Each and every vehicle in the network is equipped with OBU and Tamper-Proof device (TPD). OBU assists to make communication between RSUs and vehicles, whilst TPD implemented within in the OBU makes sure that OBU is not compromised.

25.3.1 ASSUMPTIONS

The following assumptions are made before proceeding to the proposed scheme.

- The TA can be completely trusted, that is, it can never be compromised.
- A vehicle can be compromised thereby enabling an intruder to send and receive the messages. But the secret parameters within the TPD are safe.
- The communication between TA and RSUs is done with the help of a secure channel. But all other communications are done in an insecure channel.
- The TPD present in the vehicle will have its own unique key pair assigned for it at the time of installation.

25.3.2 PROPOSED SCHEME

TABLE 25.1 Notations and Its Definitions Used in Our Scheme.

Notation	Definition
TA	Trusted Authority
RSU	Roadside Unit
V_i	*i*th Vehicle
TPD	Tamper proof device of vehicle
OBU	On Board Unit of vehicle
ID_X	Real identity of x
PWD	Vehicle's biological password
(P_X, Q_X)	Two distinct large secret odd primes of x
n_X	Number generated by x; $n_X = P_X . Q_X$
N_X	Nonce of the entity x
SA_X	Security Association denotes a set of choices that are accepted by an entity x
G	A generator point of selected elliptic curve whose order is n
Ra_X	Random value chosen by an entity x
{x}_SK	A session key and x is encrypted and integrity-protected using internal keys
(PID, R, S)	A secret certificate issued by trusted authority
KE_X	mult (G, Ra_X) is a Diffie-Hellman value
Hash (.)	A secure hash function

TABLE 25.1 *(Continued)*

Notation	Definition
PRF (.)	A pseudorandom function whose output is indistinguishable from that of a truly random function.
mult (.)	An arithmetic multiply function
Vehicle List	In this list, real identities of each vehicle along with its corresponding pseudo identities is stored known only by TA
Revocation List	It stores the pseudo identities of revoked vehicles along with revocation time

25.3.3 CERTIFICATE GENERATION BY TA

TA generates the secret certificate and issues to the vehicle $V_{i.}$ The steps carried to produce secret certificate are as follows:

1. Let $R_j = 0$ and pseudo-identity be $PID_{j.}$
2. Computes a = Hash (PID_j, R_j) and checks if $a^{(P_{ta}-1)/2} = 1(\bmod P_{ta})$ and $a^{(Q_{ta}-1)/2} = 1(\bmod Q_{ta})$.
3. If not, $R_j = R_j + 1$ and again do the calculation and verify it.
4. Computes four modular square roots $X_{1,2,3,4}$ of X^2 = a (mod N_{ta}) with the help of P_{ta} and Q_{ta} based on equation $R1,2,3,4 = \pm\alpha \cdot Q_{ta} \cdot Q^*_{ta} \pm \beta \cdot P_{ta} \cdot P^*_{ta} (\bmod N_{ta})$ where $\alpha = a^{(P_t a+1)/4)} (\bmod P_t a)$, $\beta = a^{(Q_{ta}+1)/4} (\bmod Q_{ta})$, $P^*_{ta} = P^{-1}_{ta} (\bmod Q_{ta})$, $Q^*_{ta} = Q^{-1}_{ta} (\bmod P_{ta})$ and selects the smallest square root as S_j.
5. Then outputs (PID_j, R_j, S_j) and halts.

25.3.4 VEHICLE REGISTRATION PHASE

Whenever a vehicle enters into VANET, TA identifies the vehicle and requests the vehicle's id by sending a message REQUEST ID. Vehicle reacts to the TA request by sending it's own real identity, that is, RESPOND ID. V. After receiving IDv from the vehicle, TA agrees on a Security Association SA by accepting some set of choices used for the negotiation of the cryptographic algorithms.

Based on elliptic curve cryptosystem (ECC) Diffie-Hellman Key Exchange, TA computes the public key mult (G, RaT) by selecting a random value RaT and the base point G = (x_1, y_1). A base point G in elliptic curve

$E_p(a,b)$ with order n. Selected integer RaT value should be less than n. A nonce Nt generated by TA and must be fresh to avoid replay attacks. TA computes the key exchange message by combining security association, public key, and the nonce, that is, SAt. mult (G, RaT). Nt and send it to the vehicle. Vehicle also computes the key exchange message SAt. mult (G, RaV). Nv by agreeing on the security association choosen by TA and send it to the TA. RaV value should also be less than n and Nv must be fresh.

Now both the TA and vehicle computes the session key SK = PRF (Nt. Nv. mult (mult (G, RaT), RaV)) = PRF (Nv. Nt. mult (mult (G, RaV), RaT)). Both TA and vehicle compute digital signatures {TA. {SAt. mult (G, RaT). Nt. Nv}_inv (Kt)}_SK, {V. {SAt. mult (G, RaV). Nv. Nt}_inv (Kv)}_SK signed by its private key and encrypted with the session key respectively and exchange each other. It ensures mutual authentication and also avoid attacks like man-in-the-middle attack, message modification attack and replay attack.

When authentication becomes successful, vehicle sends ID and password encrypted by the session key to the TA, that is, {(IDv, PWD)}_SK. After receiving login credentials from the vehicle, TA computes secret certificate (PID. R. S) based on the secret parameters P_{ta}, Q_{ta} chosen by TA. The secret certificate will be enclosed in a message which contains hashed login credentials and secret parameters for vehicle chosen by TA, that is, {Hash (IDv. PWD). (Pv. Qv). (PID. R. S)}_SK.

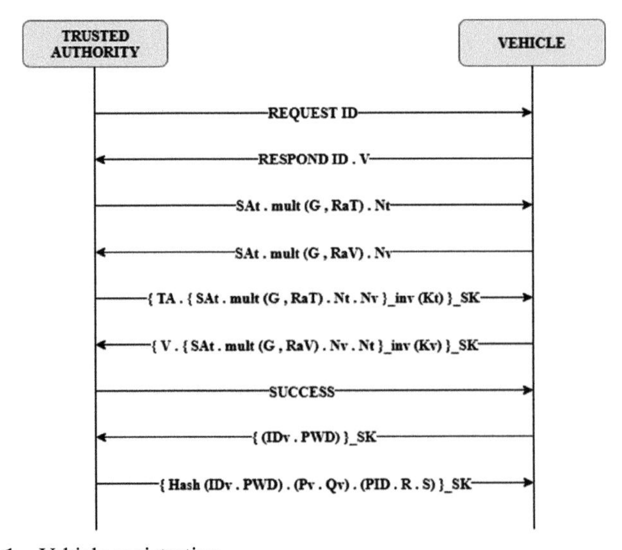

FIGURE 25.1 Vehicle registration.

25.3.5 RSU REGISTRATION PHASE

TA identifies the RSU location and requests the RSU's id by sending a message REQUEST ID. RSU reacts to the TA request by sending it's own real identity, that is, RESPOND ID. RSU. After receiving IDr from the RSU, TA agrees on a Security Association SA by accepting some set of choices used for the negotiation of the cryptographic algorithms.

Based on ECC Diffie-Hellman Key Exchange, TA computes the public key mult (G, RaT) by selecting a random value RaT and the base point G = (x_1, y_1). A base point G in elliptic curve $E_p(a,b)$ is with the order n. Selected integer RaT value should be less than n. A nonce Nt generated by TA and must be fresh to avoid replay attacks. TA computes the key exchange message by combining security association, public key, and the nonce, that is, SAt. mult (G, RaT). Nt and send it to the RSU. RSU also computes the key exchange message SAt. mult (G, RaR). Nr by agreeing on the security association choosen by TA and send it to the TA. RaR value should also be less than n and Nr must be fresh.

Now both the TA and RSU computes the session key SK = PRF (Nt. Nr. mult (mult (G, RaT), RaR)) = PRF (Nr. Nt. mult (mult (G, RaR), RaT)). Both TA and RSU compute digital signatures {TA. {SAt. mult (G, RaT). Nt. Nr}_inv (Kt)}_SK, {RSU. {SAt. mult (G, RaR). Nr. Nt}_inv (Kr)}_SK signed by its private key and encrypted with the session key, respectively, and exchange each other.

It ensures mutual authentication and also avoid attacks like man-in-the-middle attack, message modification attack, and replay attack. When authentication becomes successful, TA sends the secret parameters (Pr, Qr) to the RSU chosen by TA and encrypted with the session key, that is, {(Pr. Qr)}_SK.

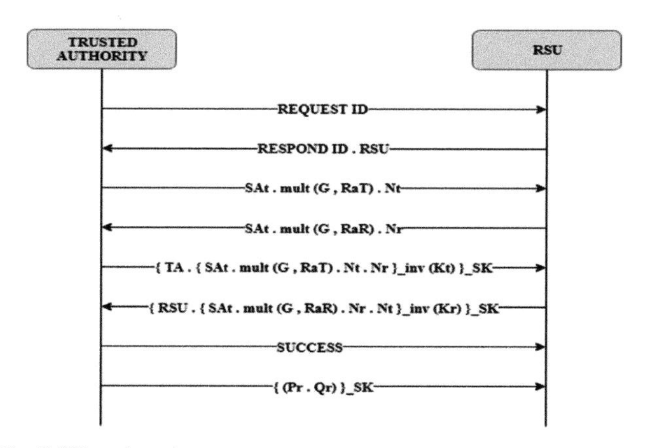

FIGURE 25.2 RSU registration.

25.3.6 INFORMAL ANALYSIS

In the informal analysis, the description of major attacks is given along with the explanation of how our proposed authentication protocol provides resistance against such security attacks.

25.3.7 MAN-IN-THE-MIDDLE ATTACK

To perform Man-in-the-Middle Attack, the intruder should capture transitions in mutual key exchange phase, that is, (SAt, mult(G, RaT), Nt) and (SAt. mult(G, RbV). Nv). Then, he will send the modified contents to both parties like (SAt, mult(G, RaT_M), Nt_M) and (SAt, mult(G, RaV_M), Nv_M). Later, both parties perform digital signature verification, which are encrypted with the session key, that is, SK = (Hash(Nt. Nv. mult(mult(G, RaT), RbV))) = (Hash(Nv. Nt. mult(mult(G, RaV), RaT))). Since, the intruder is unable to calculate the session key without the knowledge of RaT, RaV. It is computationally hard to calculate session key only with key exchange transitions. Since digital signatures are encrypted with the session key, intruders unable to modify the contents in the digital signatures. Any change in the digital signature will not match during signature verification. This avoids man in the middle attack.

25.3.8 EAVESDROPPING ATTACK (OR) NETWORK SNIFFING

In eavesdropping, the intruder try to secretly listen to the private conversation and also capture the keys, that is, (SAt, mult(G, RaT), Nt) and (SAt, mult(G, RaV), Nv) exchanged by both parties. It is computationally infeasible to calculate the session key, that is, SK = (Hash(Nt. Nv. mult(mult(G, RaT), RbV))) = (Hash(Nv. Nt. mult(mult(G, RaV), RaT))). Since, every transition after mutual key exchange is encrypted with session key, it impossible to capture what's inside the encrypted message.

25.3.9 MASQUERADE ATTACK (OR) IMPERSONATION ATTACK (OR) SPOOFING ATTACK

In order to perform this attack, the intruder needs to generate a valid message and the message should contain the valid signature (like {V.{Sat.

mult(G, RaV). Nv. Nt}_inv(Kv)}_SK) signed with a private key by the corresponding entity, it may be a vehicle (or) RSU (or) the TA. It is difficult for an adversary to produce such signature without knowing private key (like inv(Kv)). Even if the intruder try to generate the message by using it's own private key (like inv(Ki)), the entities on the other end verify the signature by decrypting it with sender's public key. Since the message is generated by an intruder, other entities like RSUs, vehicles, and TA are unaware about the intruder's public key, they fail to decrypt the signature and conclude it as a masquerading attack.

25.3.10 REPLAY ATTACK (OR) PLAYBACK ATTACK

In registration phase, any entity (vehicles or RSU or TA) before communicating with the other entities it chooses a random value called Nonce (Nx represents the random nonce value chosen by entity "x") and it acts as a timestamp. Once adversary intercepts and replays to a intercepting message, corresponding vehicle or RSU will be aware of the replay attack by verifying the current nonce Nx with the nonce value of the previously received messages. If the current nonce Nx matches with the nonce value of the previously received messages, then such a message is assumed to be a replayed message and it is discarded. Therefore, adversary is unable to clear the verification challenge due to freshness of Nx.

25.3.11 MESSAGE MODIFICATION ATTACK

It is impossible for an adversary to perform message modification attack on an encrypted message without an equivalent key to decipher them. Since, every transition after mutual key exchange is encrypted with session key (SK = (Hash(Nt. Nv. mult(mult(G, RaT), RbV))) = (Hash(Nv. Nt. mult(mult(G, RaV), RaT)))) which is computed by both the sender and the receiver, it is impossible to alter the contents in the encrypted message.

If an intruder tries to modify the messages before key exchange, the contents will not match with the digital signatures. Digital signatures are verified by both the sender and the receiver after key exchange, any modification during key exchange can be found here.

25.3.12 BRUTE FORCE ATTACK

If the intruder try to find the secret session key (SK = (Hash(Nt. Nv. mult(mult(G, RaT), RbV))) = (Hash(Nv. Nt. mult(mult(G, RaV), RaT)))), by trying with all possible combination of values (Nonce and Random values chosen by the sender and receiver). But since we are using a large generator value (G), it is highly difficult for the intruder to get to know about the session key. Even if he manages to crack the session key (SK), by the time he does so the key would be invalid because we are making use of disposable session keys which vary from session to session. Hence, brute force attack is not possible.

25.3.13 DENIAL OF SERVICE ATTACK (DOS)

In DoS attack, the attacker may prevent to establish the secure communication in between TA and Vehicles (or) TA and RSUs (or) RSUs and Vehicles. Suppose, when a vehicle tries to establish a secure session key with the TA, it sends its identity along with its nonce, by signing with its own private key (inv(Kv)) and send it as (mult(G, RaV). Nv. Nt)_inv(Kv). If an attacker tries to bombard the TA with the series of messages, TA can verify the signature of that message, by matching it with the public signature of the vehicle (Kv). Once the key pair (inv(Kv), Kv) match is found, the TA will put that vehicle in the revocation list (RL), thereby preventing the TA, and not accepting any further messages from that vehicle. Thus, it avoids DOS attack.

25.3.14 FLAW ATTACK

In flaw attack, the attacker tries to trick the other authorized entities (like vehicle, RSU, and TA) to accept the message component (instead of hash, entities are tricked to accept text (or) message (or) natural number as a message component) of one type as a message of another. Flaw attack can be successfully prevented by "tagging" types of each field of a message (like a hash component (Pseudo Random Function), is concatenated with a "Hash function" tag).

25.3.15 KEY REPLICATION ATTACK

In key replication attack, the intruder tries to replicate the private key of an entity (like vehicle, RSU, and TA), and it is used to encrypt the private conversation between other entities. It certainly takes a lot of time for the intruder to find the private key of a particular entity. Since we are making use of disposable keys like session key (SK = (Hash(Nt. Nv. mult(mult(G, RaT),

RbV))) = (Hash(Nv. Nt. mult(mult(G, RaV), RaT)))) for every new session, even if an intruder is able to find the encryption key of the previous session, it would go in vain.

25.4 EXPERIMENTAL SETUP

The simulation of our proposed scheme is performed on SPAN v1.6 on a 4GB Ubuntu system with an i3 processor. The protocol has been coded in High-Level Protocol Specification Language and then run on the SPAN[6] for the simulation.

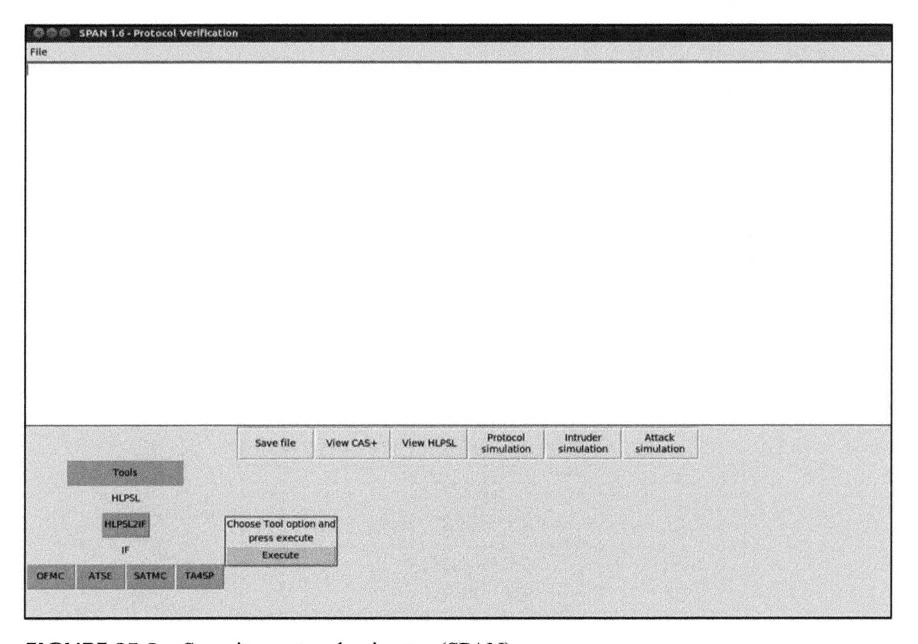

FIGURE 25.3 Security protocol animator (SPAN).

25.4.1 SIMULATION RESULTS

The simulation results of proposed lightweight authentication scheme clearly depict that the proposed protocol is safe and secure against an intruder. The simulation result is a sequence diagram that tells us that an intruder who tries to eavesdrop on the message transfer between the TA and the vehicle will be unsuccessful in his attempts.

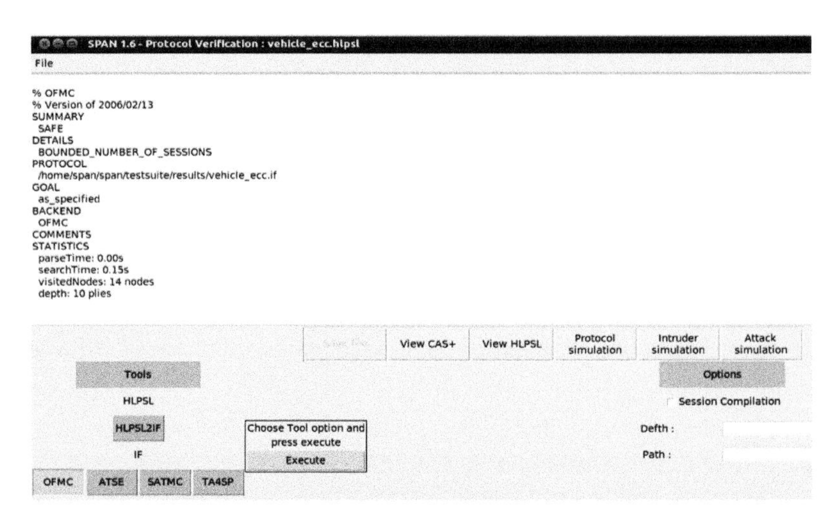

FIGURE 25.4 AVISPA output for vehicle registration.

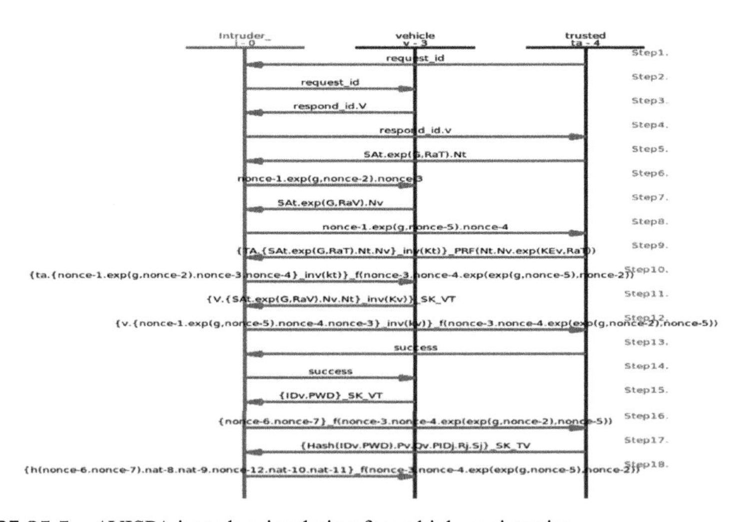

FIGURE 25.5 AVISPA intruder simulation for vehicle registration.

25.5 CONCLUSIONS

A secured lightweight key distribution scheme for VANET using Elliptic Curve Diffie-Hellman (EC-DH) has been proposed. The proposed scheme prevents eavesdropping of the messages between the TA and the vehicle in a

VANET. The informal security analysis on the proposed scheme shows the same. The simulation results in the AVISPA show that the proposed scheme makes it difficult for the intruder to eavesdrop on the messages.

ACKNOWLEDGMENTS

This part of this research work is supported by Department of Science and Technology, Science and Engineering Board (SERB), Government of India under the ECR grant (ECR/2017/000679/ES).

KEYWORDS

- **elliptic curve cryptosystem (ECC)**
- **vehicular ad-hoc network (VANET)**
- **energy efficient, security**

REFERENCES

1. Yao, L.; Wang, J.; Wang, X.; Chen, A.; Wang, Y. V2X Routing in a VANET Based on the Hidden Markov Model. *IEEE Trans. Intell. Trans. Syst.* **2018,** *19* (3).
2. Yang, X.; Yi, X.; Khalil, I.; Zheng, Y.; Huang, X.; Nepal, S.; Yang, X.; Cui, H. A Lightweight Authentication Scheme for Vehicular Ad Hoc Networks Based on MSR. *Veh. Commun.* **2019,** *15*, 16–27.
3. Jiang, D.; Taliwal, V.; Meier, A.; Holfelder, W.; Herrtwich, R. Design of 5.9 GHZ DSRC-Based Vehicular Safety Communication. *IEEE Wirel. Commun.* **2006,** *13*.
4. Williams, H. A Modification of the RSA Public-Key Encryption Procedure (Coresp.). *IEEE Trans. Inf. Theor.* **1980,** *26*, 726–729.
5. Blanchet, B. An Efficient Cryptographic Protocol Verifier Based on Prolog Rules. In *Proc. CSFW 2001*; IEEE Computer Society Press: Los Alamitos, 2001.
6. AVISPA. www.avispa-project.org IST-2001-39592 AVISPA v1.1 User Manual.
7. Mejri, M. N.; Ben-Othman, J.; Hamdi, M. Survey on VANET Security Challenges and Possible Cryptographic Solutions. *Veh. Commun.* **2014,** *1*, 53–66.
8. Raya, M.; Papadimitratos, P.; Hubaux, J.-P. Securing Vehicular communications. *IEEE Wirel. Commun.* **2006,** *13*.
9. Raya, M.; Hubaux, J.-P. Securing Vehicular Ad Hoc Networks. *J. Comput. Secur.* **2007,** *15*.

CHAPTER 26

EVALUATION AND OPTIMIZATION OF A CONGESTION CONTROL SCHEME FOR VANETS

GOPAL SINGH RAWAT and KARAN SINGH

School of Computer and Systems Sciences, Jawaharlal Nehru Unversity, New Delhi, India

ABSTRACT

The Intelligent transportation system (ITS) has changed how the transportation system was looked before. Vehicular Ad-hoc NETworks (VANETs) have become prominent communication technology for connected vehicles, assisting ITS in improving safety and traffic situations. However, the network congestion caused by the heavy data traffic can lead to excessive inaccuracy and failure of such applications. With the help of channel congestion control schemes VANETs can be made more effective and reliable. To this end, we present a swarm intelligence-based congestion control scheme named Swarm-based Intelligent Beacon Rate Adaption Scheme (S-IBRA) in this chapter. The simulation results and the comparative analysis of the proposed scheme S-IBRA, demonstrate its effectiveness in terms usage of channel (channel occupancy) and efficient beacon rate adaption.

Advanced Computer Science Applications: Recent Trends in AI, Machine Learning, and Network Security. Karan Singh, PhD, Latha Banda, PhD & Manisha Manjul, PhD (Eds.)

26.1 INTRODUCTION

The automotive industry has seen a revolution with the emergence of information and communication technologies (ICT). Transportation systems have always been an indispensable part of human activities, playing an important role in both economic and social development. The introduction of ICT in Transportation systems has led to the creation of a wide range of services, introducing the term "intelligent transportation system (ITS)". In recent years, ITSs have been developed and deployed in order to improve transportation safety and mobility, reduce environmental impacts, transportation efficiency, and productivity. ITS combines new technology and improvement in transportation systems, communication, sensors, and controllers in addition to the existing transportation infrastructure.

ITS has gained a lot of popularity over the years, both in academia and industry.[1] The main goal of ITS is improving road safety and driving conditions[2] apart from providing entertainment services. Vehicles rely on the passing of messages exchanged periodically among the various agents in the network that can include vehicles themselves or some other elements of the infrastructure. The technology enabling the communication between various agents is commonly known as Vehicular ad hoc networks (VANETs). The VANETs majorly use vehicle-to-vehicle (V2V) communication and vehicle-to-infrastructure (V2I) communication. VANETs help ITS in achieving the primary goal of traffic efficiency and road safety. Applications like cooperative vehicle safety (CVS) and traffic efficiency application are some prime example based on VANETs. Vehicles pass on short messages known as beacons and the process is called beaconing through DSRC channel. These beacons include critical information like vehicle kinematics and other such information. Vehicles broadcast beacons within their communication range with some beacon rate or beacon frequency. CVS applications rely on beacons to achieve their goals.

With the trend of an increasing number of vehicles on the roads, network congestion hinders deployment of CVS applications. Vehicles periodically send beacons which can generate communication load. The data congestion increases packet loss and communication delays, which implies a degradation of performance and Quality of Service (QoS) of VANETs. There are many techniques for improving congestion control in VANETs that have been proposed. These can be broadly classified based on[3–5]: (i) adjusting the data rate generation, (ii) transmission range adaptation of transmission channels,

(iii) hybrid methods combining both, and (iv) data packets scheduling using various channels.

The chapter discusses proposed Swarm intelligent-based congestion control scheme (S-IBRA). S-IBRA is stochastic, dynamic, and distributed in nature. Each vehicle acts as a particle of the swarm, adapts its beacon rate with cooperation from the neighboring vehicles. S-IBRA utilizes the channel capacity effectively and provides effective beacon rate adaption mechanism.

Next, the chapter is organized as follows. Section 26.2 presents the existing literature on channel congestion control in VANETs. Section 26.3 briefly discusses the fair beaconing optimization problem. Section 26.4 presents the proposed scheme S-IBRA with its methodology and operation. Section 26.5 provides the simulation results along with the performance evaluation. Finally, Section 26.6 concludes the chapter along with future scope of research in this domain.

26.2 RELATED WORK

Many schemes have been proposed addressing the channel congestion problem existing in VANETs. This section presents such schemes and their limitations.

In paper [6] a distributed fair power adjustment (D-PAV) for VANETs has been proposed. The scheme uses MaxBeaconingLoad parameter as threshold to keep beaconing traffic under control. The scheme fails to address scenarios when the threshold is violated. Huaying et al.[7] proposed a methodology based on the communication and traffic condition for adapting the power level and transmission rate. The scheme uses current speed of the vehicle, failure in attempted transmission, and success rate of beacon reception as metrics. The scheme does not considers the distributed nature of VANETs. Another scheme[8] using the vehicle information, estimates its own information and estimates its own information with the assistance of neighbors opinion. The scheme requires additional computation, which may cause communication delays.

Schmidt et al. presented a study of situation adaptive beaconing based on movement of vehicle and their neighboring vehicles. The study did not considered all possible road traffic situations and loads. In the article,[10] the concept of transmission power and transmission range control has been discussed. The strategy provides only theoretical aspect, does not considers the real scenarios. Wischhof et al.[11] proposed a congestion control scheme that is proactive in nature. The scheme is based on packet forwarding. The frequent message exchange is required leading to communication overhead.

The Tabu search approaches[12,13] have been used proposing schemes for congestion control. Schemes presented have high computation complexity adding to communication delays. Another approach of using priority in messages to tune beacon rate and transmission power presented in Ref.,[14] it increases the reliability of VANETs, however, increasing communication delay.

Toutouh and Alba used swarm intelligence for beacon rate adaption.[15,16] Schemes proposed consider channel load and distribution of the channel as the metrics for beacon rate adaption. The scheme uses greedy approach in beacon rate computation.

26.3 FAIR BEACON RATE OPTIMIZATION PROBLEM

Vehicles can adjust beacon rates efficiently use the available channels. In Refs.,[15,16] a fair beacon rate optimization problem is formulated. It computes the beacon rates while avoiding the data congestion. At some fixed interval of time the queues are monitored for a number of beacons in it, using eq 26.11 the channel occupancy is computed which further helps in computing beacon rates for specific scenarios. The FBR optimization problem defines:

- *BR (v)* is a set of beacon rates allowed for each vehicle *v*, *BR (v)* = *{br1, br2 .... brk}*
- *N (v)* as set of neighboring vehicles of *v*.
- The maximum number of beacons present in the queue without causing network overload or congestion is defined by maximum channel occupancy ($MaxQ \in Z$)
- A threshold ratio $\alpha \in [0, 1]$ defined over *MaxQ*.
- The effective channel capacity $\Omega \in [0, MaxQ]$ calculated as:

$$\Omega = a.MaxQ$$

- The channel occupancy, $Occ\ (v) \in [0,100]$, computed by each vehicle. It is using eq 26.1. Here br_v represents the beacons sent by vehicle *v* and rbr_j represents number of beacons received from neighbors.

$$Occ\ (v) = \frac{br_v + \left(\sum_j^{N(v)} rbr_j\right)}{MaxQ} \times 100\% \tag{26.1}$$

The fairness or network balance, *Fair (v)* ∈ [0, ∞], is measured using the coefficient of the variation of the beacon rates of neighboring vehicles. Computed using eq 26.2, $b\overline{r}v$ here is the average of the beacon rate of the neighbors of vehicle *v*, computed using eq 26.3.

$$Fair\ (v) = \frac{\sum_{j}^{NN(v)} \left(br_j - b\overline{r}_v\right)^2 + \left(br_v - b\overline{r}_v\right)^2}{|NN(v)|} \times \frac{1}{b\overline{r}_v} \qquad (26.2)$$

$$b\overline{r}_v = \frac{\left(\sum_{j}^{N(v)} br_j\right) + br_v}{|N(v)| + 1} \qquad (26.3)$$

In order to minimize the congestion, that is, *Occ (v)* ≤ Ω , the FBR optimization problem aims at finding the largest br_v for each vehicle v that maximizes *Occ (v)* and minimizes *Fair (v)*.

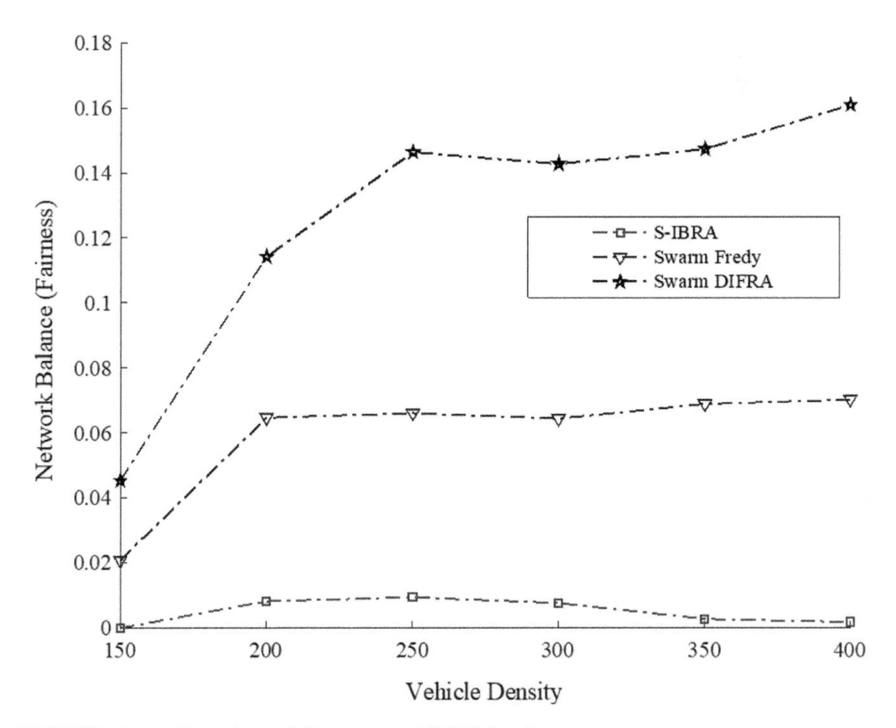

FIGURE 26.1 Flowchart of the proposed S-IBRA scheme.

26.4 PROPOSED SCHEME

The Swarm-based Intelligent Beacon Rate Adaption (S-IBRA) scheme is inspired by the existing Swarm-based control methods, that is, Swarm DIFRA[15] and Swarm FREDY.[16] S-IBRA is fully distributed as each vehicle executes it individually. S-IBRA computes the beacon rate dynamically while solving the FBR optimization problem. Each vehicle performs computations based on own experience and neighbors experience. The next sub-section presents the operation of S-IBRA.

26.4.1 S-IBRA

The congestion control methods generally perform two key operations, that is, network monitoring and reconfiguration of network components. S-IBRA monitors the network by analyzing the queues. The proposed scheme S-IBRA uses swarm intelligence-based method to improve the channel usage and beacon rate by using the information available from neighboring vehicles. S-IBRA has three chief components as illustarted by Figure 26.1:

- "Self Queue Monitoring Component (SQMC)" performs the IEEE 802.11p queue evaluation.
- "Swarm Information Exchange Component (SIEC)" assists in decoding the received beacons for useful information.
- "Intelligent Beacon Rate Adaptation Component (IBRAC)" helps in analyzing the information provided by previous two components SQMC and SIEC. It runs an algorithm to compute the final beacon rate (*fBR*), at every fixed time interval.

In S-IBRA, the components QMC and SIEC work in parallel, where the SQMC monitors the queue and computes the *fBR* after getting the neighborhood information. The first step is computation of temporary beacon rate (*tBR*) given by eq 26.4 then *fBR* is calculated as per eq 26.5. For each vehicle, the *BRBuffer* update is performed with the help of its *fBRth* component.

$$tBR = \frac{\Omega}{|N(v)+1|} \tag{26.4}$$

$$fBR = \begin{cases} br^{MIN} & \text{if } tBR < br^{MIN} \\ tBR & \text{if } br^{MIN} \le tBR \le br^{MAX} \\ br^{MAX} & \text{if } tBR > br^{MAX} \end{cases} \tag{26.5}$$

Algorithm 26.1 Intelligent Beacon Rate Adaption algorithm

```
Input: MaxQ, BRBuffer, Occ(v), fBR
Output: BR(v) , Beacon rate for reach vehicle
1: Compute OccTemp(v) for each vehicle using fBR
2: for each vehicle v do
3:   if Occ(v) <= 100 then
4:     if OccTemp(v) <= 100 then
5:       if OccTemp(v) > Occ(v) then
6:         br = Max(BRBuffer)
7:         BR(v) = index(br)
8:       end if
9:   end if
10:  else
11:    if OccTemp(v) <= 100 then
12:        br = Max(BRBuffer)
13:        BR(v) = index(br)
14:    else
15:    BR(v) = floor (fBR(v)/2)
16:    end if
17:  end if
18: end for
```

The *BRBuffer* is a vector with *k* components *BR (v)* = $[x_1 \;\; x_2 \; \ldots \; x_k]$ for each vehicle *v*. Each x_i component represents the number of requests received by the vehicle *v* . It means that vehicle v has received such number of requests to change its existing beacon rate to a new beacon rate *i* (hz or beacons per second). Let us consider the scenario where *BRBuffer (2)* = [0 0 0 33 0 5 0 0 0 0], this reflects that the vehicle id 2 has received 33 requests for 4 Hz beacon rate change and 5 requests for 6 Hz beacon rate change. *BRBuffer* gets manipulated by SQMC and SIEC. SIEC procedure analyses the received beacons for decoding fBR and BRBuffer gets updated according to stochastic Distance Discriminant procedure used in Swarm FREDY.[16]

After a fixed interval of time, IBRAC runs to compute the optimal beacon rate using Algorithm 26.1. IBRAC instead of updating the beacon rates at each fixed interval, does further comparisons of the channel occupancy with previous *BR* and with new *fBR*. The IBRAC procedure chooses the one with the maximum channel occupancy and updates the final *BR* for each vehicle after choosing the maximum value form BRBuffer after every fixed interval of time. For example, if *BRBuffer(1)* = [0 0 6 4 5 7 3 1 0 1], it means for vehicle id 1 max value of *BRBuffer* is 6 at index 3, so for vehicle id 1 the beacon rate to be updated is 3 Hz.

26.5 PERFORMANCE EVALUATION

This section presents the simulation parameters, results, and performance analysis of the simulations to evaluate the proposed Swarm-IFRA. The simulations were carried out on MATLAB 2017a Ubuntu 18 platform. The system configured with 12 GB RAM and a intel proessor i7-3.60 GHz with 4 Core(s) was used. The simulation parameters used in the experimental study are listed in Table 26.1.

TABLE 26.1 Simulation Parameters.

Parameter	Value
Simulation Period	150 s
Transmission range	250 m
MAC/PHY standard	IEEE 802.11p
Size of message	100 bytes
Data rate	6 Mbps
Beacon Interval	1 s
MaxQ	400
α	0.8
Highway Length	5 km
Number of Lanes	6 Lanes
Vehicle Density	150–400
Velocity	60–150 km

The vehicles have been placed randomly over the highway lanes. For a more realistic scenario, the outer lanes have more vehicle distributions. Each scenario is simulated 10 times for the consistency of the results. The proposed scheme S-IBRA is compared with the existing schemes that have solved the FBR optimization problem. We evaluate the existing swarm based schemes, that is, Swarm FREDY and Swarm DIFRA and optimize the FBR problem.

TABLE 26.2 Effect of Vehicle Density on Channel Usage.

Method	150 Veh.	200 Veh.	250 Veh.	300 Veh.	350 Veh.	400 Veh.
S-IBRA	74.156	75.063	75.594	75.188	74.625	73.906
Swarm FREDY	72.531	74.438	72.031	73.156	70.469	69.563
Swarm DIFRA	71.281	72.406	69	70.5	67.563	66.313

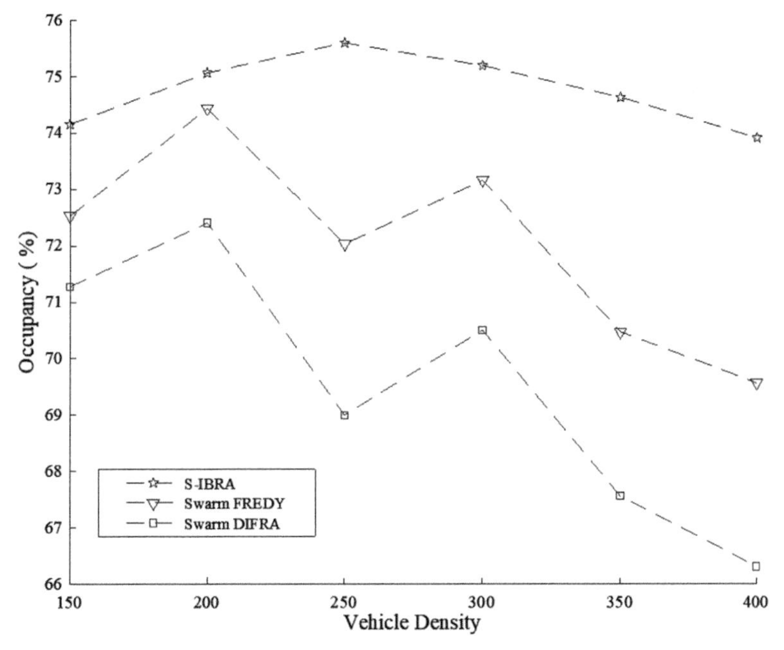

FIGURE 26.2 Channel occupancy per number of vehicles.

The scheme Swarm FREDY uses d1 and d2 as parameter. Swarm FREDY has many variants dependent on d1 and d2 values. The d1 and d2 are distances with value 50 and 100 m as they present better results compared to the other variants. So we compare the proposed scheme with Swarm FREDY (50,150) for channel usage. Table 26.2 presents the channel usage of each scheme. The data is obtained through median values of 10 simulations for each scenario. The proposed scheme S-IBRA has better performance compared with existing control method Swarm FREDY and Swarm DIFRA as seen from Table 26.2. The channel usage has no particular trend with respect to vehicle density. However, the comparisons can be seen for each scheme (see Fig. 26.2).

For each scenario, the median of the individual beacon rate of vehicles has been computed after running the simulation 10 times. The best variant of Swarm FREDY, that is, SF (50,100) has been used for comparison along with Swarm DIFRA and the proposed scheme. Table 26.3 provides the results obtained after simulations. As the density of vehicles increases the beacon rates start decreasing. This trend reflects how the beacon rate is directly dependent on vehicle density. The proposed scheme S-IBRA uses the IBRAC component and runs Algorithm 26.1 to compute the optimal beacon rate. Results from Table 26.3 demonstrate that the proposed scheme improves the beacon rate when compared with Swarm FREDY and DIFRA (see Fig. 26.3).

TABLE 26.3 Effect of Vehicle Density on Beacon Rate.

Method	150 Veh.	200 Veh.	250 Veh.	300 Veh.	350 Veh.	400 Veh.
S-IBRA	10	8	6	5	4	4
Swarm FREDY	9	7	5	4	3.875	3
Swarm DIFRA	8.75	6.125	4.5	3.875	3	2.75

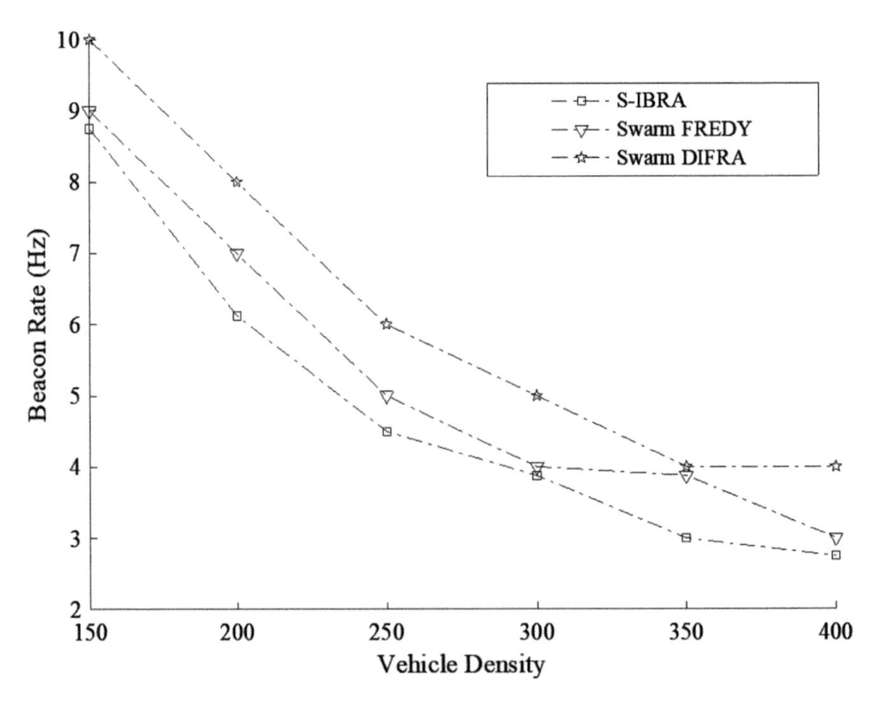

FIGURE 26.3 Beacon rate per number of vehicles.

Similarly, we evaluated the proposed scheme for network fairness or balance metrics. Figure 26.4, shows that the Swarm DIFRA scheme performs best and the proposed scheme S-IBRA needs further improvement. Swarm DIFRA considers all the neighbor vehicles while computing optimal beacon rates, hence the network fairness is high. However, in Swarm FREDY and S-IBRA, the beacon rate computation involves a smaller number of neighbors, which explains the results obtained (see Fig. 26.4).

We have performed the experimental analysis with the help of simulations. The simulation helped in evaluating the existing swarm-based congestion control schemes. The proposed scheme is compared with the basline scheme and the results demonstrate a improved performance of both baseline schemes, that is, Swarm FREDY and DIFRA.

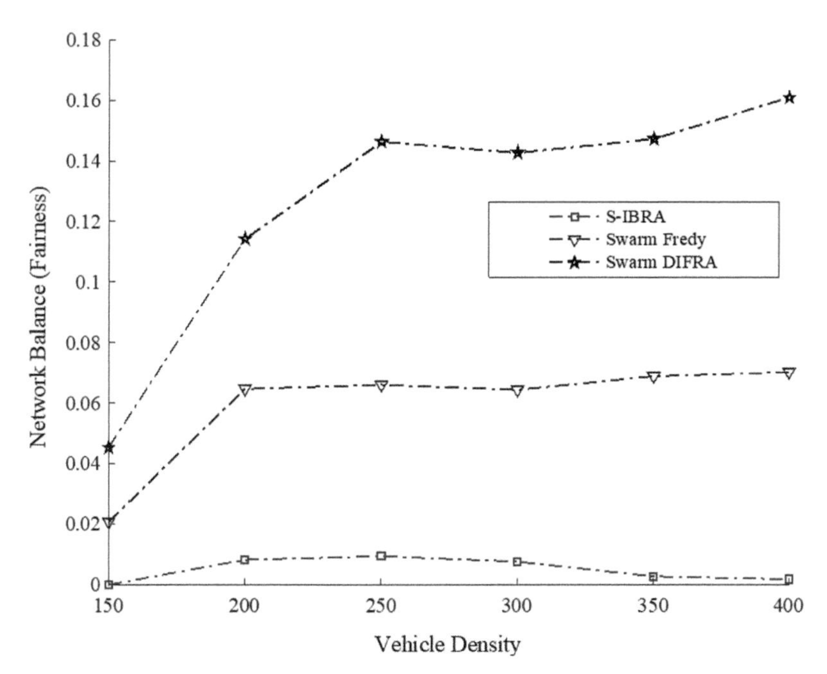

FIGURE 26.4 Network balance per number of vehicles.

26.6 CONCLUSIONS

The network congestion in VANET causes loss of packets and communication delays, thus degrading the performance of the VANET and its QoS. This chapter has evaluated the existing Swarm-based congestion control schemes based on beaconing. A Swarm-based intelligent beacon rate adaption scheme (S-IBRA) is proposed in this chapter, optimizing the FBR problem. We have evaluated the other schemes like Swarm FREDY and DIFRA. The proposed scheme is compared with these schemes. The experimental evaluation demonstrates a significant improvement in congestion control by the proposed scheme S-IBRA, improving Occupancy and individual beacon rates of vehicles. It demonstrated competitive performance in comparison with DIFRA and FREDY. The proposed scheme with intelligent and fair beacon rate adaption improves the channel occupancy, while ensuring that the network congestion is avoided. However, the network balance or fairness is better for Swarm DIFRA and the proposed scheme needs some improvement. The future research directions may involve improvement in network balance and evaluation of the scheme using realistic urban scenarios. It will be an interesting research direction, solving the FBR optimization problem using additional QoS metrics.

KEYWORDS

- **VANETs**
- **ITS**
- **beaconing**
- **swarm intelligence**

REFERENCES

1. Alam, M.; Ferreira, J.; Fonseca, J. Introduction to Intelligent Transportation Systems. In *Intelligent Transportation Systems*; Springer: Cham, 2016; pp 1–17.
2. Mejri, M. N.; Ben-Othman, J.; Hamdi, M. Survey on VANET Security Challenges and Possible Cryptographic Solutions. *Veh. Commun.* **2014,** *1* (2), 53–66.
3. Pereñiguez, F.; Lozano, J. S.; Fernández, P. J.; Bernal, F.; Skarmeta, A. F.; Ernst, T. Vehicular Ad Hoc Networks: Standards. *Solutions Res*. 2015, 253–282.
4. Sattari, M. R. J.; Md Noor, R.; Keshavarz, H. A Taxonomy for Congestion Control Algorithms in Vehicular Ad Hoc Networks. In *2012 IEEE International Conference on Communication, Networks and Satellite (ComNetSat)*; IEEE, 2012; pp 44–49.
5. Zhang, W.; Festag, A.; Baldessari, R.; Le, L. Congestion Control for Safety Messages in VANETs: Concepts and Framework. In *2008 8th International Conference on ITS Telecommunications*; IEEE, 2008; pp 199–203.

6. Torrent-Moreno, M.; Mittag, J.; Santi, P.; Hartenstein, H. Vehicle-to-Vehicle Communication: Fair Transmit Power Control for Safety-Critical Information. *IEEE Trans. Veh. Technol.* **2009,** *58* (7), 3684–3703.
7. Xu, H.; Barth, M. A Transmission-Interval and Power-Level Modulation Methodology for Optimizing Inter-Vehicle Communications. In *Proceedings of the 1st ACM International Workshop on Vehicular ad Hoc Networks*; 2004; pp 97–98.
8. Rezaei, S.; Sengupta, R.; Krishnan, H. Reducing the Communication Required by DSRC-Based Vehicle Safety Systems. In *2007 IEEE Intelligent Transportation Systems Conference*; IEEE, 2007; pp 361–366.
9. Schmidt, R. K.; Leinmuller, T.; Schoch, E.; Kargl, F.; Schafer, G. Exploration of Adaptive Beaconing for Efficient Intervehicle Safety Communication. *IEEE Netw.* 2010, *24* (1), 14–19.
10. Tielert, T.; Jiang, H. H.;Delgrossi, L. Joint Power/Rate Congestion Control Optimizing Packet Reception in Vehicle Safety Communications. In *Proceeding of the Tenth ACM International Workshop on Vehicular Inter-Networking, Systems, and Applications*, 2013; pp 51–60.
11. Wischhof, L.; Rohling, H. Congestion Control in Vehicular Ad Hoc Networks. In *IEEE International Conference on Vehicular Electronics and Safety*; IEEE, 2005; pp 58–63.
12. Taherkhani, N.; Pierre, S. Congestion Control in Vehicular Ad Hoc Networks Using Meta-Heuristic Techniques. In *Proceedings of the Second ACM International Symposium on Design and Analysis of Intelligent Vehicular Networks and Applications*; 2012; pp 47–54.
13. Taherkhani, N.; Pierre, S. Improving Dynamic and Distributed Congestion Control in Vehicular Ad Hoc Networks. *Ad Hoc Netw.* **2015,** *33*, 112–125.
14. Djahel, S.; Ghamri-Doudane, Y. A Robust Congestion Control Scheme for Fast and Reliable Dissemination of Safety Messages in VANETs. In *2012 IEEE Wireless Communications and Networking Conference (WCNC)*; IEEE, 2012; pp 2264–2269.
15. Toutouh, J.; Alba, E. Distributed Fair Rate Congestion Control for Vehicular Networks. In *Distributed Computing and Artificial Intelligence, 13th International Conference*; Springer: Cham, 2016; pp 433–442.
16. Toutouh, J.; Alba, E. A Swarm Algorithm for Collaborative Traffic in Vehicular Networks. *Veh. Commun.* **2018,** *12*, 127–137.

CHAPTER 27

DISPUTE BETWEEN COUNTRIES, A CORRESPONDING ATTACK ON CYBERSPACE: THE NEW NATIONAL SECURITY CHALLENGE

KOLLABATHINI SIDDHARDHA

Doctoral Candidate, Centre for European Studies, School of International Studies, Jawaharlal Nehru University, New Delhi

ABSTRACT

Today, cyberspace is a fact of daily life, and cyberspace's impact has not bypassed states' national security. Cyberspace, a manmade technological advancement over the past decades, transformed the way economies work around the world, reshaping social interactions, and a paradigm shift in politics. Cyberspace being boundary less, omnipresent across multiple domains, and anarchic, have been considered to attack whenever there are any disputes between two countries. In the context described above, a pressing question arises: Cyberspace is not a domain like land, water, and air, and it is an environment inhabited by information and knowledge, existing in electronic form. If cyberspace is a mere inhabitation of information and knowledge, why do states want to consider cyberspace as an arena for confrontation in any dispute between countries? This chapter proposes discussing this new phenomenon, looking into the evaluation, and analyzing aspects of the recent phenomenon.

Advanced Computer Science Applications: Recent Trends in AI, Machine Learning, and Network Security. Karan Singh, PhD, Latha Banda, PhD & Manisha Manjul, PhD (Eds.)

27.1 INTRODUCTION

Over the past decades, cyberspace, a manmade technological advancement, transformed the way economies work worldwide, reshaping social interactions, and a paradigm shift in politics. Today cyberspace is a fact of daily life, and it cuts across multiple domains. Cyberspace plays a vital role in atomic energy, space, communications, defense, education, agriculture, manufacture, services, entertainment, and employment generation and in addressing national priorities. Similarly, cyberspace became indispensable in the area of national security and defense. The protection of cyberspace has become a significant challenge to states as cyberspace is intertwined with other warfare domains, namely land, water, air, and space.[1]

Interestingly in the realm of the computer networks, state actors are not less in exploiting the incognito, precession impact, cost-effective, and minimal human resources requirement characters of cyberspace, which is an informative environment, to achieve national interests.[2] The attacks and threats to national security that are pervasive offline have begun to penetrate the world online. Thus cyberspace has become a new arena of confrontation leading to cyber-insecurity. Such an attack took place in 2007 in Estonia, where Estonia was subjected to systematic distributed attacks for 3 weeks. The cyberattack crippled the critical information infrastructure of financial centers, banks, parliament, ministries, security, and public transport. The cyberattack on Estonia is the first "documented proper cyberattack" and is the beginning of cyber warfare.[3]

Similarly, the attack on Georgia's cyberspace in 2008 changed the threat landscape for all the states that rely on cyberspace. One distinctive characteristic of the cyberattack on Georgia in the above context is—the outbreak of physical hostilities between Georgia and Russia over Abkhazia and South Ossetia landed up in the cyber domain.[4] "Europe" that became a battlefield for World War One and World War Two, coincidentally became a theatre for confrontation in the cyber domain.

Till the cyberattacks against Estonia 2007 and Georgia 2008, the cyberspace does not have strategic security attire. Pre Estonia 2007 and Georgia 2008 cyberattacks, cyberspace was viewed as a 21st-century technological infrastructure, a platform for sociocultural concepts, and a predominant support structure for economic activities. The Estonia and Georgia cyberattacks led to the conclusion that the cyberspace meant to conduct commerce, communicate with the citizens, and interface with the critical infrastructure via electronic means can be a battle space and has become a central security concern for governments across the world.[5]

The phenomenon—"whenever there are any disputes between two countries, a corresponding attack on the digital space has been seen" become more common in recent times. The 2010 Stuxnet cyberattack to fail Iran's nuclear enrichment program; Operation Nitro Zeus 2015 was an elaborate plan developed by the US for a cyberattack on Iran, in case the diplomatic efforts to limit Iran's nuclear program failed and led to a military conflict; and more recently in 2020 the India-China border clash at Galwan valley led to heightened cyberattacks on India by China, evinced the aforementioned phenomenon. In this context, a pressing question arises: Cyberspace is not a domain like land, water, and air. It even does not exist like space. In the words of Wing Commander M K Sharma (Indian Air Force): cyberspace is a bio-electronic environment that is literally universal, it exists where there are telephone wires, coaxial cables, fiber-optic lines, or electromagnetic waves. This environment is inhabited by information and knowledge existing in electronic form.[6] If cyberspace is a mere inhabitation of information and knowledge, why do states want to consider cyberspace as an arena for confrontation in any dispute between countries or as an area of strategic importance? The answers, intuitively, lie in studying cyber security in national security.

27.2 REVIEW OF LITERATURE

27.2.1 CYBER SECURITY AND NATIONAL SECURITY

There is a fair consensus in the literature that our societies are cyber dependent, and cyber security is a growing matter of national security concern. Therefore *Cyberspace* is crucial in studying from the perspective of security and international relations. Cyberspace constitutes an environment significantly different from other realms of internationally regulated activity.[7–9] Czosseck and Geers are of the opinion that each era brings with it new techniques and methods of waging war; while military scholars and experts have mastered land, sea, air, and space warfare, the time has come that they have to study the art of cyber war also. They felt that cyberspace is narrowly defined, and the concepts of attack, defense, and security remain unchanged, as do the threats posed by adversary propaganda, espionage, and attack on critical infrastructure.[10] Nye contrasts with Czosseck and Geers and defines that the characteristics of cyberspace reduced some of the power differentials among actors, and thus provides a good example of the diffusion of power that typifies the global politics in this century.[11]

Czosseck and Geers were complemented by M.K. Sharma and Reveron by saying that "the concept of cyber capabilities in war are slowly emerging" and sites the incident Russian cyber warriors entering the Georgian Ministry of Defense critical infrastructure in strengthening their argument. Sharma and Reveron by their books *Cyber Warfare: The Power of the Unseen; Cyber Warfare and National Security: Is securing Military Networks Enough?;* and *Cyber Space and National Security: Threats, Opportunities, and Power in Virtual world,* established a coherent framework for understanding how cyberspace fits within the national security.[12,13]

Rid, in his comprehensive work *"Cyber war will not take place,"* argued that cyber war has never happened in the past, it does not occur in the present, and it is highly unlikely that it will disturb our future. Further, he argued that most of the writers on cyberspace in the context of national security distracted from the real significance of cybersecurity: cyberattacks are not creating more vectors of violent interaction; instead they make previously violent interactions less violent. Rid, with his analysis, opens a fresh viewpoint that cyberspace is not a domain of military activity; instead, the use of computer networks permeates all other domains of military conflict, land, sea, air, and space.[14] Contrarily to Rid, Yates, Lieutenant Commander of US Navy and writer of *Cyber Warfare: An Evolution in Warfare not just War Theory*, asserted that cyberattack in conjugation with the military would rise to the level of national security concerns.[15]

Singer and Friedman altogether brought a new dimension of the debate cyberspace and national security by means of extensively discussing how it all works in cyberspace, why cyberspace matters and what anyone can do in cyberspace.[16] Green presented a multidisciplinary analysis of cyber war,[17] and Steed implanted Greens' analysis by illustrating the strategic implications of cyber war.[18] In their writings, Eun and Abmann opinioned that cyberspace must be taken into consideration more seriously to have enriched analytical and theoretical understanding of international politics in the digital age. Furthermore, they argued that cyberspace does not have changed the very nature of the war, but cyber warfare indeed will reshape the way in which war begins or is carried out in the near future.[19]

27.3 RESEARCH METHODOLOGY

The proposed qualitative research will be inductive. The research analyses perceptions of cyber security issues through the lens of realism. The research shall be based on both primary and secondary sources. Primary sources will include the documents and reports related to cyber security from the United Nations, India, European Union, and NATO. Likewise, the study also considers the primary sources like the national security policy of India, the US, EU, and NATO. The study will also make use of the secondary sources that include books, articles, and newspapers clipping for this research.

27.4 DISCUSSION

27.4.1 CYBER SECURITY IN NATIONAL SECURITY

Cyberspace is such a term that is not still completely defined, yet it has become virtually an inseparable element of our existence. Science fiction writer William Gibson coined the word 'cyberspace' in a short story published in 1982, and with the advent of the internet in the 1990s, cyberspace entered the real world. It created new space for information and communications, interactions, conducting businesses, and creating social media, among many other activities. Thus, it created also a new platform for conflicts. There is a healthy debate that all technological advancements are valuable in their own right, but new technology brings a new set of challenges. Indeed, cyberspace being boundary less and anarchic, which meant for economic, social, scientific, and military purposes, has been considered a state entity and states developed strategies, weapons, and stratagem in the cyber domain to safeguard their critical infrastructures and to defend the national security. The impact of cyberspace has not bypassed the national security of states.[20]

Cyberspace constitutes an environment significantly different from other realms of internationally regulated activity. The advent of cyberspace accelerated the military use of cyber capabilities, and simultaneously, the militarization of cyberspace took place due to the lack of convention on cyberspace,[21] and the Western world is the epicenter of these changes. Drek.S.Reveron, in his 2012 work on "*Cyber Space and National Security: Threats, Opportunities, and Power in a Virtual World,*" postulated that, by developing a computer language code, one could be capable enough of cyberattacking the systems anywhere in the world across almost all domains that are connected to computers or networks, and it is highly unlikely to attribute the attack. The following summarized cyberattack incidents might help in better understanding the complexity of cyberspace and the concerns surrounding it.

In 2007, a diplomatic row reputed between Russia and Estonia when Estonian authorities moved a monument "Red Army" from the center of the capital city, Tallinn, to the outskirts of the town. Estonia, an internet-reliant country, was cyber-attacked after the initial unrest. The cyberattack brought down the vast computerized infrastructure of Estonia by what experts in cyber security termed a coordinated "denial of services attack." The devastation was such that Ene Ergma, the Speaker of the Estonian Parliament and a nuclear physics scientist, has made the comparison: "When I look at a nuclear explosion and the explosion that happened in our country in May, I see the same thing." As with nuclear radiation, cyber war can destroy a modem state without drawing blood. At the time, Russia was suspected of the attacks, and Moscow has denied allegations of Russian involvement. The Estonian government denounced the attacks as an unprovoked act of aggression and was unsuccessful in establishing the origin of the cyberattack. However, reports and observations on the Estonian cyberattack incident pointed out that while nationalist fervor on the Russian side certainly played a part in rallying independent hackers, there is a possibility that Russia was involved.[22] Scott Shackelford, Cyber security Program chair, Indiana University, specified the Estonia cyberattack as "the first large scale incident of a cyber assault on a state."

Identically, a computer attack on Georgian websites had started slowly in 2008, weeks before the military confrontation on a territorial dispute over Abkhazia and South Ossetia with Russia. Georgia's prominent websites were defaced, for instance, that of Georgia's National Bank and the Ministry of Foreign Affairs.[23] Noticeably, the Georgia cyberattack was the first case in the cyber security history, where an independent cyberattack has taken place in sync with a conventional military operation.

There is another side to the story; over the years, states have also increased their use of cyber operations to further their national interests. In 2015, for example, the United States developed an elaborate plan code-named Nitro Zeus aimed at Iran under the President Obama administration. The project was a strategy to be launched after the Stuxnet 2010 cyber incident, to disable Iran's air defenses, communications systems, and crucial parts of its power grid in case the nuclear talks between Iran and *P5 plus one* (UN security council permanent members and Germany) fails. The project was shelved in July 2015 after the nuclear deal struck between Iran and six other nations. The bloodless, cost-effective, precision impact and incognito characters of cyberattack attained a military perspective. A February 2016 report in "The New York Times" by David E. Sanger and Mark Mazzetti categorically acknowledged that the states started considering cyberattack as an alternative.[24]

Withal, recently in 2020, the India-China border clash at Galwan valley led to heightened frictions between both countries. The Indian establishment responded by banning Chinese mobile apps, legitimately by invoking the provisions mentioned under 69 A of its Information Technological Act. However, the Ministry of Information and Technology, Govt. of India, repudiated the Chinese mobile app ban action in any association with the ongoing tension along the Himalayan border.

Year	Country	Dispute	Cyberattack
2007	Estonia	Diplomatic row b/w Estonia and Russia over a monument—"Red Army"	Coordinated Distributed Denial of Service Attack.
2008	Georgia	Physical hostility b/w Georgia and Russia over Abkhazia and South Ossetia.	Distributed Denial of Service Attack.
2010	Iran	Conflict b/w US and Iran over Iran's Natanz uranium enrichment plant—a key part of the nuclear power generation process.	A 500 kilobyte computer worm *"Stuxnet,"* the world's documented first cyber/digital weapon.
2015	Iran	Negotiations b/w P5+1 (US, UK, Russia, France, China, plus Germany) and Iran over Iran's Nuclear program.	Comprehensive cyberattack code name *"Nitro Zeus."* An elaborate plan developed by US for a cyberattack on Iran.
2020	India	India-China border clash at Galwan valley.	Distributed Denial of Service Attack, and Internet Protocol Hijack.

Furthermore, the government of India explicitly stated that the unprecedented decision to prohibit Chinese mobile apps was based on the reports by the security agencies that China has been engaged in massive data mining in India and likely has stolen the personal information of Indian citizens. Post the embargo; China has made more than 40,000 cyberattacks on Indian cyber space.[25] India-China cyberattack incident provides further evidence that cyberspace is integrated with or in conjugation with a military operation, and the cyberattack gives an edge.

27.5 CONCLUSION

Cyberspace in recent decades has become part and parcel of our lives, and in this day and age, it became an integral part in national security in this 21st century. Arnold Wolfers, in his 1952 article on *National Security as an Ambiguous Symbol,* wrote that national security is the absence of threat to a society's core values.[26] As the states and societies are increasingly becoming information societies, and following Wolfers' argument, the threat to information can be seen as a threat to the core of these societies. With Wolfers's reasoning, cyber security in national security must be taken more seriously in understanding various perceptions underlying security in this cyber era. Eventually, one can safely conclude from the aforementioned illustration that a corresponding attack on cyberspace has been seen whenever there are any disputes between countries. In point of fact, one can safely conclude from the aforementioned illustration that whenever there are any disputes between countries, a corresponding attack on the cyber space has been seen.

KEYWORDS

- **cyberspace**
- **cyber security**
- **national security**
- **security challenges**

REFERENCES

1. Gupta, A. Securing Cyberspace: "A National Security Perspective". In *Securing Cyberspace: International and Asian Perspective*; Samuel, C., Sharma, M., Eds.; Pentagon Press: New Delhi, 2015.
2. Singer, P. W.; Friedman, A. *Cybersecurity and Cyberwar: What Everyone Needs to Know*; Oxford University Press: New York, 2014.
3. Stiennon, R. *There Will Be Cyberwar: How the Move to Network Centric War Fighting Has Set the Stage for Cyberwar*; IT-Harvest Press: Birmingham, 2015.
4. Diebert, R. J.; Rohozinski, R.; Crete-Nishihata, M. Cyclones in Cyberspace Information Shaping and Denial in the 2008 Russia–Georgia War. *Security Dialogue* **2012,** *43* (1), 3–24.
5. Dinu, M. S. *The 5 Operational Domain and the Evolution of NATO'S Cyber Defence Concpet*, Ph.D Thesis, Bucharest: National Defence University of Romania "Carol I", 2017.

6. Sharma, M. K. *Cyber Warfare: The Power of the Unseen*; KW Publishers; New Delhi, 2011.
7. Clemente, D. *Cyber Security and Global Interdependence: What Is Critical?* Chatham House: London, 2013.
8. Meyer, P. *Global Cyber Security Norms: A Proliferation Problem?* ICT for Peace Foundation: Geneva, 2015.
9. Kuru, H.; Ocak, M. A. Determination of Cyber Security Awareness of Public Employees and Consciousness Rising Suggestions. *J. Learn. Teach. Digital Age* **2016,** *1* (2), 57–65.
10. Czosseck, C.; Geers, K. *The Virtual Battlefield: Perspectives on Cyber Warfare*; IOS Press BV: Amsterdam, 2009.
11. Nye, J. S., Jr *Cyber Power.* President and Fellows of Harvard College: Cambridge, 2010.
12. Reveron, D. S. *Cyber Space and National Security: Threats, Opportunities, and Power in a Virtual World*; Georgetown University Press: Washington, DC, 2012.
13. Sharma, M. K. *Cyber Warfare: The Power of the Unseen*; KW Publishers: New Delhi, 2011.
14. Rid, T. *Cyber War Will Not Take Place*; Oxford University Press: New York, 2013.
15. Yates, LCDR J.. A *Cyber Warfare: An Evolution in Warfare not Just War Theory*, Master's Thesis, Virginia: Unites States Marine Corps Command and Staff College, Marine Corps University, 2013.
16. Singer, P. W.; Friedman, A. *Cybersecurity and Cyberwar: What Everyone Needs to Know*; Oxford University Press: New York, 2014.
17. Green, J. A. *Cyber Warfare: A Multidisciplinary Analysis*; Routledge: Oxon, 2015.
18. Steed, D. The Strategic Implications of Cyber Warfare. In *Cyber Warfare: A Multidisciplinary Analysis*; Green, J. A., Eds.; Routledge: Oxon, 2015.
19. Eun, Y. S.; Abmann, J. S. Cyberwar: Taking Stock of Security and Warfare in the Digital Age. *Int. Stud. Persp.* **2016,** *17*, 343–360.
20. Choucri, N. Co-Evolution of Cyberspace and International Relations: New Challenges for the Social Sciences. Lecture delivered on 13 October 2013 at the World Social Science Forum, Montreal: Canada, 2013.
21. Mazzucchi, N.; Alix, D. *Web Wars: Preparing for the Next Cyber Crisis*; Carnegie Europe: Brussels, 2019.
22. Sharma, M. K. *Cyber Warfare: The Power of the Unseen*; KW Publishers: New Delhi, 2011.
23. Rid, T. *Cyber War Will Not Take Place*; Oxford University Press: New York.
24. Sanger, E. D.; Mazzetto, M. *U.S. Had Cyberattack Plan If Iran Nuclear Dispute Led to Conflict*, 2016. https://www.nytimes.com/2016/02/17/world/middleeast/us-had-cyberattack-planned-if-iran-nuclear-negotiations-failed.html (accessed 24 June 2020).
25. Koser, M.; Thaver, M. *In 5 days, over 40,000 Chinese Searches for Vulnerabilities in Indian Cyber Space*, 2020. https://epaper.indianexpress.com/c/52967213 (accessed 24 June 2020).
26. Wolfers, A. National Security as an Ambiguous Symbol. *Polit. Sci. Quart.* **1952,** *67* (4), 481–502.

INDEX

D

E

T

U

V